Cases on Research Ethics

Janice E. Jones
Carroll University, USA

Mette L. Baran
https://orcid.org/0000-0002-7262-9698
Troy University, USA

IGI Global Scientific Publishing
Publishing Tomorrow's Research Today

Vice President of Editorial	Melissa Wagner
Director of Acquisitions	Mikaela Felty
Director of Book Development	Jocelynn Hessler
Production Manager	Mike Brehm
Cover Design	Jose Rosado

Published in the United States of America by
IGI Global Scientific Publishing
701 East Chocolate Avenue
Hershey, PA, 17033, USA
Tel: 717-533-8845 | Fax: 717-533-7115
Website: https://www.igi-global.com E-mail: cust@igi-global.com

Library of Congress Cataloging-in-Publication Data

Names: Jones, Janice E., 1959- editor | Baran, Mette, 1965- editor
Title: Cases on research ethics / edited by Janice E. Jones, Mette L. Baran.
Description: Hershey, PA : IGI Global Scientific Publishing, 2025. | Includes bibliographical references and index. | Summary: "This book explores both historical cases and current issues that are plaguing the research community. It provides both the history of research ethics along with current examples that will allow aspiring researchers to move from dependent students to independent scholars"-- Provided by publisher.
Identifiers: LCCN 2025036202 (print) | LCCN 2025036203 (ebook) | ISBN 9798369383025 hardcover | ISBN 9798369383032 paperback | ISBN 9798369383049 ebook
Subjects: LCSH: Social sciences--Research--Moral and ethical aspects--Case studies | Artificial intelligence--Moral and ethical aspects--Case studies
Classification: LCC H62 .C344185 2025 (print) | LCC H62 (ebook) | DDC 300.72--dc23/eng/20251117
LC record available at https://lccn.loc.gov/2025036202
LC ebook record available at https://lccn.loc.gov/2025036203

British Cataloguing in Publication Data
A Cataloguing in Publication record for this book is available from the British Library.

Table of Contents

Detailed Table of Contents

Institutional Review Boards (IRBs) did not emerge from abstract ethical theory or administrative preference. They developed in direct response to repeated and well-documented failures to protect human research participants (Faden & Beauchamp, 1986; Katz, 1972). Throughout the twentieth century, researchers across multiple national contexts conducted studies that exposed individuals to harm, deception, and exploitation—often selecting participants precisely because they were marginalized, institutionalized, or lacked the power to refuse (Brandt, 1978). This chapter examines three landmark cases from Germany, the United States, and Sweden. Although differing in political and cultural context, each case demonstrates similar ethical breakdowns: absence of informed consent, exploitation of vulnerable populations, disproportionate risk without benefit, and lack of independent accountability (Belmont Report, 1979).

The ideas of "êthikê theôria" were likely originally discussed by Aristotle, who also studied and provided standards for evaluating human behavior. Formal rules were developed in response to a history of unethical human testing, including the Nuremberg Trials and the Tuskegee Syphilis Study. Informed consent, beneficence, and justice were among the concepts created by these rules, which included the Nuremberg Code, the Declaration of Helsinki, and the Belmont Report, to safeguard study subjects. In order to assure ethical conduct in research, institutional review boards are now essential. The idea of using humans in studies was starting to gain traction by the turn of the 20th century, provided that large-scale investigations were initially carried out on animals. Humans and animals were employed in studies much more frequently as the understanding of bacteria advanced and pharmaceutical corporations grew.

Human subjects research history has always tried to strike a balance between science & safety of research subject. The way of carrying out most experiments in the traditional manner, development of test in most cases were tragic thus needed a lot of morality. Research in contemporary times is guided by principles of autonomy, justice & beneficence that create dignity, inclusion & protection of research subjects. This is no longer the case as it is now the people as the research partners & not the objects of inquiry. However, rapid technological growth & global collaborations present new moral challenges. Emerging domains (genomics, artificial intelligence & data science) have difficult questions that present privacy, consent & equity problems. Those issues imply that there is a need to have hard & yet flexible laws that would enable people to be secured & inventions to be improved. Ethical research does not just establish a knowledge, but also trust, dignity, & ethical responsibility of the society in such a way that science is built in a responsible manner to serve people & society.

Plagiarism in academia is a notorious and emerging phenomenon that violates the essence of values in the academic fraternity, such as intellectual honesty, trust and originality. It is a theoretically, empirically, and technologically informed chapter that provides a deep analysis and starts by providing a contextual picture of the historical and philosophical basis of plagiarism and intellectual property. It uses ethical theories, including ethical deontology, consequentialism, and virtue ethical theories, to understand moral reasoning, and discusses cultural and educational factors that affect how people perceive plagiarism. There are challenges in establishing intent and accountability that are reflected in global case studies. Academic pressure, cultural misperceptions, and lack of policy are some root causes explored in the chapter, as well as their consequences to individuals and institutions. Finally, it states that fighting plagiarism not only needs enforcement but also needs a culture of academic integrity that is durable to absorb the transformation of technology.

In the dynamic world of research, conflicts of interest whether personal, financial, or institutional can quietly undermine credibility, objectivity, and public trust. This chapter explores the complex ethical terrain of competing interests in academic and applied research environments. Through real-world case studies and theoretical insights, it sheds light on how subtle biases, hidden incentives, and institutional pressures can compromise research integrity. Personal conflicts often stem from relationships or individual ambitions, subtly influencing research decisions. Financial conflicts, including funding sources, stock ownership, and consultancy roles, pose significant risks of data manipulation or selective reporting. Meanwhile, institutional conflicts arise when universities or organizations prioritize prestige, funding, or political interests over academic rigor. This chapter presents a comparative analysis of each conflict type, supported by landmark cases and recent controversies from diverse disciplines.

Sports and exercise science research presents unique ethical challenges extending beyond traditional biomedical frameworks. This chapter examines moral considerations in studying human performance within competitive athletic contexts. Fundamental principles of respect for persons, beneficence, justice, and integrity require careful adaptation for athletic populations facing distinctive pressures and vulnerabilities. Critical issues include informed consent procedures for diverse populations, risk assessment in high intensity protocols, confidentiality protection in elite populations, and data privacy with emerging technologies. Research design challenges encompass participant selection, randomization ethics, and emergency preparedness. Intervention studies raise concerns regarding ergogenic aids, fatigue boundaries, and placebo controls. The chapter emphasizes research integrity through proper data handling, conflict management, and transparent reporting.

Important scientific discoveries have been followed using animal research particularly in the field of medicine and biology. Nevertheless, animals have to be treated in an ethical manner. Ethics in animal research follow the 3S model whereby: Replacement (where possible, using alternatives), Reduction (involving the use of fewer animals) and Refinement (whereby procedures are altered to cause minimal suffering). The ethics committees must approve their studies on animals in order to make sure that research is necessary and that they take good care of the animals. Good welfare would entail adequate food, hygienic surroundings, medicine and the reduction of pain or stress. Almost all the countries have legislative acts that guard the research animals, but the enactments differ. Uniform care of animals There should be international standards and collaboration that bring about uniformity in taking care of animals. Education of researchers, regulation of laws, and education of alternatives present the important steps towards improved animal welfare in research all over the world.

The concept of ethical integrity permeates descriptive accounting research, because, although this type of research is observational in its design, it nonetheless holds considerable responsibility because it relies on live financial information. Inaccuracy or prejudice to any degree can lose stakeholder credibility, as well as taint reputations and reputation of the research and the profession. In this chapter, the philosophy-based moral values including professional and philosophical theories of ethics are discussed with reference to the principles of integrity, objectivity, accountability, and confidentiality. It examines the real-life data managing, methodological Openness,

The chapter explores the transformative evolution of research ethics in the 21st century, shaped by rapid technological innovation, global collaboration, and shifting societal values. It examines how traditional ethical principles such as respect, beneficence, justice, and nonmaleficence must adapt to address emerging challenges in artificial intelligence, automation, immersive technologies, and big data. Key concerns including algorithmic accountability, informed consent in digital environments, participant vulnerability, data governance, and intellectual property are critically analyzed. The discussion highlights tensions between global harmonization and cultural relativism in ethical governance, while emphasizing the need for anticipatory and adaptive frameworks. Special attention is given to ecological responsibility, participant care, and the growing role of ethics education in ensuring human dignity, scientific integrity, and social responsibility.

Preface

Scientific research depends on the participation and trust of individuals and communities. Every dataset, clinical trial, survey, or interview ultimately reflects the willingness of people to share information, experiences, and sometimes personal risk in the service of knowledge. When conducted responsibly, research improves lives, informs policy, advances medicine, and deepens our understanding of the world. Yet the history of research also demonstrates that the pursuit of knowledge can cause harm when ethical responsibility fails to keep pace with scientific ambition.

The modern system of research ethics did not emerge spontaneously. It developed in response to a series of deeply troubling historical episodes in which research participants were deceived, exploited, or exposed to unnecessary risk. These cases revealed a critical flaw in the early research enterprise: researchers and institutions were often left to judge the ethics of their own work. Without independent oversight, conflicts of interest, professional pressures, and institutional incentives sometimes overshadowed the rights and welfare of research participants.

Institutional Review Boards (IRBs) emerged as a response to these failures. Designed to provide independent ethical review of research involving human participants, IRBs serve as an essential safeguard within the research process. By requiring informed consent, evaluating potential risks and benefits, protecting vulnerable populations, and promoting transparency in research design, IRBs help ensure that scientific progress does not come at the expense of human dignity.

This book grew out of a shared recognition that many students, researchers, and practitioners encounter research ethics primarily as a regulatory requirement rather than as a historically grounded and intellectually meaningful component of responsible scholarship. IRB processes are often experienced as procedural steps to complete rather than as mechanisms designed to protect participants and strengthen the integrity of research itself.

Our goal in this volume is to place ethical oversight back within its broader historical, institutional, and intellectual context. Through historical analysis, contemporary cases, and discussions of emerging ethical challenges, the chapters that follow examine how research ethics developed, how IRBs function today, and how ethical oversight must continue to evolve in response to new technologies, global research collaborations, and increasingly complex forms of data collection.

The book is written for a wide audience, including students learning about research methods, scholars conducting human subjects research, faculty members mentoring new researchers, and administrators or policymakers responsible for ethical oversight. By combining historical reflection with case-based analysis, we hope to encourage readers to see research ethics not merely as a set of rules to follow, but as a shared responsibility central to the credibility and legitimacy of scholarly inquiry.

ORGANIZATION OF THE BOOK

Chapter 1: Unethical Research Cases and the Lessons That Shaped Modern Research Ethics Making the Case for Institutional Review Boards: A Historical Overview

Institutional Review Boards (IRBs) did not emerge from abstract ethical theory or administrative preference. They developed in direct response to repeated and well-documented failures to protect human research participants (Faden & Beauchamp, 1986; Katz, 1972). Throughout the twentieth century, researchers across multiple national contexts conducted studies that exposed individuals to harm, deception, and exploitation—often selecting participants precisely because they were marginalized, institutionalized, or lacked the power to refuse (Brandt, 1978). This chapter examines three landmark cases from Germany, the United States, and Sweden. Although differing in political and cultural context, each case demonstrates similar ethical breakdowns: absence of informed consent, exploitation of vulnerable populations, disproportionate risk without benefit, and lack of independent accountability (Belmont Report, 1979).

Chapter 2: History of Ethics in Research

The ideas of "êthikê theôria" were likely originally discussed by Aristotle, who also studied and provided standards for evaluating human behavior. Formal rules were developed in response to a history of unethical human testing, including the Nuremberg Trials and the Tuskegee Syphilis Study. Informed consent, beneficence, and justice were among the concepts created by these rules, which included the Nuremberg Code, the Declaration of Helsinki, and the Belmont Report, to safeguard study subjects. In order to assure ethical conduct in research, institutional review boards are now essential. The idea of using humans in studies was starting to gain traction by the turn of the 20th century, provided that large-scale investigations were initially carried out on animals. Humans and animals were employed in studies much more frequently as the understanding of bacteria advanced and pharmaceutical corporations grew.

Chapter 3: Protection of human subject

Human subjects research history has always tried to strike a balance between science & safety of research subject. The way of carrying out most experiments in the traditional manner, development of test in most cases were tragic thus needed a lot of morality. Research in contemporary times is guided by principles of autonomy, justice & beneficence that create dignity, inclusion & protection of research subjects. This is no longer the case as it is now the people as the research partners & not the objects of inquiry. However, rapid technological growth & global collaborations present new moral challenges. Emerging domains (genomics, artificial intelligence & data science) have difficult questions that present privacy, consent & equity problems. Those issues imply that there is a need to have hard & yet flexible laws that would enable people to be secured & inventions to be improved. Ethical research does not just establish a knowledge, but also trust, dignity, & ethical responsibility of the society in such a way that science is built in a responsible manner to serve people & society.

Chapter 4: Borrowed Words, Broken Ethics: Addressing Plagiarism In Academia

Plagiarism in academia is a notorious and emerging phenomenon that violates the essence of values in the academic fraternity, such as intellectual honesty, trust and originality. It is a theoretically, empirically, and technologically informed chapter that provides a deep analysis and starts by providing a contextual picture of the historical and philosophical basis of plagiarism and intellectual property. It uses ethical theories, including ethical deontology, consequentialism, and virtue ethical theories, to understand moral reasoning, and discusses cultural and educational factors that affect how people perceive plagiarism. There are challenges in establishing intent and accountability that are reflected in global case studies. Academic pressure, cultural misperceptions, and lack of policy are some root causes explored in the chapter, as well as their consequences to individuals and institutions. Finally, it states that fighting plagiarism not only needs enforcement but also needs a culture of academic integrity that is durable to absorb the transformation of technology.

Chapter 5: Untangling Conflicts Personal, Financial, and Institutional Dilemmas in Research Ethics

In the dynamic world of research, conflicts of interest whether personal, financial, or institutional can quietly undermine credibility, objectivity, and public trust. This chapter explores the complex ethical terrain of competing interests in academic and applied research environments. Through real-world case studies and theoretical insights, it sheds light on how subtle biases, hidden incentives, and institutional pressures can compromise research integrity. Personal conflicts often stem from relationships or individual ambitions, subtly influencing research decisions. Financial conflicts, including funding sources, stock ownership, and consultancy roles, pose significant risks of data manipulation or selective reporting. Meanwhile, institutional conflicts arise when universities or organizations prioritize prestige, funding, or political interests over academic rigor. This chapter presents a comparative analysis of each conflict type, supported by landmark cases and recent controversies from diverse disciplines.

Chapter 6: Navigating Ethical Challenges in Sports and Exercise Science Research

Sports and exercise science research presents unique ethical challenges extending beyond traditional biomedical frameworks. This chapter examines moral considerations in studying human performance within competitive athletic contexts. Fundamental principles of respect for persons, beneficence, justice, and integrity require careful adaptation for athletic populations facing distinctive pressures and vulnerabilities. Critical issues include informed consent procedures for diverse populations, risk assessment in high intensity protocols, confidentiality protection in elite populations, and data privacy with emerging technologies. Research design challenges encompass participant selection, randomization ethics, and emergency preparedness. Intervention studies raise concerns regarding ergogenic aids, fatigue boundaries, and placebo controls. The chapter emphasizes research integrity through proper data handling, conflict management, and transparent reporting.

Chapter 7: Animal Research Ethics and Welfare Standards

Important scientific discoveries have been followed using animal research particularly in the field of medicine and biology. Nevertheless, animals have to be treated in an ethical manner. Ethics in animal research follow the 3S model whereby: Replacement (where possible, using alternatives), Reduction (involving the use of fewer animals) and Refinement (whereby procedures are altered to cause minimal suffering). The ethics committees must approve their studies on animals in order to make sure that research is necessary and that they take good care of the animals. Good welfare would entail adequate food, hygienic surroundings, medicine and the reduction of pain or stress. Almost all the countries have legislative acts that guard the research animals, but the enactments differ. Uniform care of animals There should be international standards and collaboration that bring about uniformity in taking care of animals. Education of researchers, regulation of laws, and education of alternatives present the important steps towards improved animal welfare in research all over the world.

Chapter 8: Ethical Guidelines in Descriptive Accounting Research

The concept of ethical integrity permeates descriptive accounting research, because, although this type of research is observational in its design, it nonetheless holds considerable responsibility because it relies on live financial information. Inaccuracy or prejudice to any degree can lose stakeholder credibility, as well as taint reputations and reputation of the research and the profession. In this chapter, the philosophy-based moral values including professional and philosophical theories of ethics are discussed with reference to the principles of integrity, objectivity, accountability, and confidentiality. It examines the real-life data managing, methodological Openness, and responsible reporting and deals with contemporary technological issues. The moral questions that could appear to researchers during their work find an illustration in a case study, confirming theoretical research and practical moral choice.

Chapter 9: Ethical Dilemmas in Social Science Research: Case Studies and Solutions

This chapter discusses social science researchers' ethical dilemmas as scientists and social agents. Social scientists must provide correct data and ethical issues arising from their involvement in social movements or politically sensitive situations. Insufficient studies have examined how researchers manage ethical issues in practice, especially when their findings contradict existing structures or affect disadvantaged people. This article examines how deontological, utilitarian, and virtue ethics apply to real-world ethical situations. Our findings show that research ethics must balance scientific rigor and ethical responsibility. The chapter suggests that researchers need stronger frameworks and training to overcome ethical issues and ensure research integrity and social relevance. These findings join conversations on ethical research practices and their effects on study validity and researcher and participant well-being.

Chapter 10: Research Ethics in Islam (Principles, Practices, and Contemporary Case Reflections)

Research ethics is essential for responsible knowledge production in today's globalized academic environment. While largely shaped by secular Western traditions, Islamic scholarship provides a deep ethical framework grounded in the Qur'an, Hadith, and jurisprudence. Core principles include truthfulness (sidq), justice (adl), intention (niyyah), public good (maslahah), and accountability (amanah). This chapter examines how these values guide research in theory and practice, using case studies from Muslim-majority contexts and global scholarship. It calls for an inclusive model that integrates Islamic ethics with international standards without compromising science. Addressing biomedical research, social sciences, and emerging technologies, it offers practical guidance for researchers and institutions worldwide.

Chapter 11: Ethics of Using Artificial Intelligence and Digital Tools in Research

The rapid advancement of artificial intelligence and digital technologies has transformed the research landscape, offering powerful tools for data collection, analysis, and interpretation. Alongside these opportunities, new ethical challenges have emerged, extending beyond traditional research ethics. Issues such as algorithmic bias, data privacy, ownership, transparency, and accountability have become central to maintaining integrity in scientific inquiry. This chapter explores key ethical principles, including respect for persons, beneficence, non maleficence, and justice, in the context of AI assisted research. It examines risks of overreliance on automated decision making, threats to authenticity in digital environments, and complexities of intellectual property. Emphasis is placed on responsible innovation, human oversight, global guidelines, and frameworks to balance technological potential with ethical responsibility.

Chapter 12: Ethical Dimensions of AI-Powered Business Decisions Ensuring Transparency and Responsibility

Artificial Intelligence (AI) is reshaping business decision-making by enabling predictive analytics, automation, and real-time insights across domains like finance, marketing, HR, and operations. While enhancing efficiency, AI also raises ethical challenges of transparency, fairness, privacy, and responsibility—especially when opaque algorithms affect vulnerable groups. This chapter explores the ethical dimensions of AI adoption in business through case-based learning, applying principles of justice, autonomy, beneficence, and non-maleficence. Real-world cases—such as biased loan approvals and facial recognition in retail—illustrate dilemmas of bias, consent, and surveillance. Students engage in reflection exercises, dilemma maps, and role-based analysis, guided by ethical frameworks including OECD, IEEE, the EU AI Act, and a proposed model, Ethical Reflexivity in AI (ER-AI). By combining global perspectives, emerging challenges like generative AI, and the intersection of research and business ethics, the chapter equips future scholars and practitioners to navigate AI responsibly.

Chapter 13: Role of Artificial Intelligence in Public Administration Research Ethics

The use of artificial intelligence (AI) in public administration research provides transformative opportunities and complex ethical challenges. In this chapter, the focus is on the role of AI in public administration research ethics. Despite the prevalent use of the term AI, there is no generally accepted definition of the concept. However, in this chapter, AI was used to mean technology that enables machines to imitate different complex human competencies related to public administration research. The discussion is focused on how AI impacts key ethical principles, specifically, informed consent, autonomy, privacy, confidentiality, equity, bias, transparency, accountability, scientific validity, and public interest. In this chapter a cases study on issues and dilemmas associated with the application of AI in public administration research was included to assists in explaining that ethics in research are indispensable because the pursuit of knowledge should not come at the expense of public trust and equitable governance outcomes.

Chapter 14: Safeguarding Ethics and Academic Integrity in the Age of Generative Artificial Intelligence: A Case Study on Plagiarism Prevention

This chapter examines the evolving challenges to academic integrity posed by the integration of generative artificial intelligence (AI) in higher education. Focusing on a case study of plagiarism prevention at a university, the chapter explores how AI's dual role as both a tool for enhancing research and a potential source of academic dishonesty necessitates a robust ethical framework. It discusses the importance of institutional support, advanced detection software, and continuous education on ethical research practices. By highlighting the collaboration between ethics committees, libraries, and research networks, the chapter underscores the need for universities to rigorously enforce academic integrity policies while adapting to the rapidly changing digital landscape. Ultimately, this chapter provides insights into the balance required between leveraging AI's capabilities and preserving the core principles of academic honesty and originality in scholarly work.

Chapter 15: Future Trends in Research Ethics: Navigating New Ethical Frontiers

The chapter explores the transformative evolution of research ethics in the 21st century, shaped by rapid technological innovation, global collaboration, and shifting societal values. It examines how traditional ethical principles such as respect, beneficence, justice, and nonmaleficence must adapt to address emerging challenges in artificial intelligence, automation, immersive technologies, and big data. Key concerns including algorithmic accountability, informed consent in digital environments, participant vulnerability, data governance, and intellectual property are critically analyzed. The discussion highlights tensions between global harmonization and cultural relativism in ethical governance, while emphasizing the need for anticipatory and adaptive frameworks. Special attention is given to ecological responsibility, participant care, and the growing role of ethics education in ensuring human dignity, scientific integrity, and social responsibility.

Acknowledgment

We would like to recognize the following reviewers and express our sincere appreciation for their contributions. Your expertise and professionalism are greatly valued.

Naseem Akhter

Roger J. Baran

Janice Chretien

Meijing Cui

Evan Diehnelt

Ying Gao

Brandon A. Griffin

Junnatun Naym

Christine Newman

Rami Saleh

Introduction

WHY ETHICAL OVERSIGHT MATTERS

In the name of scientific progress, researchers have infected patients with disease, withheld effective treatments, exposed vulnerable populations to harm, and conducted experiments without the knowledge or consent of those involved. These actions were not the work of isolated individuals operating outside the scientific community. They were carried out by respected researchers, supported by reputable institutions, and often justified as necessary sacrifices for the advancement of knowledge. History therefore offers a sobering lesson: when the pursuit of discovery proceeds without ethical restraint, the enterprise intended to improve human life can instead produce profound injustice and lasting harm.

Modern systems of ethical oversight emerged in response to these failures. Institutional Review Boards (IRBs) were created to ensure that research involving human participants is conducted with respect, transparency, and accountability. By requiring informed consent, assessing risks and benefits, safeguarding vulnerable populations, and providing independent review, IRBs help align scientific ambition with ethical responsibility.

This volume examines how and why these systems of oversight developed, how they function in contemporary research environments, and how they must evolve in response to emerging challenges. Ethical governance is not static. Advances in digital technology, global research collaboration, large-scale data analytics, and artificial intelligence are reshaping the ways knowledge is produced. As research methods evolve, ethical oversight must adapt to address new risks related to privacy, algorithmic decision-making, data ownership, and accountability.

The book is organized into five sections that explore the development and application of research ethics from historical foundations to future challenges.

Section I: Making the Case for IRB — A Historical Overview

Section I examines the historical context that gave rise to modern ethical oversight, focusing on landmark cases that revealed the need for independent review of research involving human participants. These cases illustrate how the absence of ethical safeguards allowed harm to occur and demonstrate why principles such as informed consent and participant protection became central to modern research governance.

Section II: The Role of the Institutional Review Board

Section II focuses on the contemporary role of the Institutional Review Board. Through case studies involving human subject protections, plagiarism, conflicts of interest, and institutional responsibility, this section explores how ethical oversight operates in practice and how ethical failures can still arise even within regulated research environments.

Section III: Special Interest and Discipline-Specific Cases

Section III examines discipline-specific and culturally informed ethical challenges. Cases from sports and exercise science, animal research, accounting, social science, and Islamic perspectives on research ethics illustrate how universal ethical principles must be interpreted within particular methodological, professional, and cultural contexts.

Section IV: The Role of Artificial Intelligence in Ethical Research

The fourth section addresses one of the most significant developments shaping contemporary research: the rise of artificial intelligence and digital research tools. The chapters examine ethical issues related to AI-assisted research design, data analysis, decision-making, confidentiality, anonymity, and academic integrity.

Rather than presenting AI as a rupture from past ethical concerns, the cases demonstrate how emerging technologies amplify longstanding ethical challenges related to transparency, accountability, and oversight. AI becomes an ethical force multiplier—capable of strengthening responsible research practices while simultaneously magnifying risks.

Section V: Looking Forward — Future Directions in Research Ethics

Section IV turns to one of the most significant developments affecting modern research: the increasing role of artificial intelligence and digital technologies. These chapters examine ethical questions related to AI-assisted research design, automated data analysis, confidentiality, academic integrity, and algorithmic decision-making.

Section V: Looking Forward — Future Directions in Research Ethics

Section V looks forward to the future of research ethics. As scholarship becomes increasingly global, interdisciplinary, and technologically mediated, ethical governance must evolve accordingly. Addressing emerging ethical challenges will require not only regulatory frameworks but also a renewed commitment to ethical competence among researchers and institutions alike.

Taken together, the chapters in this volume advance a central argument: ethical oversight is not a barrier to innovation but a necessary foundation for trustworthy research. By protecting participants, promoting transparency, and encouraging ethical reflection, IRBs and related oversight mechanisms help ensure that the pursuit of knowledge remains aligned with the values of responsibility, justice, and respect for human dignity.

Chapter 1

Unethical Research Cases and the Lessons That Shaped Modern Research Ethics Making the Case for Institutional Review Boards:

A Historical Overview

Mette L. Baran
http://orcid.org/0000-0002-7262-9698
Troy University, USA

EXECUTIVE SUMMARY

Institutional Review Boards (IRBs) did not emerge from abstract ethical theory or administrative preference. They developed in direct response to repeated and well-documented failures to protect human research participants (Faden & Beauchamp, 1986; Katz, 1972). Throughout the twentieth century, researchers across multiple national contexts conducted studies that exposed individuals to harm, deception, and exploitation—often selecting participants precisely because they were marginalized, institutionalized, or lacked the power to refuse (Brandt, 1978). This chapter examines three landmark cases from Germany, the United States, and Sweden. Although differing in political and cultural context, each case demonstrates similar ethical breakdowns: absence of informed consent, exploitation of vulnerable populations, disproportionate risk without benefit, and lack of independent accountability (Belmont Report, 1979).

DOI: 10.4018/979-8-3693-8302-5.ch001

UNETHICAL RESEARCH CASES AND THE LESSONS THAT SHAPED MODERN RESEARCH ETHICS

Making the Case for Institutional Review Boards: A Historical Overview

Institutional Review Boards (IRBs) did not emerge from abstract ethical theory or administrative preference. They developed in direct response to repeated and well-documented failures to protect human research participants (Faden & Beauchamp, 1986; Katz, 1972). Throughout the twentieth century, researchers across multiple national contexts conducted studies that exposed individuals to harm, deception, and exploitation—often selecting participants precisely because they were marginalized, institutionalized, or lacked the power to refuse (Brandt, 1978). These cases revealed a consistent lesson: when scientific ambition proceeds without independent oversight, ethical principles are easily overridden.

Professional norms and individual goodwill proved insufficient to safeguard participants. In the absence of enforceable standards, ethical decision-making was shaped by institutional pressure, nationalism, public health urgency, and scientific prestige (Faden & Beauchamp, 1986). IRBs emerged as a corrective to these systemic failures, translating ethical ideals into formalized, enforceable mechanisms of review (National Research Act, 1974). Understanding this history is essential not only for explaining why IRBs exist, but for recognizing the human cost of research conducted without them.

This chapter examines three landmark cases from Germany, the United States, and Sweden. Although differing in political and cultural context, each case demonstrates similar ethical breakdowns: absence of informed consent, exploitation of vulnerable populations, disproportionate risk without benefit, and lack of independent accountability (Belmont Report, 1979). Together, they clarify why modern research ethics requires structural oversight rather than reliance on individual virtue.

Germany: Nazi Human Experiments (1930s–1940s)

During World War II, Nazi physicians and scientists conducted brutal experiments on prisoners in concentration camps, including Jews, Roma, people with disabilities, and political detainees (Nuremberg Code, 1947). These studies involved hypothermia and freezing experiments, high-altitude simulations, exposure to infectious diseases, forced sterilization, and invasive procedures performed without anesthesia (Katz, 1972). Participation was entirely coerced, and many subjects were permanently injured or killed.

The ethical violations were total. There was no consent, no concern for participant welfare, and no distinction between research and punishment. Human beings were treated as expendable objects, justified by an ideology that denied their moral worth (Faden & Beauchamp, 1986). These experiments represent the most extreme consequences of divorcing science from ethics and permitting state power to override individual rights.

The exposure of these atrocities after the war led directly to the **Nuremberg Code (1947)**, which established voluntary informed consent, risk minimization, and the right to withdraw as essential requirements for ethical research (Nuremberg Code, 1947). Although initially dismissed by some as applicable only to wartime crimes, the Code introduced the foundational principle that ethical research requires explicit standards rather than implicit trust—an idea central to later IRB systems (Faden & Beauchamp, 1986).

United States: The Tuskegee Syphilis Study (1932–1972)

The Tuskegee Syphilis Study, conducted by the U.S. Public Health Service, followed hundreds of Black men with syphilis in rural Alabama in order to observe the natural progression of the disease (Brandt, 1978). Participants were misled about the nature of the study and were denied effective treatment even after penicillin became the standard cure (Brandt, 1978). Many experienced severe complications, and some died as a direct consequence of the study design.

Tuskegee constitutes a profound ethical failure rooted in deception, racism, and exploitation. Participants were selected because of their social vulnerability and limited access to healthcare (Brandt, 1978). The study violated what would later become the core ethical principles of research: respect for persons was denied through deception and lack of informed consent; beneficence was ignored through the withholding of treatment; and justice was violated through the systematic targeting of a marginalized population (Belmont Report, 1979).

Public exposure of the study in the early 1970s resulted in widespread outrage and directly led to the **National Research Act of 1974**, which mandated the creation of Institutional Review Boards (National Research Act, 1974). The Act also established the National Commission, whose work culminated in the **Belmont Report (1979)**. Belmont articulated the principles of respect for persons, beneficence, and justice, providing the conceptual framework that continues to guide IRB review (Belmont Report, 1979).

Sweden: The Vipeholm Dental Experiments (1945–1955)

Between 1945 and 1955, researchers in Sweden conducted the Vipeholm dental experiments at a state institution for individuals with intellectual and developmental disabilities (Svensson & Broberg, 2013). Residents were deliberately fed extremely high-sugar diets to study the development of dental caries. Participants were unable to provide informed consent, and guardians were often inadequately informed or not consulted (Svensson & Broberg, 2013).

Subjects were exposed for years to sticky candies and frequent sugar intake specifically designed to maximize tooth decay. Even after severe dental damage became evident, the study continued without therapeutic intervention (Svensson & Broberg, 2013). The research was supported by government agencies and academic institutions, making it an example of officially sanctioned ethical failure rather than rogue misconduct.

The Vipeholm experiments challenge the assumption that unethical research occurs only in authoritarian or overtly racist systems. Conducted within a welfare-state context and justified by public health goals, the study demonstrates how utilitarian reasoning and institutional convenience can override ethical judgment (Svensson & Broberg, 2013). Its legacy contributed to the development of European Research Ethics Committees (RECs), which function analogously to IRBs, and it remains a central case in international ethics education (Declaration of Helsinki, 1964/2013). While European Research Ethics Committees (RECs) differ administratively from U.S. IRBs, they serve the same core function: independent ethical oversight prior to and during human subjects research.

Theoretical Framework: Principlism and Systems Ethics

This chapter is grounded in an integrated theoretical framework that combines **principlism** with **systems ethics** to explain both the normative foundations of research ethics and the institutional necessity of ethical oversight. Principlism articulates the moral standards governing human subjects research, while systems ethics accounts for the structural conditions under which ethical failures recur despite professional norms and individual good intentions.

Principlism, as articulated in the Belmont Report (1979), rests on three core principles: **respect for persons**, **beneficence**, and **justice**. Respect for persons requires recognition of individual autonomy through informed consent and mandates additional protections for those with diminished capacity. Beneficence obligates researchers to minimize harm and maximize potential benefit through systematic risk–benefit analysis. Justice requires the fair selection of research participants and equitable distribution of research burdens and benefits. Each historical case examined in this chapter—Nazi human experimentation, the Tuskegee Syphilis Study, and the Vipeholm dental experiments—represents a profound violation of all three principles.

While principlism provides essential normative guidance, it does not fully explain how ethical violations persist within established scientific and medical institutions. As Faden and Beauchamp (1986) note, ethical principles are vulnerable when their application depends solely on investigator discretion, professional self-regulation, or institutional goodwill. This limitation highlights the need for a complementary framework capable of addressing organizational dynamics and power asymmetries.

Systems ethics shifts the focus from individual moral failure to the institutional structures that shape ethical decision-making. From this perspective, unethical research emerges when systems reward scientific productivity, national interest, or public health outcomes without independent mechanisms for ethical accountability. Across the cases examined, researchers operated within environments that normalized harm, minimized participant autonomy, and systematically exploited vulnerable populations. These outcomes were not anomalies but predictable consequences of unchecked authority and structural incentives. "Historical cases demonstrate how vulnerable populations can be exploited when ethical oversight is absent, a concern extensively analyzed in the ethics literature" (Levine, 1988, p. 88).

Institutional Review Boards represent a systems-level response to these recurring failures. By requiring independent review, collective deliberation, documented risk–benefit assessment, informed consent procedures, and continuing oversight, IRBs redistribute ethical responsibility from individual investigators to institutional processes. This structural intervention mitigates conflicts of interest and embeds ethical evaluation into routine research practice.

Together, principlism and systems ethics offer a comprehensive framework for understanding both the moral failures of past research and the institutional mechanisms developed in response. Principlism defines the ethical standards by which research must be judged, while systems ethics explains why those standards require formalized, enforceable oversight. "Ethical principles are operationalized not only through principlism but also through structured protections such as IRBs, which enforce risk–benefit assessment, informed consent, and equitable participant selection" (Emanuel, Wendler, & Grady, 2000, pp. 2701–2703).

Within this integrated framework, IRBs function as the operational bridge between ethical theory and research practice, ensuring that ethical principles are applied consistently even under conditions of scientific, political, or social pressure.

Why IRBs Became Necessary

Across these cases, common patterns emerge. Researchers approved their own studies, institutions prioritized scientific advancement and national interest, and no independent bodies existed to evaluate risk or protect participants (Katz, 1972). Informed consent was not culturally or legally entrenched, and vulnerable populations were routinely selected because they lacked the capacity to refuse (Faden & Beauchamp, 1986).

Ethical frameworks developed reactively in response to public exposure of abuse. The **Nuremberg Code** followed Nazi atrocities (Nuremberg Code, 1947); the **Declaration of Helsinki** responded to medical abuses (World Medical Association, 2013); and the **Belmont Report** emerged from Tuskegee (Belmont Report, 1979). These developments demonstrate that ethical principles without institutional enforcement are insufficient to prevent harm. The experiments represent the most extreme consequences of divorcing scientific inquiry from ethical constraint and independent oversight.

The Transformative Role of Institutional Review Boards

IRBs fundamentally altered the conduct of research involving human subjects by introducing independent review, systematic risk–benefit analysis, mandatory informed consent, special protections for vulnerable populations, and continuing oversight (Office for Human Research Protections, 2018). Ethical responsibility shifted from individual investigators to institutional structures designed to identify and mitigate conflicts of interest.

IRBs are not obstacles to scientific progress; they are safeguards shaped by historical failure. They exist to ensure that research advances without repeating the ethical violations that necessitated their creation (Belmont Report, 1979).

CONCLUSION: ENDURING LESSONS FOR CONTEMPORARY RESEARCH

The historical cases examined in this chapter demonstrate that unethical research is rarely accidental. Rather, it emerges when power goes unchecked, participants are reduced to means rather than ends, and ethical oversight is absent (Faden & Beauchamp, 1986). Across disparate political, cultural, and institutional contexts, the same structural failures recurred: vulnerable populations were exploited, risks were externalized, and scientific goals were prioritized over human dignity.

Modern systems of ethical review exist precisely because these failures occurred. Institutional Review Boards were developed not as administrative formalities, but as structural safeguards designed to prevent the normalization of harm and to counteract institutional pressures that historically undermined ethical judgment. By embedding independent review, risk–benefit assessment, informed consent, and continuing oversight into the research process, IRBs transform ethical principles into enforceable practice.

As research expands through globalization, emerging technologies, and data-intensive methods, the lessons of Nuremberg, Tuskegee, and Vipeholm remain critically relevant (World Medical Association, 2013). Remembering these cases is not merely an exercise in historical reflection; it is an essential component of ethical vigilance. The true measure of scientific progress lies not only in what is discovered, but in the manner in which knowledge is pursued.

REFERENCES

Brandt, A. M. (1978). Racism and research: The case of the Tuskegee Syphilis Study. *The Hastings Center Report*, *8*(6), 21–29. DOI: 10.2307/3561468 PMID: 721302

Code, N. (1947). Permissible medical experiments. In *Trials of war criminals before the Nuremberg Military Tribunals under Control Council Law No. 10* (Vol. 2, pp. 181–182). U.S. Government Printing Office.

Declaration of Helsinki. (2013). *Ethical principles for medical research involving human subjects*. World Medical Association. (Original work published 1964)

Emanuel, E. J., Wendler, D., & Grady, C. (2000). What makes clinical research ethical? *Journal of the American Medical Association*, *283*(20), 2701–2711. DOI: 10.1001/jama.283.20.2701 PMID: 10819955

Faden, R. R., & Beauchamp, T. L. (1986). *A history and theory of informed consent*. Oxford University Press.

Katz, J. (1972). *Experimentation with human beings: The authority of the investigator, subject, professions, and state*. Russell Sage Foundation. DOI: 10.2307/1227842

Levine, R. J. (1988). *Ethics and regulation of clinical research* (2nd ed.). Urban & Schwarzenberg.

National Research Act, Pub. L. No. 93-348, 88 Stat. 342 (1974).

Office for Human Research Protections. (2018). *Federal policy for the protection of human subjects ("Common Rule")*. U.S. Department of Health and Human Services.

Report, B. (1979). *Ethical principles and guidelines for the protection of human subjects of research. National Commission for the Protection of Human Subjects of Biomedical and Behavioral Research*. U.S. Department of Health, Education, and Welfare.

Svensson, R., & Broberg, G. (2013). Ethics and public health: The Vipeholm experiments. In *The history of medical ethics* (pp. 303–322). Springer.

World Medical Association. (2013). Declaration of Helsinki: Ethical principles for medical research involving human subjects. *Journal of the American Medical Association, 310*(20), 2191–2194. DOI: 10.1001/jama.2013.281053 PMID: 24141714

Chapter 2
History of Ethics in Research

Meenu Sharma
http://orcid.org/0000-0003-0493-556X
The Assam Royal Global University, India

Supriya David
http://orcid.org/0009-0009-6004-878X
RTM Nagpur University, India

EXECUTIVE SUMMARY

The ideas of "êthikê theôria" were likely originally discussed by Aristotle, who also studied and provided standards for evaluating human behavior. Formal rules were developed in response to a history of unethical human testing, including the Nuremberg Trials and the Tuskegee Syphilis Study. Informed consent, beneficence, and justice were among the concepts created by these rules, which included the Nuremberg Code, the Declaration of Helsinki, and the Belmont Report, to safeguard study subjects. In order to assure ethical conduct in research, institutional review boards are now essential. The idea of using humans in studies was starting to gain traction by the turn of the 20th century, provided that large-scale investigations were initially carried out on animals. Humans and animals were employed in studies much more frequently as the understanding of bacteria advanced and pharmaceutical corporations grew.

INTRODUCTION

Ethical principles influence researchers' behavior. It has an impact on both the process of scientific discovery and its applications and consequences. Instances

DOI: 10.4018/979-8-3693-8302-5.ch002

of ethical considerations in research encompass human rights, data management, resource allocation, treatment of animal and human subjects, integrity, societal responsibility, honesty, and publication of research conclusions. Primary objective of ethics in scientific research is to keep welfare of society and individuals from being compromised in the sake of knowledge growth.It can thrive. Research ethics are heavily influenced by prior experiences, when a lack of ethical concern resulted in undesirable implications. (Miteu, 2024).

The history of research ethics is a history of events, mostly tragedies, inspiring the articulation of principles and guidelines in the domain of research ethics. Since Nuremberg Code post World War II, the ethics in research has been gradually forming into a multileveled structure, at least until we were blessed with the declaration of Helsinski and policy and principles introduced and reviewed in the Belmont report, where the values of autonomy, beneficence, justice, and respect for persons take each its own central stage (IV, 2006). A retrospective glance at this evolution illustrates how the past serves not only as a portent warning of what must be circumvented evermore, but also reminds us why ethical vigilance as a staple of modern research practices is so vitally important.

From Risk-Benefit to Precaution — A Shift in Research Ethics Paradigm Over the Decades Concerning Human Participation in Science

The field of research ethics emerged (and evolved) into a distinct discipline, in the worst of ways, through shocking examples of unethical experimentation in the 20th century, including well-known texts on Nazi medicine and the Tuskegee Syphilis Study. The Nuremberg code introduced the notion of voluntary informed consent, and the Declaration of Helsinki provided a common ground for biomedicine on an international level. A report from the National Commission for the Protection of Human Subjects of Biomedical and Behavioral Research, known as the Belmont Report, and the US Common Rule enshrined these ethical principles, such as respect for persons, beneficence, and justice. As research grew more challenging, research ethics evolved to meet difficult issues such as genetic research, data security, biobanking, and global health research (Coppola L, 2019). General principles from international guidelines and human rights-based frameworks (eg.Recognizing this trajectory is critical to the ethical standards that inform present-day research practices, as well as to avoiding irresponsible innovation in the future.

Table 1. Historical Development of Research Ethics

Historical Development of Research Ethics			
Year/Period	**Event / Document**	**Context & Key Issues**	**Ethical Principles Introduced / Reinforced**
1900	Prussian Directive on Human Experimentation	Response to unethical medical experiments in Germany	Voluntary consent, physician responsibility
1932–1972	Tuskegee Syphilis Study (USA)	Long-term study on untreated syphilis without patient knowledge	Led to demand for informed consent & justice in research
1947	**Nuremberg Code**	After Nazi doctors' trials for inhumane experiments in WWII	Voluntary informed consent, beneficence, right to withdraw
1964	**Declaration of Helsinki** (World Medical Association)	Set international ethical guidelines for biomedical research	Informed consent, independent review, scientific validity
1974	National Research Act (USA)	Response to Tuskegee scandal	Established IRBs (Institutional Review Boards)
1979	**Belmont Report** (USA)	Ethical principles for human subject research	Respect for persons, beneficence, justice
1991	**Common Rule** (USA)	Federal policy governing human research	Requirements for IRBs, informed consent, vulnerable populations
2000s–Present	CIOMS Guidelines,	Global harmonization	Protection of
	ICH-GCP, UNESCO Declarations	of research ethics	vulnerable groups, cultural sensitivity, data protection, integrity in research

Source: Compiled from data from National Institute of Environmental Health Sciences.

REVIEW OF LITERATURE

Historical occurrences, especially Tuskegee Syphilis Study and Nuremberg Code, have had a significant influence on development of research ethics. The formation of fundamental criteria that put participant wellbeing and informed consent first resulted from these incidents, which brought attention to the need for Research ethicsinvolving humans.

Nuremberg Code has been developed in response to horrors of World War II, established guidelines for moral medical research with a focus on informed agreement and the need to prevent needless suffering. (Udo Schuklenk, 2013).

It established the foundation for upcoming ethical frameworks and signaled a significant change in the recognition of research participants' rights.(Muto, 2022).

The Study of Tuskegee Syphilis: This study, which was carried out between 1932 and 1972, involved the unethical treatment of African American men who have been denied syphilis treatment and misinformed about their health status (megítélésére, 2008). The Nuremberg physicians' trials had a significant impact and marked the start of a new era in the ethical evaluation of human experimentation. Belmont Report, which defined ethical standards such as beneficence, justice, and respect for humans in research, was created as a result of the public outcry after its disclosure. (Harrington, 2020).

Even though these historical occurrences sparked important developments in research ethics, moral conundrums still arise in modern research, requiring constant attention to detail and modification of ethical guidelines to adequately safeguard human beings.

The ethics of research have been greatly influenced by the works of Immanuel Kant and John Rawls, who emphasized the value of moral thinking and fairness. Kant's ethical system, which is based on the ideas of human dignity and universal moral rules, offers a fundamental viewpoint that shapes current ethical discussions. Building on Kantian ideas, Rawls presents the concept of "justice as fairness," promoting a society in which the well-being of the most disadvantaged is given priority. When combined, their ideologies highlight importance of ethical considerations to research procedures.

Most significant and most distinctive products of the European Enlightenment is Immanuel Kant's (1724–1804) ethical theory, which continues to have an impact on moral philosophers as well as the larger moral and political culture. One of the best and most defining products of the European Enlightenment is Immanuel Kant's (1724–1804) ethical theory. It continues to have an impact on moral philosophers as well as the larger moral and political culture. The core tenet of Kantian ethics is that since we can see them as laws established by our logical ability to control our lives, the same fundamental moral principles are objectively binding on all of us. The idea that every human being has equal dignity as a member of the moral community, forming an ideal "realm of ends," is equally significant to the Kantian vision. Ethical principles are thought of as those that would compel people to join forces in the common pursuit of a system of ends that upholds the dignity (or absolute worth) of every individual and equally provides for their happiness.

John Rawls is a representative of moderate liberalism and one of the best thinkers in moral philosophy. Justice as fairness, or more precisely, the creation of a society in which everyone's social security is guaranteed, with an emphasis on treatment of the less fortunate (social principle), is the fundamental idea of his 1971 book A Theory of Justice. He contrasts the utilitarian (teleological) and deontological morals, and he relied heavily on Kant and the social contract tradition. Since he incorporates several Judeo-Christian virtues into his theory, theologians have been debating it a lot. These encompass the initial position, signifying an ideal state (the Garden of Eden), the social principle (the poor Lazarus), the focus on general agreement (Babylon or Pentecost), fairness in choosing the principles of a just society, and the inherent dignity of all humans. This suggests, on the one hand, a particular quest within liberalism itself (almost fearful of itself). But liberalism's methodology and underlying assumptions preclude the recognition of specific virtues and their justifications. The absence of metaphysics and existence of these virtues in the secularized form of Rawls' liberal theory suggest some ambiguity or perhaps contradiction.

The Ethical Framework of Kant: According to Kant, morality and respect for human dignity are the sources of moral principles that apply to everyone (Bouzanis, 2023). According to Bouzanis (2023), he imagines a "realm of ends" in which people work together to achieve common moral objectives while valuing equality and respect.

The Impact of Rawls on Ethical Theory: By incorporating Kantian ethics into his framework, Rawls presents a conception of justice that guarantees social stability for everyone, especially the underprivileged (Jamnik A., 2021). Ethical discourse is being revitalized by his work, which questions positivism and linguistic analysis while regaining the importance of Kant and other historical moral thinkers (Nussbaum, 1999).Although Kant and Rawls offer strong foundations for ethical inquiry, some opponents contend that these theories could ignore the difficulties of practical applications, indicating the necessity for more useful methods of making moral decisions.

Particularly in the wake of past abuses in human experimentation, the idea of informed consent has been crucial in defining ethical norms in research. Informed consent, which dates back to the Nuremberg Code of 1947, has developed via several legal and ethical frameworks, highlighting the importance of participant autonomy and understanding when participating in research. The rights and welfare of research volunteers were significantly impacted by this evolution, which has ensured they are well-informed and capable of making their own decisions about participating voluntarily.After WWII atrocities, the Nuremberg Code (1947) laid the groundwork for informed consent.The Declaration of Helsinki (1964) improved moral standards for human subjects' medical research.Informed consent was emphasized as a process rather than a one-time occurrence in the Belmont Report (1979), which introduced the fundamental ethical concepts of beneficence, fairness, and respect for humans. (Nagai H, 2022).

According to O'Sullivan et al. (2020), informed consent increases participants' autonomy in research settings by enabling them to make knowledgeable decisions. (O'Sullivan, 2020). According to Holm (2006), informed consent is now required by law to guarantee that participants are competent to give their consent free from coercion and are fully informed. (Holm, 2015). Problems still exist, including making sure that different populations understand one another and modifying consent procedures in rapidly changing study fields. Even though informed consent has significantly increased participant protection, it is still difficult to guarantee that everyone, especially members of vulnerable populations, completely understands the consequences of their involvement.This emphasizes how important it is to have ethical awareness in research procedures.

ETHICS AS A FOUNDATION OF HUMAN CIVILIZATION

A civilization without ethics cannot be called a civilization. This is one reason that human beings have organized themselves with codes of conduct, going back to the earliest law codes such as Hammurabi's, the teachings of philosophers such as Confucius, Aristotle, and Kant, and at every level of government, indicating that societies are not prepared to coexist without ethical norms. Ethics provides the framework within which trust, cooperation, and justice can flourish. Without ethical principles, human interaction would collapse into conflict, exploitation, and chaos. At its core, ethics functions as a 'moral compass'. It regulates desires, curbs destructive impulses, and encourages the cultivation of virtues (Soniya Billore, 2023). For instance, Aristotle emphasized that ethics is not merely about rules but about developing character—becoming a person who embodies virtues like courage, honesty, and justice. Similarly, Confucian ethics focused on harmonious relationships and the importance of respect, loyalty, and benevolence. These insights reinforce that ethics is not merely an external thing, but an internal thing, according to who we ought to be. Therefore, any global conclusion about ethics needs to recognize ethics as the building block of human civilization. Remove ethics and laws have no legitimacy, politics tumbles into tyranny, and every interpersonal relationship regresses into subterfuge and treason. Ethics is the invisible architecture that holds human life together.

Theories and Traditions: Multiple Pathways Toward the Good

Ethics is not monolithic. Across cultures and centuries, multiple traditions and theories have offered different answers to the question: *What should I do?*

Deontological ethics, represented by Immanuel Kant, insists on morality and respecting the inherent dignity of every person.Derived from Aristotle, virtue ethics emphasizes character and moral development. It demands not only to know what is the right action but also what type of person you should be (Kranak, 2019).

More recently articulated, care ethics centers on empathy, compassion, and the responsibility of interpersonal relations, in contrast to the abstraction of rule- or outcome-based systems.

The plurality of ethical theories shows that morality is multi-faceted. The insights into human moral experience gained through the four lenses of duty, consequences, virtue, and care are indispensable to effective moral analysis. No one framework fully contains the richness of ethics, and in practice, moral reasoning often draws on consonance from various traditions.

Finally, as far as ethics are concerned, one must embrace this plurality rather than boil ethics down to a single fixed formula. In a pluralistic world (and one hopes, it is pluralistic), ethics is designed dialectically, where multiple theories talk amongst themselves, as they must in order to face the diversity of experience and social moment, but also the commonalities in universality, justice, equality, fairness, dignity, etc.

HISTORICAL TRAJECTORY: PIVOTAL EVENTS AND TURNING POINTS IN RESEARCH ETHICS

For ages, human subjects have been used in medical research. The experimental method's widespread use in science and medicine during the nineteenth century led to important advancements in human subjects research. However, as using animals for research became commonplace, some scientists—mostly doctors—started to wonder if research involving human subjects was vital. There were several ethical discussions over the suitability of this kind of research among the scientific community.

Long before formalized codes of research ethics existed, physicians were guided by professional oaths and traditions, most notably the Hippocratic Oath. Originating in ancient Greece around the 5th century BCE, the Hippocratic tradition emphasized principles such as beneficence ("do good") and nonmaleficence ("not harm"). These were intended to guide clinical practice, ensuring that physicians acted in the best interests of patients. However, these norms were largely patient-centered and clinical in focus, rather than directed toward research.

Experiment and Inventions Involving Ethics in Research

1. Akbar the Great, the Mughal emperor, experimented in 1500 to find out if children raised in a silent atmosphere would pick up language. He directed silent nurses who used sign language to communicate with one another to nurture twelve infants. Upon his return, he found that the twelve kids had learned sign

language instead of an audible language. Other monarchs have conducted similar experiments, many of which were aimed at identifying the "original" language.

2. In contrast to the Catholic Church's belief that the Earth is fixed and the Sun revolves around it, scientist and astronomer Galileo Galilei (1564–1642) made the case for a heliocentric interpretation of the solar system in his Discourse on Two New Universe Systems.
3. Galileo went before the Catholic Church Inquisition in 1633. He was placed under home arrest for the rest of his life and told to change his opinions. The Church forbade his book. Pope John Paul II issued a solemn apology for the way Rome treated Galileo in 1992, 359 years after Galileo was arrested.
4. Medical experimentation in pre-modern Europe often walked the line between therapeutic and inquisition. New interventions were tested by physicians directly on patients, with little separation between treatment and knowledge production objectives. There was little room for ethical contemplation, and informed consent did not exist. Early experiments (for example, the 17th-century studies on blood circulation by William Harvey or the late-18th-century trials of a smallpox vaccination by Edward Jenner) existed without ethical oversight as we would recognize it today, but often appealed to persuasion or trust. Edward Jenner (1749–1823), an English physician, is known as the "father" of immunology and tested a smallpox vaccination on James Phipps, age 8, in 1796. Dairymaids were not susceptible to smallpox, as Jenner and others had noted. Jenner speculated that their exposure to cowpox was the cause of this. Jenner exposed Phipps to smallpox after inoculating him with cowpox to verify this theory. Smallpox did not infect Phipps. Ten to twenty percent of the world's population perished from smallpox annually during Jenner's time.
5. To ascertain an electron's charge, Robert Millikan performed oil drop experiments in 1909. Millikan was awarded the Nobel Prize in 1923 for such research. Millikan kept 33 of the 149 oil drop observations he categorized as "fair" and "poor," and he excluded his pupil, Harvey Fletcher, from the author of the publication describing these findings. Fletcher significantly contributed to the experimental design by recommending that Millikan utilize oil droplets rather than water droplets, as noted by journalists and historians examining Millikan's notebooks.
6. When museum curator Charles Dawson discovered skull in a gravel bed close to Surrey, UK, in 1912, it was assumed to be fossilized remains creature that was halfway between humans and apes (often referred to as "a missing link").
7. For decades, the skull was the subject of discussion, and many scientists thought it was a fake. These concerns were supported by chemical investigations conducted in 1953, which revealed that the skull was a combination of an orangutan

jaw and a human skull that had been chemically altered to give the appearance of age.

8. The United States undertook the Manhattan Project from 1942 to 1945 to create an atomic weapon. The Project was headed by General Leslie Groves, and scientific project has been managed by physicist Robert Oppenheimer. Other notable scientists who contributed to the Project included Edward Teller, Richard Feynman, Enrico Fermi, and Hans Beth.The Project's scientists were exposed to exceptionally high radiation doses from handling uranium and plutonium during a period when the harmful health consequences of radiation were not generally understood.
9. The US Department of Energy funded covert studies on the effects of radiation on humans from 1944 1980s. The subjects have not been informed that they were taking part in the studies. Military people, pregnant women, and cancer patients were all subjects of experiments. A total of several hundred radiation discharges on human volunteers were part of these experiments. They were frequently carried out to test safety gear or weapons. The Human Radiation Experiments Advisory Committee, established in 1994, found that the government's failure to ventilate the mines adequately contributed to the lung cancer deaths of at least several hundred uranium miners.
10. The historic incident also forced the people to think about ethics in research, as the Cold War started when the Soviet Union tested an atomic weapon in 1949.
11. Francis Crick & James Watson proposed the structure of DNA in 1953, which they were awarded Nobel Prize in 1962 alongside Maurice Wilkins. The idea of Watson and Crick was confirmed by Rosalind Franklin's (1920–1958) X-ray direct image of DNA (Photo 51). Wilkins presented Photo 51 to Watson and Crick without Franklin's consent. The 1953 edition of Nature included the papers by Watson and Crick, as well as those by Wilkins and Franklin. Wilkins and Franklin were not writers in Watson-Crick. Franklin was not given Nobel Prize since she died at the age of 37 in 1958 cause to ovarian cancer.
12. Through his "electric shock" tests, Stanley Milgram (1933–1984) showed that when people obey an authoritative figure, they are willing to perform actions that they believe to be immoral. A student, an instructor, and a researcher participated in the studies, which came in several versions. Electrodes were attached to the student. Even when the machine's dial was set to "hazardous" levels, and the students were screaming in agony and pleading for the experiments to end, teachers were still willing to carry out this task. No shocks were administered. The trials were designed to assess the subjects' obedience to an authority figure. These tests have since been duplicated by other researchers, who have found comparable outcomes.

13. The first in vitro fertilization-conceived child in history, Louise Brown, was born alive and well in the United Kingdom in 1978.

Books and articles publications involving ethics in research

1. In his book The Novum Organon, Francis Bacon (1561–1626), a philosopher and scientist, makes the case that scientific knowledge should be founded on trial and observation rather than custom and authority and that it should advance humankind.
2. The Origin of Species, published in 1856 by Charles Darwin and Alfred Wallace, puts forth a theory of natural selection-based evolution for living organisms. The book causes much controversy since it suggests that humans are descended from apes. Darwin gathered the majority of data for theory. Because he anticipated fierce opposition to his ideas and wanted to be sure he could support them with arguments and proof, he waited more than 20 years before publishing them.
3. The father of cybernetics, Norbert Wiener, wrote an article titled "A Scientist Rebels" in the Atlantic Monthly in 1947, stating that he would not perform research for the military.
4. To achieve Bacon's scientific objectives, the first scientific organization, Royal Society of London for Advancement of Natural Knowledge, was established in 1962.The Royal Society of London Philosophical Transactions, oldest scientific publication in existence, was published by the Royal Society in 1665.
5. In her 1961 book Silent Spring, Rachel Carson (1907–1964) advises against the harmful impacts of pesticides, especially DDT, on public health and the environment. Her book initiates the ecology movement.
6. In Asilomar, California, leaders in the nascent field of recombinant DNA research gathered to talk about biosafety concerns and create biosafety procedures. In 1974, Paul Berg as well as other prominent researchers, called for voluntary suspension of recombinant DNA research till these issues were fully recognized and appropriate safety protocols were put in place.An article outlining some biosafety guidelines and suggestions for recombinant DNA studies was published in 1975 by Berg and associates.
7. Peter Singer offers a philosophical defense of animal rights movement in his 1975 publication Animal Liberation. According to Singer, it is wrong to use animals in many socially acceptable ways.
8. In his book Sociobiology, Wilson (1929–2021) rekindles the age-old "nature vs. nurture" controversy. In his work, he suggests biological and evolutionary theories to explain human culture and behavior.

Medical Experiment involving ethics in research

1. Mary Rafferty, 30, was a mentally challenged patient being treated by Roberts Bartholow (1831–1904) in 1874. She had a 2-inch hole in her head from malignant ulcer. In order to investigate how electrical stimulation affected her brain, he placed electrodes into the hole. Rafferty went through convulsions, seizures, discomfort, and distress before going into a coma and passing away a few days later.
2. Two days after Joseph Meister, then nine years old, had been mauled by an infected dog, Louis Pasteur tested a rabies vaccine on him in 1885. Pasteur was able to heal hundreds of patients by 1886 as a result of the vaccine he used. The vaccination was examined on dogs, not humans. Given that he was not a doctor, Pasteur might have been sued for vaccinating people.
3. In 1897, Giuseppe Sanarelli forced five patients to get injections of microbes against their will to examine his theory that germs cause yellow fever.
4. Three patients passed away, and all of them experienced yellow fever symptoms. Walter Reed, however, subsequently disproved the theory. Numerous medical professionals harshly condemned Sanarelli's experiments as unethical.
5. By 1900, the U.S. Early in the 20th century, medical investigations in Cuba by Army physician Walter Reed (1851–1902) revealed that Aedes aegypti mosquitoes spread yellow fever. In the Caribbean and Central America, yellow fever had been a serious public health issue that killed thousands of people annually and endangered US military operations and trade. When the participant had never had yellow fever, Reed asked them to either have a blood injection from a yellow fever patient or let mosquitoes that had fed on yellow fever patients bite them. The participants attested to their understanding of the experiment's hazards, including the potential for death, by signing informed consent forms, which are thought to be the first known use of this documentation. Spanish was used to translate the forms. In addition to $100 in gold, participants received free medical treatment for those who got yellow fever.
6. Additionally, $100 has been given to families of participants who passed away. James Carroll and Jesse Lazear, two of Reed's associates, offered their services for the Project. Lazear passed away after they both contracted yellow fever. Because of his advanced age (the sickness was more dangerous for individuals over 40, including Reed), Carroll talked Reed out of volunteering for the experiment. Yellow fever claimed six lives.
7. Unit 731's Japanese scientists conducted immoral trials on hundreds of Chinese prisoners of war between 1932-1945. These experiments included vivisections, vaccinations, wound-healing and surgical research, and tests using biological and chemical weapons. In return for the scientists' data from chemical and bi-

ological weapons research, the US government agreed not to bring war crimes charges against them.

8. German scientists carried out unethical research on captives in concentration camps between 1939 and 1945. These tests included wound-healing and surgical procedures, as well as exposure to ionizing radiation, low air pressure, electricity, freezing temperatures, and infectious disorders. In Nuremberg Trials, German scientists were charged with war crimes by the Allies. The scientists were prosecuted based on the Nuremberg Code.
9. The CIA launches a clandestine program called MKUltra in the 1950s and 1960s to create mind control methods. Unknowingly, victims were tortured psychologically by MKUltra researchers using electric shocks and the hallucinogenic chemical LSD. Additionally, Nazi and Japanese physicians who had conducted immoral tests on living humans were hired as part of the scheme. The public first learned about the MKUltra program during hearings given by Congress between 1975 and 1977.
10. Between 1956 and 1980, Joan Giles, Saul Krugman, and other researchers performed tests on mentally challenged children with hepatitis at Willowbrook State School.To create a vaccine against a more severe strain of hepatitis, which was endemic at Willowbrook, they purposefully infected individuals with a mild version of the illness. Children were exposed to hepatitis by injection as well as by being forced to consume chocolate milk laced with the excrement of hepatitis patients. In 1965, Robert F. Kennedy referred to the institution as a "snake trap" in a speech to Congress.

Legal and political decisions involving ethics in research

1. In McDonald v. Massachusetts General Hospital, decided between 1876 and 1957, the Massachusetts Supreme Court held that a nonprofit hospital is not responsible for the conduct of its staff. Because courts upheld its logic, hospitals were protected from liability for medical experiments performed by their employees, making this a significant case for human research ethics. Mary Schloendorff may sue her doctors for performing surgery against her will, but not the hospital, according to the state's Court of Appellate's decision in Schloendorff v. Society of New York Hospital (1914).Such a legal concept, called the Schloendorff rule, was in force until it was declared unlawful by the state's Court of Appeals in the 1957 case of Bing v. Thunig.
2. President Roosevelt received a letter from physicist Albert Einstein (1879–1955) alerting him about Germany's threat. Leó Szilárd wrote the letter after consulting with Eugene Wigner and Edward Teller, and Einstein signed it. The US should build a nuclear weapons program, the letter recommended.

3. An estimated 200,000 civilians had been killed when US dropped atomic bombs on Nagasaki&Hiroshima, Japan, between 6&9 August 1945.Oppenheimer, with his fellows, spearheaded "Atoms for Peace" movement following bombings Endless Frontier is a report written by Vannevar Bush (1890–1974) for President Franklin D. Roosevelt, promoted funding for research and technology to advance both economic growth and national security.
4. A law shifting control of atomic energy development from the military to the civilian sector was passed by the US Congress in 1946. Atomic Energy Commission has been established by law to encourage peaceful use of nuclear energy.
5. US Atomic Energy Commission canceled Oppenheimer's security clearance in 1954when it was determined that his connections to Communist Party members made him a security risk and untrustworthy.
6. In 1957, the Soviet Union launched first satellite,Sputnik, into orbit, and the US government invested more in science and technology to compete in space.
7. President John F. Kennedy promised that by end of 1961, the US would send a man to the moon.
8. Dozens of nations joined the Partial Nuclear Test Ban Treaty in 1963, which forbids the testing of nuclear weapons other than underground, out of concern for the long-term effects of nuclear fallout.
9. A guide on the use and care of laboratory animals was released by the Public Health Service, outlining guidelines, protocols, as well as criteria for using animals in investigations while ensuring their welfare. Eight revisions have been made to the Guide, the most recent being in 2011.
10. In 1965, the Laboratory Animal Care Assessment and Accreditation Association was introducedto accredit institutions that engage lab animals for research.
11. 1966 saw the US. The Animal Welfare Act, passed by Congress, protects study animals, except for rodents and birds. The Act was passed in part because of concerns about dogs being stolen for scientific purposes. Several states enacted or amended animal cruelty laws protecting domestic, wild, laboratory, and agricultural animals in the 1960s.
12. One hundred ninety nations have signed Treaty on Nonproliferation of nuclear weaponsas of 1968. But still, many countries have not signed on to it.
13. The Biological Weapons and Toxins Convention, which was signed in 1972 by numerous nations, forbids the creation, manufacture, transfer, acquisition, and utilization of biological and toxic weapons, yet allows for research into defensive solutions.
14. To offer direction for recombinant DNA experiments that National Institutes of Health sponsors, Recombinant DNA Advisory Committee was established. Establishment of institutional biosafety committees to supervise recombinant

DNA operations is one of the requirements outlined in the 1976 Recombinant DNA Advisory Committee recommendations for recombinant DNA research.

15. An extreme animal rights organization called the Animal Liberation Front was founded in 1978 to save animals from experiments. Luc Montagnier charges Robert Gallo with research misconduct and stealing an HIV strain after they collaborated to extract HIV from biological samples supplied by people living with AIDS between 1984 and 1993; Gallo is acquitted. Gallo and Montagnier also disagree on who possesses the right to patent an HIV test and who should be credited with discovering the virus. The governments of US and France agree to settle conflict.
16. The Animal Welfare Information Center was established in 1985 as a result of an amendment to the Animal Welfare Act. Its goals are to increase public access to information about animal welfare and to advance more compassionate practices in animal research. Additionally, the amendment mandates that Animal Care and Use Committees assess government-funded animal research.
17. In 1986, Roger Boisjoly, an aerospace expert, warned NASA of the possibility of an O-ring failure during the Space Shuttle Challenger launch because of cold temperatures. While rubber O-rings have been designed to operate effectively at 32° F (0°C) or lower temperatures, anticipated air temperature at the time of launch has been 26° F (-3°C). All seven crew members perished in the Challenger's explosion after Boisjoly met with NASA authorities to address the matter, and they agreed to move forward with the launch. According to a special committee looking into the tragedy, the cold temperature caused O-rings to fail to seal correctly, allowing rocket propellant to leak and explode. Richard Feyman, a physicist who won the Nobel Prize, played a crucial role in persuading the committee that O-ring failure caused the explosion.

Use of Falsified and Fabricated Data: Academic Misconduct

1. In his 1830 book Reactions on the Fall of Science in England, and a Few of its Causes, English mathematician and inventor Charles Babbage (1791-1871) makes case that many of his contemporaries employed unethical research techniques, such as data fudging, falsification, and clipping. Babbage invented a programmable mechanical computer.
2. Twenty-two unethical experiments published in the medical literature are described by Henry Beecher (1904–1976) in the New England Journal of Medicine. These papers, out of 186 he found in medical journals, were chosen by Beecher, who said they demonstrated a pattern of unethical behavior.

3. At Sloan Kettering Cancer Institute, William Summerlin acknowledged in 1974 that he had fabricated data by creating black blotches on white mice with a marker. He was making a method for skin graft transplantation.
4. The Technology Transfer Act of 1986 changed Bayh-Dole Act of 1980, allow researchers to patent ideas created with funding from government. In 1980, the US Supreme Court declared in Diamond v. Chakrabarty that a genetically modified bacteria is patentable as it is an outcome of human intellect. This helps to create strong intellectual property protection for the emerging biotechnology sector and establishes a precedent for patents on other living species.
5. The Whitehead Institute was founded at MIT in 1981. Following this lead, other academic institutions start establishing intricate alliances with businesses. 1981 saw significant changes to the federal laws governing research involving human beings.
6. Harvard postdoctoral associate John Darsee was charged with data fabrication in 1981. Seventeen of his publications were taken down.
7. Betrayers of Truth, book written by Nicholas Wade and William Broad, was published in 1982. According to the book, there is more scientific misconduct than scientists would want to acknowledge, and it implies that well-known scientists like Robert Millikan, Gregor Mendel, and Isaac Newton were not entirely truthful with their findings. They assist in ushering in a new era of scientific "fraud-busting" with their book.
8. A panel from the National Institute of Mental Health concluded in 1987 that Steven Breuning had falsified and fabricated data in twenty-four papers. In 1988, Breuning has been found guilty of cheating federal government. Martin Luther King was charged with plagiarism in his doctoral dissertation in 1987. In a 1986 publication submitted by six of her colleagues in journal Cell, Margot O'Toole, postdoctoral researcher at Whitehead Institute from 1987 to 1996, expresses multiple concerns regarding data.She charges Imanishi-Kari with data fabrication and falsification. The resulting probe prompts investigations. One of the co-authors of the contentious study is David Baltimore, a Nobel Prize recipient. Baltimore left his position as president of Rockefeller University without facing any charges of wrongdoing. He called New York Times-reported investigation "witch hunt." Imanishi-Kari was ultimately cleared by a Department of Health & Human Services appeals tribunal, admitting solely to keeping bad records. In 1988, a genetically modified mouse was patented for cancer research by Harvard and Dow Chemical.
9. Public Health Service established the Office of Scientific Integrity and Office of Scientific Integrity Review in 1989 to conduct investigations into cases of scientific misconduct and provide guidance to organizations.It also changes the definition of misbehavior. In 1992, two organizations combined to form Office

of Research Integrity. In 1989, National Institutes of Health required training funds for all graduate students to include instruction in appropriate research practices. These standards are expanded in later years to include intramural researchers and postdoctoral trainees. The Responsible Conduct of Research training guidelines were introduced by the National Sanitation Foundation in 2010. In 1989, Stanley Pons and Martin Fleischmann declared their discovery of a method to achieve nuclear fusion at room temperature during news conference at University of Utah. When laboratories globally fail to replicate their findings, they face allegations of fraud, negligence, and self-deceit.

As long as large-scale studies were first carried out on animals, the concept of using humans in studies was gaining traction by the turn of 20th century. The number of individuals and animals utilized in research expanded dramatically with the growth of pharmaceutical businesses and the advent of bacteriology, the study of bacteria.Before enactment of Pure Food & Drug Act in 1906, there have been no regulations concerninghuman subjectsin research. Some unethical tactics were used in bacteriological research toward the close of 19th and the start of 20th centuries. For instance, without their knowledge or agreement, contagious agents were injected into convicts, mentally ill people, and orphans. There have been several other documented trials in which susceptible people were shocked with electricity. Regulation of human testing had been attempted. For instance, a guideline on human experimentation was distributed in 1900 by the Prussian Minister of Religion, Education, and Health. In 1931, novel therapies and human experimentation were regulated by German Reich Ministry of Interior. But these rules were mostly disregarded. Although medical and scientific groups denounced the acts above, no professional, disciplinary, or criminal proceedings were brought against them. Such accusations were made only after World War II and Nuremberg trials.

MAJOR HISTORICAL CASES OF ETHICS IN RESEARCH

Moore v. Regents of University of California (1990)

California Supreme Court determined in **Moore v. Regents of University of California (1990)** that while Moore had no property rights over his tissue, researchers maintained rights to a cell line derived from it.Additionally, Court finds that researchers' failure to inform Moore of their financial interests in his tissue sample constituted a violation of his right to informed consent. By this finding, the majority of courts have held that patients give up their property rights to tissues when they donate them for research purposes or when they are left over (or "abandoned") after surgery or other medical operations. (Kalkstein, 2000).

The Nuremberg Code (1947)

The Nuremberg Doctors' Trial (1946-1947), an offshoot of the larger post-war Nuremberg trials, uncovered gruesome information about Nazi medical experiments inside concentration camps. On December 9, 1946, 23 prominent German administrators and doctors have been charged with offenses crimes against humanity and war crimes committed by an American military court. German doctors are accused of conducting medical examinations on thousands of prisoners in concentration camps without obtaining their consent. The majority of individuals passed away or had permanent disability. The Nuremberg Code has been founded in 1948 as a direct consequence of trial. It emphasizes that "The voluntary permission of the human subject is vitally fundamental," meaning that participants must provide their consent advantages of the research must justify risks. Despite having no legal force behind it, Nuremberg Code was the first international agreement to encourage informed consent and voluntary participation. (Code, 1947).

The Code encompassed ten principles, with the following being the most influential:

- Voluntary consent of the human subject is essential.
- Research must yield fruitful results for the good of society.
- Anticipated benefits must justify the risks.
- Investigators must terminate experiments if harm becomes likely.

Although the Nuremberg Code lacked statutory authority and was initially dismissed by some physicians as relevant only to Nazi crimes, it provided a moral baseline. Its insistence on voluntary consent marked a radical shift from earlier norms, embedding the idea of individual autonomy at the heart of research ethics. Over time, its principles influenced both domestic policies and subsequent international declarations.

Thalidomide

1957–1962: Thalidomide was first sold in West Germany in 1957 as a medicine to treat pregnancy-related morning sickness. Due to exposure to this medicine, around 10,000 babies are born with severe birth abnormalities, primarily in West Germany. Exposure to thalidomide kills 2,000 kids. (Kim JH, 2011). The FDA's drug reviewer, Frances Kathleen Oldham Kelsey (1914-2015), rejected the medicine in

1960. Countries all around the world will soon outlaw the substance. Kelsey has been awarded 1962 President's Award for Outstanding Federal Civilian Service.

Thalidomide received approval as a sedative in Europe in late 1950s. The drug was initially utilized for treating nausea and insomnia in pregnant women; however, it was subsequently discovered that its use during pregnancy resulted in serious fetal defects. A significant number of patients did not give informed permission and were unaware of the fact that they were providing a medicine that was not approved by the FDA for use.

Following Senate hearings, Drug, Food, as well as Cosmetic Act's so-called "Kefauver Amendments" had been enacted in 1962 to guarantee increased efficacy anddrug safety.Pharmaceutical companies were required to show the efficacy of their products to the FDA prior to market approval for the first time. Thalidomide received approval as a sedative in Europe during late 1950s; however, FDA did not authorize its use in US.

The Helsinki Declaration

In 1964, World Medical Association recognized standards for doctors who use human beings in biomedical research. International research ethics are governed by Declaration of Helsinki, which was revised in 1975, 1983, 1989, as well as 1996. It forms basis of contemporary Good Clinical Practices. Guidelines for "non-therapeutic research" and "research linked with clinical treatment" are also established. The Declaration of Helsinki addressed following issues: participants must provide their informed consent; research mustbe conductedthroughpersons with the necessary medical or scientific training; risks should not outweigh benefits; an independent committee should review research methods before they are initiated; and human research must bebased on results of laboratory or animal research. (Association, 2013).

Helsinki was about how to practice medicine every day, all around the world.

Key contributions of the original Declaration included:

Scientific validity: Only scientifically sound studies were ethically justifiable.

Over the decades, the Declaration has undergone multiple revisions (1975, 1983, 1989, 1996, 2000, 2008, and 2013), reflecting evolving debates. Later revisions addressed:

Vulnerable populations: Additional protections for children, prisoners, and economically disadvantaged groups.

Post-trial access: Obligations to ensure subjects benefit from successful interventions.

Data transparency: Requirements for study registration and publication of results.

The Declaration remains a living document, embodying the medical profession's collective conscience while adapting to new challenges such as globalized trials, genetic research, and data sharing.

Study of Tuskegee Syphilis (1932–1972)

U.S. conducted Tuskegee Syphilis Study from 1932-1972. The Department of Education, Health & Welfare was established in 1932. Research examined impact of untreated syphilis in 400 African American males from Tuskegee, Alabama area. Researchers did not inform participants that they had been part of a trial. Researchers kept participants from receiving treatment, even after penicillin, an effective medication for illness, became widely accessible in 1950s. As a result, most Tuskegee clinic participants believed they were receiving treatment for "bad blood." Experiment concluded in 1972 following an Associated Press article that informed Congress and public about the ethical issues associated with research.A lawsuit filed by participants and their families was settled by the US government. (Thomas SB, 1991). Only when its presence was made known and it turned into a political embarrassment did it become the Welfare, Education, and Health Department. In 1997, President Clinton issued an apology to the research participants and their families under growing pressure.

Stanford Prison Experiment

The 1971 **Stanford Prison Experiment**, conducted by psychologist Philip Zimbardo, is another well-known example of unethical activity with negative results. The investigation, which looked into psychological impact of perceived power, caused participants to experience emotional anguish. Underestimating the risk of psychological harm while not providing enough protection for human participants is unethical in psychological research. This scenario illustrated the importance of ethical guidelines that prioritize emotional and mental well-being of participants, particularly in psychological research (McLeod, 2025). It caused in more strict review processes and the development of methods to prevent psychological harm in research. It prompted the American Psychological Association and other organizations to raise their ethical standards to protect participants' mental and emotional well-being.

The Belmont Report

The Belmont Report: In 1979, National Commission for Protection of Human Subjects of Biomedical as well as Behavioral Research established Ethical Research

Principles involving Human Subjects. It also served as the conceptual foundation for a fundamental change to federal research limitations in the United States in 1981.

The Report outlines fundamental ethical standards and concepts that help address the moral dilemmas associated with conducting research involving human subjects. (Research, 1979). The following are the three fundamental ideas and how they are applied:

The idea

- Respect for Persons: People ought to be regarded as independent beings. Individuals who have less autonomy should be protected.
- Beneficence: No harm should be done to human beings. Research should minimize potential risks and enhance potential rewards.
- Justice: Research risks and rewards must be shared equitably.

Utilization

- Informed consent: To the extent that they are able, subjects should be given a chance to choose what will or won't happen to them. The consent process requires the presence of three components: information, understanding, and voluntariness.
- Evaluation of the advantages and risks
- A methodical evaluation of the types and extent of risks and benefits is required.

Subject selection

- Fair methods and results must be used when choosing research participants.

The 1974 National Research Act

Tuskegee Syphilis Study created so much publicity that National Research Act of 1974 has been approved. National Research Act established National Commission for Protection of Human Subjects in Behavioral and Biomedical Research. This group was tasked with determining basic ethical principles that should guide behavioral and biological research with human subjects. It was additionally charged with formulating criteria to ensure that such research occurs by these ethical norms. The committee produced a landmark report known as the Belmont Report that established the ethics of human subjects' research in US. Tuskegee Syphilis Study created so much publicity that National Research Act of 1974 has been approved. National Research Act established National Commission for Protection of Human

Subjects of Behavioral as well as Biomedical Research, responsible for determining fundamental ethical standards concerning biomedical as well as behavioral research concerning human subjects, or developing standards to ensure that researchers follow these standards. The Belmont Report, which the Commission published, served as a model for US research ethics involving human subjects.

Abuse of Human Participants in Radiation Experiments

During World War II and early stages of the Cold War, US officials conducted tests on hospital patients, pregnant women, and children with developmental disabilities. They recruited military people to examine the effects of radiation. This is another instance of the mistreatment of human subjects. The majority of experiment participants were unaware that they were being exposed to radioactive materials, and just a small percentage provided informed consent. To create health or safety regulations for thousands of employees at atomic weapon plants, Manhattan Project officials approved wartime tests. On grounds of national security, policymakers defended a more thorough investigation of radiation impacts after war as the Cold War grew more intense. In the 1990s, the government apologized and provided financial compensation to some victims of human radiation testing after legislative inquiries, many official reports, academic studies, and lawsuits. (Popper SE, 1995).

SOCIAL-BEHAVIORAL CONTROVERSIES

The social critique reached more than just biomedical researchers. The obedience experiments of Stanley Milgram (1961–63) inflicted very high levels of psychological stress by deceiving participants into thinking they were delivering painful shocks—the Milgram paradigm. In the same vein, the study of "Tearoom Trade" (1970) by Laud Humphreys, which researched secret homosexual encounters, was criticized for issues of privacy, lies, and exploitation. Then, those controversies expanded moral disagreement to include psychological damage, privacy, and knowledge consent in behavioral science research. In tandem, these scandals buzzed up enough public outrage demand more accountability to help spur the creation of new oversight regimes and codified norms. (Goode, 1996).

GOOD CLINICAL PRACTICE

Good Clinical Practice guidelines were developed by the International Council for Harmonisation of Technical Requirements for Pharmaceuticals for Human Use

(ICH) in the 1990s. These standards harmonized requirements for clinical trial design, conduct, and documentation from the major regulatory agencies (U.S. FDA, European EMA, Japanese PMDA).Good Clinical Practice ascertains scientific validity of the results as well as ethical protection of subjects through practices such as informed consent, integrity, and monitoring systems. (Vijayananthan A, 2008).

REGULATIONS BASED ON THE BELMONT REPORT

In 1981, the Food and Drug Administration and the Department of Health and Human Services published regulations based on the Belmont Report. The Department of Health and Human Services published Title 45 (public welfare) and Part 46 (protection of human subjects) of Code of Federal Regulations. The FDA released Code of Federal Regulations Parts 50 (protection of human subjects), 56 (institutional review boards), and Title 21 (food and pharmaceuticals). In 1991, the fundamental regulations of the Department of Health and Human Services (45 Code of Federal Regulations Part 46, Subpart A) have beenofficially accepted by over a dozen more authorities and departments involved in or supporting research concerning human subjects, thereby establishing Federal Policy for Protection of Human Subjects, widely referred to as "Common Rule."Department of Veterans Affairs established this regulation in 1991 at 38 Code of Federal Regulations Part 16.1991 edition of "Federal Policy" remains commonly utilized by federal organizations and departments today.

The primary components of the Common Rule are as follows: research institutions are required to ensure compliance; guidelines for gathering and maintaining informed consentfrom researchers; the Institutional Review Board's membership, function, and operations are regulated; research review or documentation are regulated; and certain vulnerable study participants (children, prisoners, and pregnant women) receive additional protections.

Additionally, under 21 Code of Federal Regulations Parts 50-56, the FDA has established limitations on a significant amount of privately-financed and specific federally sponsored research (FDA Rules and Regulations, 2024).Regarding electronic items, food additives, drugs, devices, biologics, color additives,and other test objects that have been regulated by FDA, the FDA is in responsible for regulating usage of human subjects in research.Standards of Common Rule and FDA regulations are comparable, despite the existence of some considerable differences. Study subjects are protected by FDA regulations and Common Rule.

OTHER BOARDS AND BODIES THAT ENFORCE STANDARDS

In addition to the Institutional Review Board, other national and international regulatory organizations and agencies review research ethics norms. These contain:

Office for Human Research Protections is responsible for enforcement of Federal Policy for the Protection of Human Subjects in US. Global biomedical research ethics guidelines are provided by the World Health Organization. International Committee of Medical Journal Editors establishes ethical standards for biomedical research publications. Together with the Institutional Review Board, these groups form a broad network that ensures scientific research is conducted ethically. They protect integrity of research by utilizing lessons learned and reflections from past.

CHALLENGES AND TENSIONS IN ETHICAL LIFE

Although ethics guides us, it is not at all easy to live an ethical life. An ethical life is riddled with tension and conflict. Justice versus mercy, freedom versus equality, loyalty versus truth, short-term benefit versus long-term sustainability: All of these are conflicts that people and societies often face when it comes to competing values. Then, the nature of ethical consensus, the cultural diversity, further complicates it. That which is obligatory in one society is optional or even immoral in another. However, the problem lies in the ability to discern genuine cultural pluralism, which enhances moral understanding from harmful practices that violate fundamental human rights.

Then, there is the sword of moral relativism driven by a charge that ethics are inherently subjective or culturally determined. But as much as it is true that there are ethics that vary, the extreme case of relativism makes universal human rights humanly impossible, and moral accountability difficult or impossible. A fair conclusion about ethics cannot be that cultural diversity is not real, but it needs a claim that some aspects, such as human dignity, justice, and compassion, are relevant to everybody.

Finally, we are living in modern times, and that brings different challenges — consumerism, technology, and environmental disasters literally shake (not even just rattle) the foundations of morality. For instance, the fast evolution of artificial intelligence brings with it deep questions about autonomy, privacy, responsibility, and even personhood. As we face these hurdles, ethics should also change channeling strength, functionality, perhaps even feasibility, while remaining firmly rooted in essential values.

EMERGING ISSUES

In recent decades, ethical discourse has expanded to address new challenges:

Ancillary care obligations: Responsibilities researchers may have to provide care unrelated to the study.

Post-trial responsibilities: Ensuring participants and host communities have access to successful interventions after research concludes.

Benefit sharing: Fair distribution of knowledge, products, and resources generated by research, especially in international collaborations.

Data ethics: Protection of privacy, responsible sharing of genomic and digital data, and management of artificial intelligence in research.

THE FUTURE OF ETHICS: TOWARD A MORE HUMANE WORLD

In the future, ethics will become an even more significant aspect in determining the future course of mankind, which the world has never witnessed before in the form of opportunities and dangers. With the power and potential of scientific and technological advances still yet to come, the promise of extending life, enhancing capabilities, and connecting minds around the globe represents something we aspire to. But these very same advances risk exploitation, dehumanization, and ecological destruction. Ethics will decide whether technology will help or harm mankind.

In the same way that globalization has compressed distance and made cultures closer, there have also been wars, disparities, and migrations that are leaving millions of people fighting for their cultural survival. This, in turn, calls for a dynamic approach in global ethics which must be based on human rights principles, ecological responsibility, and the need for inter-cultural and interreligious dialogue. The great ethical challenge of the future will be to develop a planetary cognizant that makes no distinction, that affirms that we are all connected as a life force and have a collective responsibility as a human race. Central to this future will be education. Ethics is about critical moral reasoning, empathy, and responsibility, not indoctrination. Ethical training helps a person make wise and considerate decisions and to oppose injustice, corruption, and exploitation. Because when it comes to ethics, two things that we cannot really separate are the future of ethics and the future of human civilization. To achieve a more just, peaceful, and sustainable world, our choices, big and small, at personal, social, professional, and global levels must be ethical.

CONCLUSION

Ethics is not an ancillary issue but itself the essence of the human. It tells us where human action should be directed, is the glue that binds societies together, and is the light showing us the way to justice, peace, and the ultimate flourishing of all humanity. Ethics is eternal and temporal: at its core, it rests on timeless values such as justice, compassion, and honesty, yet evolves to meet new challenges. Ethics points out that morality is not about being perfect; it is about striving: striving to be better people, better communities, and a better world. It cannot even aspire to be so unless it is backed by ethics that dares us to rise above self-interest and parochial interests, to act with honour, even at a loss, and to be responsible, not just for ourselves but for others and generations yet to come. Ethics, a philosophy, religion, and cultural practice that has been around for 2000 years, has always been a response to the deepest questions about what kind of life is worth living, what kind of society is worth building, and what legacy we should leave behind. Ethics is not a conclusion, but an open conversation that transcends generations, cultures, and experiences. It is a commitment to live with integrity, respect others' dignity, pursue justice, and nurture the fragile bonds that sustain life on this planet. Ethics is not a conclusion to be reached, but a way of life to be continually practiced.

Nuremberg Code has been established in response to the Nazi atrocities conducted during World War II, which highlighted absence of international guidelines for human subject's research. Pharmaceutical companies are required to demonstrate the efficacy of their drugs to the FDA before production through the passage of "Kefauver Amendments" to Food, Drug, and Cosmetic Act, which had been introduced in response to thalidomide tragedy.Today's Good Clinical Practices are based on the Declaration of Helsinki. The Tuskegee Syphilis Study is possibly most severe instance of unethical human subjects' research in American history. National Research Act established need to protect human subjects in research and established groundwork for publication of Belmont Report.

The comprehensive ethical standards that currently govern scientific research were significantly influenced by these historical instances of ethical lapses. They serve as a sobering reminder of the continuous necessity of prioritizing welfare and rights of participants in all research projects and potential repercussions of unethical behavior.

The history of research ethics is a narrative of reactionary reform and slow-moving institutionalization. Each of these milestones—from the clinical, not-so-moral path of the Hippocratic Oath, to the horrors that begat the Nuremberg Code, to the practical guidance reflected in the Declaration of Helsinki also represented the moral failures and aspirations of the times. Ethical norms are not fixed and must respond to ongoing challenges; the history of research ethics highlights this point. It is a path that in the end is telling of our human tradition of searching for knowledge but not loss of the instinct of achieving human dignity.

REFERENCES

Association, W. M. (2013). Declaration of Helsinki: Ethical Principles for Medical Research Involving Human Subjects. *Journal of the American Medical Association.* PMID: 24141714

Bouzanis, C. (2023). Kant, Immanuel. *The International Encyclopedia of Ethics*, 1–12.

Code, T. N. (1947). *Trials of War Criminals before the Nuremberg Military Tribunals under Control Council Law No. 10.* U.S. Government Printing Office.

Coppola L, C. A. (2019). Biobanking in health care: evolution and future directions. *Journal of Translational Medicine*, 172. (2024). *FDA Rules and Regulations.* US Food and Drug Administration.

Goode, E. (1996). The ethics of deception in social research: A case study. *Qualitative Sociology*, *19*(1), 11–33. DOI: 10.1007/BF02393246

Harrington, M. E. (2020). Research Ethics. In *The Design of Experiments in Neuroscience* (pp. 13–29). Cambridge University Press. DOI: 10.1017/9781108592468.003

Holm, S. (2015). Informed Consent: Ethical and Legal Issues. *eLS* .

IV, B. A. (2006). A Summary of Important Documents in the Field of Research Ethics. *Schizophrenia Bulletin*, 69–80. PMID: 16192409

Jamnik, A. (2021). Rawls' Theory of Justice as Fairness. *Bogoslovska Smotra*, 1037–1058.

Kalkstein, J. A. (2000). Moore v. Regents of the University of California Revisited. *Yale Symposium on Law and Technology.*

Kim, J. H., & Scialli, A. R. (2011). Thalidomide: The tragedy of birth defects and the effective treatment of disease. *Toxicological Sciences*, *122*(1), 1–6. DOI: 10.1093/toxsci/kfr088 PMID: 21507989

Kranak, J. (2019). Kantian Deontology. In *Introduction to Philosophy: Ethics.* Rebus Press.

McLeod, S. (2025). Stanford Prison Experiment. *Simply Psychology*. megítélésére, K.-F. S. (2008). Influence of the Nuremberg physicians' trials—beginning a new era in the ethical judging of human experiments. *Orvosi Hetilap*, 195–201.

Miteu, G. D. (2024). Ethics in scientific research: A lens into its importance, history, and future. *Annals of Medicine and Surgery (London)*, *86*(5), 2395–2398. DOI: 10.1097/MS9.0000000000001959 PMID: 38694399

Muto, H. (2022). *The Holocaust as an Inflection Point in the Development of Research Ethics.* Springer eBooks.

Nagai, H. N. E., Nakazawa, E., & Akabayashi, A. (2022). The creation of the Belmont Report and its effect on ethical principles: A historical study. *Monash Bioethics Review, 40*(2), 157–170. DOI: 10.1007/s40592-022-00165-5 PMID: 36357708

Nussbaum, M. (1999). Virtue Ethics: The Misleading Category. *The Journal of Ethics, 3*(3), 163–201. DOI: 10.1023/A:1009877217694

O'Sullivan, L. O., Crowley, R., McAuliffe, É., & Doran, P. (2020). Contributory factors to the evolution of the concept and practice of informed consent in clinical research: A narrative review. *Contemporary Clinical Trials, 19*, 100634. DOI: 10.1016/j.conctc.2020.100634 PMID: 33024880

Popper, S. E., & McCloskey, K. (1995). Ethics in human experimentation: Historical perspectives. *Military Medicine, 160*(1), 7–11. DOI: 10.1093/milmed/160.1.7 PMID: 7746436

Research, T. N. (1979). *The Belmont Report.* U.S. Department of Health, Education, and Welfare.

Soniya Billore, T. A. (2023). Self-regulation and goal-directed behavior: A systematic literature review, public policy recommendations, and research agenda. *Journal of Business Research.*

Thomas SB, Q. S. (1991). The Tuskegee Syphilis Study, 1932 to 1972: implications for HIV education and AIDS risk education programs in the black community. *Am J Public Health.*

Udo Schuklenk, S. P. (2013). *International Research Ethics. International Encyclopedia of Ethics.* Wiley.

Vijayananthan, A. N. O., & Nawawi, O. (2008). The importance of Good Clinical Practice guidelines and their role in clinical trials. *Biomedical Imaging and Intervention Journal, 4*(1). Advance online publication. DOI: 10.2349/biij.4.1.e5 PMID: 21614316

Chapter 3
Protection of Human Subject

Krutika Patre
http://orcid.org/0009-0001-7969-4409
Ajeenkya D.Y. Patil University, India

Sunil Sankathala
http://orcid.org/0009-0004-7809-7647
Ajeenkya D.Y. Patil University, India

Kritikka Kunte
Ajeenkya D.Y. Patil University, India

Aryan Vasa
http://orcid.org/0009-0000-1837-8587
Ajeenkya D.Y. Patil University, India

Devanshi Soni
Ajeenkya D.Y. Patil University, India

Susanta Das
http://orcid.org/0000-0002-9314-3988
Ajeenkya D.Y. Patil University, India

EXECUTIVE SUMMARY

Human subjects research history has always tried to strike a balance between science & safety of research subject. The way of carrying out most experiments in the traditional manner, development of test in most cases were tragic thus needed a lot of morality. Research in contemporary times is guided by principles of autonomy, justice & beneficence that create dignity, inclusion & protection of research subjects. This is no longer the case as it is now the people as the research partners & not the objects of inquiry. However, rapid technological growth & global collaborations present new moral challenges. Emerging domains (genomics, artificial intelligence & data science) have difficult questions that present privacy, consent & equity problems. Those issues imply that there is a need to have hard & yet flexible laws that would enable people to be secured & inventions to be improved. Ethical research does not just establish a knowledge, but also trust, dignity, & ethical responsibility of the society in such a way that science is built in a responsible manner to serve people & society.

DOI: 10.4018/979-8-3693-8302-5.ch003

1. INTRODUCTION

In medical research, there would not be medical advancement without human beings. It is the largest percentage of the medicines, treatment, procedures undergoing which had been previously experimental because of the experimentation on the human population. Vaccine, imaging technology and bone marrow transplants are some of the examples of the innovations that begin as an experiment and become the normalcy. But there is the dark history of inhumane experiments on human beings that do hold those exploited, abused or even de-dignified in the name of science. The consequences of ethical neglect in cases of failure to be extremely alert of ethical practices in research are quite punitive as indicated in the case of the Nazi medical test, and the Tuskegee Syphilis Study. It is these atrocities that brought about the universal tenets of respect, beneficence, justice and autonomy as the principle of research ethics (Bertholf, 2001; Manton et al., 2014; Xu et al., 2020; Sng et al., 2016; Felzmann et al., 2014).

Medical research did not get under the extremely restrictive scrutiny until later, when the time had passed, the World War II had elapsed. With the demonstration of the free consent, the force is barred by the Nuremberg Code of 1947. This was to be supplemented by the reference of Helsinki (1964) that entails the culpability of the doctors, the well-being of the patient, and the reduction of harm than good. These theories became incorporated in Belmont Report (1979) of the U.S and became the foundation of the institute review boards (IRBs). The theories transformed the paradigm of researches to the participant centred approach founded on rights and well-being (Levine et al., 2004).

It has the issue of dealing with the vulnerable groups. The vulnerable individuals might be characterized as children and prisoners who were economically disadvantaged in the past and such an approach would be likely to lead to the development of the problem of stereotyping. The challenge of the current generation of authors and ethics boards is to discover the trade-off between the indisputable autonomy and the extreme generalization (Levine et al., 2004). One of the most important protective issues can be referred to as informed consent since the participants of the study were the subjects of the study in a voluntary manner at the moment when the concept of the purpose and risk, advantages, and other opportunities were elucidated to them. This must be existent and culture aware and possessed. Nonetheless, the intervention of the vulnerable groups is an ethical peculiarity in that an emergency does not always give the parties the opportunity to negotiate (Manti & Licari, 2018; Parija & Mandal, 2014; Gupta, 2013).

The question of whether science should be held back or hard ethics is not new. The other ones narrate that life-saving procedures were gained following the risky experiment and the other ones narrate that the professional duty can guarantee the moral and scientific excellence of work (Wendler, 2012). The safeguarding of the vulnerable is a very fine line to take in regard to the lawfulness of the adequate degrees of protection which are based on the premise of the fair access (Gostin, 1991).

Governments have come up with regulations (which are in their latest version as the Common Rule 1991) to alleviate the burden on the administration, and steps, policies, procedures and requirements of the protection of the human subjects. Some opponents also believe that the processes of effecting changes are working out of the effectiveness of safety. In the international arena, its equivalent to the national policies is CIOMS (Council of International Organizations of Medical Sciences) Ethical Guidelines (Largent, 2016).

The technologies are new in terms of ethical issues. There are such concerns as genomics, biobanking, AI, and electronic health records ownership, consent and breach of privacy. The second because that has led into the consent element of this behaviour has been the adoption of the social network research that renders it difficult to gather information with consent of the third party (Klovdahl, 2005).

The minority groups in society have their interests. This can be negated through force of prisoners, children would be obliged to agree with parents, old people would need surrogate decision-makers and the crippled would demand special methods of consent (Vanclay et al., 2013; Singer & Levine, 2003; Stineman & Musick, 2001). As such, protection should not be without inclusion.

And the last but not the least, the IRBs, RECs and the researchers must also ensure that it also has the culture of moral responsibility which is not wholly reliant on the bureaucracy. It is needed especially in the developing nations where the local values can be at odds with the experiments of integration (Hrebert, 2015). Community learning and Community level acquisition of faith have been achieved at the community level and it has been effectively carried out in Cambodia and Laos (Kue et al., 2017).

It is an innovation/regulation battle that is won whatever the will of fate. And most of them are bad, insofar as they might tend to revive the evils of the old days; and, at the same time, an extremity of liberality of government will revive the evils of the old days (Lynn, 2004; Roets, 2017; Pisani et al., 2016). The most difficult part of any modern research ethics is to strike a balance.

The chapter examines the nature of human subject protection and its history and some of the key milestones in ethical progress and regulatory developments. It then discusses the primary ethical principles and control measures, in particular, the informed consent and the regulatory review procedure. The chapter also looks at the ethical issues which surround vulnerable groups, the protection and inclusion dilemma and proceeds to give an answer to the emerging ethical issues posed by technology and data-driven research. This is then followed by an international perspective on ethics and good governance committees and then implemented and management implications of research ethics. The chapter has also introduced a bibliometric analysis as a demonstration of the trends of the research and the paper has concluded by saying that the matter of whether to be an ethical researcher or a scientific one has to be balanced.

2. ETHICAL FRAMEWORKS AND EVOLVING CHALLENGES IN HUMAN SUBJECT PROTECTION

Building on the foundational context established in the introduction, the following literature review critically examines the key dimensions of human subject protection as discussed in existing scholarship. In this section we going to discuss about following topics.

2.1. Historical Foundations and Evolution of Human Subject Protection

The protection of human subjects in research emerged as a response to a long history of unethical experimentation in biomedical and clinical studies. Early medical advancements often relied on human experimentation conducted without adequate knowledge, consent, or concern for participant welfare. While such research contributed to major breakthroughs such as vaccines, imaging technologies, and bone marrow transplantation it also involved serious ethical violations that exploited vulnerable populations. These abuses highlighted the urgent need for ethical safeguards to ensure that scientific progress does not come at the cost of human dignity (Bertholf, 2001).

Some of the most notorious examples of unethical research include the Nazi medical experiments during World War II and the Tuskegee Syphilis Study in the United States. In these cases, participants were subjected to extreme harm, deception, and denial of treatment, revealing the catastrophic consequences of conducting research without ethical oversight. These events served as a turning point in global awareness of research ethics and directly influenced the establishment of universal ethical principles such as respect, beneficence, justice, and autonomy (Bertholf, 2001).

The first formal international response to these atrocities was the Nuremberg Code (1947), which introduced voluntary informed consent as an absolute requirement for research involving human subjects. The Code emphasized that participants must freely consent to research with full understanding of its risks and benefits. This marked a fundamental shift in research practice, placing the rights and autonomy of participant above scientific ambition. However, despite its importance, the Nuremberg Code was initially applied inconsistently and lacked enforcement mechanisms (Levine et al., 2004).

Subsequently, the Declaration of Helsinki (1964) expanded ethical responsibilities by explicitly addressing the role of physicians in research and emphasizing participant welfare over scientific or societal interests. It introduced principles such as independent ethical review, risk minimization, and the obligation to protect vulnerable populations. This declaration significantly influenced medical research worldwide and reinforced the idea that ethical responsibility extends beyond consent alone (Levine et al., 2004).

In the United States, these international principles were later consolidated in the Belmont Report (1979), which articulated three foundational ethical principles: respect for persons, beneficence, and justice. Respect for persons emphasized autonomy and informed consent, beneficence focused on minimizing harm and maximizing benefits, and justice addressed fair selection and treatment of research participants. The Belmont Report became the cornerstone of modern research ethics and directly led to the establishment of Institutional Review Boards (IRBs), which oversee and regulate human subject research (Bertholf, 2001; Levine et al., 2004).

Over time, these ethical frameworks were incorporated into federal regulations such as the Common Rule, which standardized human subject protections across research institutions. Together, these historical developments reflect a paradigm shift from researcher-centred experimentation to participant-centred ethical governance. The evolution of human subject protection demonstrates that ethical vigilance is essential to maintaining public trust, safeguarding human rights, and ensuring responsible scientific advancement (Gostin, 1991; Wendler, 2012).

2.2. Core Ethical Principles and Regulatory Oversight Mechanisms

The ethical governance of research involving human subjects is grounded in three core principles respect for persons, beneficence, and justice which have been refined through decades of ethical reflection and regulatory development. These principles were formally articulated in the Belmont Report (1979) and continue to guide modern research ethics across biomedical, clinical, and social science research. Respect for persons emphasizes individual autonomy and the right to make informed decisions about participation, while also acknowledging the need for additional protections for individuals with diminished autonomy. Beneficence requires researchers to minimize potential harm and maximize anticipated benefits, whereas justice focuses on the fair distribution of research burdens and benefits across populations (Bertholf, 2001; Gostin, 1991).

These ethical principles are operationalized through regulatory frameworks designed to protect human subjects and ensure accountability. In the United States, the Common Rule was established to standardize ethical review and oversight across federally funded research. The regulation mandates prior ethical review, informed consent requirements, and ongoing monitoring of research involving human participants. The creation of Institutional Review Boards (IRBs) provided a structured mechanism to evaluate risks, benefits, and ethical compliance before research begins. These oversight bodies serve as a safeguard against exploitation and unethical research practices (Levine et al., 2004).

Internationally, ethical governance has been supported by guidelines such as those developed by the Council for International Organizations of Medical Sciences (CIOMS), which complement national regulations and address ethical challenges in low- and middle-income countries. These guidelines emphasize fairness, transparency, and protection of vulnerable populations, particularly in settings where regulatory infrastructure may be weaker. The literature notes that international collaboration has increased the complexity of ethical oversight, requiring harmonization across diverse legal and cultural contexts (Largent, 2016).

Despite their importance, regulatory systems have faced criticism for being overly bureaucratic and inconsistent across institutions. Revisions to the Common Rule in 2018 aimed to reduce administrative burden, streamline low-risk research, and improve clarity in informed consent requirements. While these changes were welcomed by researchers, critics argue that they may dilute participant protections by prioritizing efficiency over ethical rigor. This tension reflects an ongoing debate within the literature about how to balance innovation with responsibility (Tsan, 2025).

Overall, the literature emphasizes that ethical principles alone are insufficient without effective regulatory oversight. At the same time, excessive regulation may hinder valuable research. Scholars conclude that ethical governance must remain flexible, context-sensitive, and grounded in the foundational principles of respect, beneficence, and justice to ensure both participant protection and scientific progress (Wendler, 2012).

2.3. Informed Consent as a Central Ethical Safeguard

Informed consent is widely regarded as the cornerstone of ethical research involving human subjects. Rooted in the principle of respect for persons, informed consent ensures that participants voluntarily agree to participate in research with a clear understanding of its purpose, procedures, risks, benefits, and alternatives. Early ethical frameworks treated consent as a formal requirement; however, contemporary literature emphasizes that informed consent is a process rather than a one-time event that must be maintained throughout the research lifecycle (Parija & Mandal, 2014).

The literature highlights several essential components of valid informed consent, including voluntariness, comprehension, and disclosure. Researchers are responsible for presenting information in a manner that is understandable and culturally appropriate. This is particularly important in settings involving low literacy, language barriers, or differing cultural norms. Studies emphasize that consent must be free from coercion or undue influence, especially when incentives or authority relationships are involved (Gupta, 2013).

Challenges to informed consent arise in specific research contexts, such as emergency research, paediatric studies, and research involving cognitively impaired individuals. In such cases, obtaining traditional consent may be impractical or impossible. The literature discusses ethically justified consent waivers and surrogate decision-making mechanisms, while stressing the need for heightened ethical scrutiny and participant protection. These exceptions demonstrate the complexity of applying consent principles across diverse research designs (Manti & Licari, 2018).

The evolution of research methodologies has further complicated informed consent practices. The rise of large-scale data collection, electronic health records, and biobanking raises concerns about secondary data use and long-term data storage. Scholars argue that consent processes must adapt to these developments by emphasizing transparency, ongoing communication, and participant trust. Consent is increasingly viewed as a relational process built on mutual respect rather than a legal formality (Parija & Mandal, 2014).

Overall, the literature concludes that informed consent remains a fundamental ethical safeguard but acknowledges its limitations. While it cannot eliminate all risks, effective consent practices reinforce autonomy, promote trust, and enhance ethical integrity. Continuous refinement of consent processes is necessary to address evolving research contexts and protect participant dignity (Manti & Licari, 2018; Gupta, 2013).

2.4. Vulnerable Populations, Inclusion, and Ethical Balance:

Research involving vulnerable populations has long been recognized as one of the most complex areas of human subject protection. Vulnerability in research refers to a reduced capacity to protect one's own interests due to factors such as limited autonomy, power imbalance, cognitive impairment, social marginalization, or economic disadvantage. Early ethical frameworks often approached vulnerability by excluding such populations entirely from research participation to prevent exploitation. However, contemporary literature strongly critiques this exclusionary approach, arguing that it may itself constitute injustice by denying vulnerable groups access to the potential benefits of research (Levine et al., 2004).

Children are among the most frequently discussed vulnerable populations in ethical literature. Their limited decision-making capacity necessitates parental or guardian consent, along with the child's assent when developmentally appropriate. Ethical scholars emphasize that children should not be automatically excluded from research, as doing so can hinder advances in paediatric care. Instead, research designs must incorporate child-centred protections, age-appropriate information, and minimal risk exposure. Overprotection, the literature argues, can paradoxically harm children by limiting evidence-based treatments tailored to their needs (Felzmann et al., 2014).

Prisoners represent another group historically subjected to severe research abuses, leading to stringent regulatory controls. Their constrained autonomy and dependence on institutional authority create a high risk of coercion or undue influence. Ethical literature highlights the tension between protecting prisoners from exploitation and ensuring fair access to potentially beneficial research, especially in studies addressing health disparities prevalent in incarcerated populations. Institutional Review Boards are therefore required to apply heightened scrutiny to prisoner research, ensuring voluntariness, minimal risk, and clear societal value (Vanclay et al., 2013).

Older adults also present ethical challenges due to factors such as cognitive decline, sensory impairment, and reliance on caregivers. The literature notes that older individuals are often underrepresented in clinical research, resulting in treatments that may not be adequately tested for this population. Ethical scholars advocate for adaptive consent processes, simplified information materials, and inclusion of caregivers or legally authorized representatives when necessary. Importantly, assumptions about incapacity must be avoided, as many older adults retain full decision-making ability (Singer & Levine, 2003).

Persons with disabilities face similar challenges, particularly when disabilities affect communication or cognition. The literature emphasizes that disability alone should not be equated with incapacity. Instead, researchers are encouraged to employ alternative communication methods, accessible consent formats, and individualized assessments of decision-making ability. Ethical protections should focus on enabling participation rather than reinforcing exclusion (Stineman & Musick, 2001).

Across all vulnerable populations, contemporary scholarship argues that vulnerability should be understood as contextual rather than categorical. Individuals may be vulnerable in one research context but not in another. Broad labelling risks stereotyping and overgeneralization, which can undermine autonomy and justice. Ethical protection, therefore, must be proportionate, situation-specific, and responsive to actual risks rather than assumed weaknesses (Levine et al., 2004; Gostin, 1991).

Ultimately, the literature emphasizes that ethical research must strike a balance between protection and inclusion. Safeguards such as tailored consent processes, enhanced monitoring, and community involvement can reduce risks while ensuring that vulnerable populations are not unjustly excluded from scientific progress.

2.5. Emerging Ethical Challenges in Technology and Data-Driven Research

Rapid technological advancement has significantly transformed the landscape of human subject research, introducing new ethical challenges that traditional frameworks struggle to address. The literature highlights those emerging technologies such as genomics, biobanking, artificial intelligence, electronic health records, and large-scale data analytics have expanded the scope of research beyond direct participant interaction. As a result, issues of privacy, consent, data ownership, and confidentiality have become increasingly complex (Pisani et al., 2016).

One major ethical concern arises from the collection and reuse of biological and digital data. Biobanking and genomic research often involve long-term storage of biospecimens for future, unspecified research purposes. The literature notes that traditional informed consent models may be inadequate in such contexts, as participants cannot fully anticipate how their data will be used. This raises concerns about autonomy, transparency, and trust. Scholars emphasize the need for governance mechanisms that ensure accountability, ongoing communication, and respect for participant preferences (Largent, 2016).

Social network research presents unique ethical dilemmas related to third-party data and confidentiality. Studies in this area often collect information about individuals who have not directly consented to participation. While anonymization is commonly used to mitigate risk, the literature argues that anonymization alone does not eliminate ethical responsibility. Researchers must carefully balance the feasibility and scientific value of network research with the obligation to protect individuals whose data may be indirectly involved (Klovdahl, 2005).

Digital health technologies and electronic health records further complicate consent and privacy protections. Data collected for clinical purposes may later be repurposed for research, sometimes without explicit participant awareness. Ethical scholars highlight the risk of erosion of trust when participants feel they have lost control over their personal information. The literature stresses the importance of transparency, robust data security, and clear communication regarding data use (Pisani et al., 2016).

Artificial intelligence and algorithmic analysis introduce additional concerns related to bias, accountability, and informed consent. While AI-driven research promises efficiency and innovation, it can also perpetuate inequalities if training data reflect existing social biases. The literature warns that ethical oversight must extend beyond data collection to include algorithm design, interpretation, and application to ensure fairness and justice (Wendler, 2012).

Overall, the literature underscores that technological innovation must be accompanied by ethical evolution. Existing principles of respect, beneficence, and justice remain relevant but require reinterpretation in digital contexts. Scholars advocate adaptive ethical frameworks that emphasize participant trust, continuous oversight, and shared responsibility between researchers, institutions, and policymakers to ensure that technological progress does not compromise human dignity (Pisani et al., 2016; Largent, 2016).

2.6. Global Perspectives, Ethics Committees, and the Future of Human Subject Protection

The protection of human subjects increasingly operates within a globalized research environment, where studies often span multiple countries, cultures, and regulatory systems. Bibliometric analyses presented in the literature demonstrate that research on human subject protection is dominated by high-income countries, particularly the United States and Europe, where ethical oversight structures are well established. However, the expansion of research into low- and middle-income countries has raised concerns about ethical consistency, power imbalances, and exploitation (Roets, 2017).

Ethics committees and Institutional Review Boards play a central role in safeguarding participants by reviewing research protocols, assessing risks and benefits, and ensuring compliance with ethical standards. While these bodies provide essential oversight, the literature notes growing dissatisfaction among researchers due to bureaucratic delays, inconsistent interpretations of guidelines, and administrative burden. Reforms such as the revised Common Rule aimed to address these concerns, yet scholars caution that efficiency should not come at the expense of participant protection (Tsan, 2025).

International guidelines, such as those developed by CIOMS, attempt to bridge ethical gaps between countries by providing context-sensitive guidance, particularly for research conducted in resource-limited settings. These guidelines emphasize fair benefit sharing, community engagement, and cultural sensitivity. The literature stresses that ethical standards must be adaptable to local contexts without compromising fundamental principles of autonomy, beneficence, and justice (Largent, 2016).

Community engagement is increasingly recognized as a critical component of ethical research, especially in marginalized or culturally diverse populations. Studies highlight that involving communities in the research process enhances trust, improves consent quality, and reduces ethical conflicts. Culturally relevant ethics training programs have been shown to strengthen understanding of human subject protections among community researchers and participants alike (Kue et al., 2017).

Open-access publishing has further influenced the global dissemination of ethical knowledge. Bibliometric findings indicate a significant rise in open-access literature on human subject protection, facilitating wider access to ethical standards, policies, and best practices. This increased accessibility supports transparency and global dialogue on ethical research conduct (White, 2020).

Looking forward, the literature concludes that the future of human subject protection depends on adaptive governance, interdisciplinary collaboration, and a shift from purely rule-based compliance toward a culture of ethical responsibility. Ethics committees, researchers, and institutions must work together to respond to emerging challenges while preserving the central goal of research ethics: protecting human dignity while enabling meaningful scientific advancement (Hébert et al., 2015; Wendler, 2012).

2.7. Applied Case Illustration in Human Subject Protection

Although the chapter extensively discusses ethical principles, regulatory mechanisms, and protections for vulnerable populations, it lacks a concrete applied case study demonstrating how these principles function in real research contexts. This absence is particularly notable given the book's focus on *Cases on Research Ethics*. The literature included in the chapter already provides rich material for applied analysis, especially in discussions of children, prisoners, older adults, and persons with disabilities, where ethical tensions between protection and inclusion are well documented (Vanclay et al., 2013; Singer & Levine, 2003).

Additionally, while Institutional Review Boards (IRBs), ethics committees, and community engagement are mentioned, their practical decision-making processes are not illustrated through a unified example. The literature emphasizes that ethical protection must be contextual, proportionate, and responsive to real-world constraints rather than purely theoretical (Levine et al., 2004; Gostin, 1991).

2.8. Informed consent in Human Subject Research

The concept of informed consent is one of the essential conditions of ethical research with human subjects and the implication of the principle of respect of persons expressed in the most renowned ethical theories like the Belmont Report (National Commission for the Protection of Human Subjects of Biomedical and behavioural Research, 1979). As indicated in this chapter, the informed consent is meant to make sure that the participants in the research will have a free will to either join or not to join the research as a result of sufficient information about the purpose of the study, procedures, risks, and anticipated benefits (Levine et al., 2004). Notably, informed consent is not a discrete administrative act but rather a process that has to be maintained through to the end of research (Manti & Licari, 2018).

Meaningful informed consent is a very difficult issue to achieve in practice. In research that deals with technical methods or unpredictable results, the researchers need to present complicated information in a way that is reasonably understood by the participants of the research (Wendler, 2012). In spite of the fact that formal consent forms are offered, the participants might not be thoroughly informed about the consequences of their involvement, which demonstrates the difference between procedural adherence and informed choice (Levine et al., 2004). These issues are made more acute in studies that deal with the vulnerable, including children and prisoners, as explained elsewhere in the chapter. Under those circumstances, the question of decision-making capacity, surrogate consent and the possibility of coercion or undue influence become ethically relevant (Levine et al., 2004). These instances show that no uniform application of informed consent can be used in all research situations.

These issues on consent provide a convenient guide to practice by the Belmont Report principles, which include respect to persons, beneficence and justice. Respect of person focuses on voluntariness and understanding, beneficence focuses on risk disclosure and minimization of harm, and justice ascertains equitable choice of participants (National Commission, 1979). Combined, these principles make informed consent an applied ethical duty and not a merely procedural thing.

2.9. Managerial, Policy and Applied Implications of Research Ethics

Although it is significant that the historical development of the research ethics and the bibliometric trends offer certain background, it is the application of ethics in the actual research settings that makes ethics relevant. In accordance with the ethical principles, including those expressed in the Belmont Report, institutional policies regarding research in biomedical and behavioural sciences must be translated into specifics and day-to-day research activities (National Commission for the Protection of Human Subjects of Biomedical and behavioural Research, 1979). The Emerging role of institutional Review Boards and ethics committees is to perform the central managerial role of interpretation of these principles, determining regulatory compliance versus ethical judgment depending on the situation (Levine et al., 2004).

Ethical oversight, in the policy aspect, is not only about following the formal rules. The researchers are also bound by practical limitations, administrative regulations, and changing regulatory pressures without violating the participants and reducing the harm (Wendler, 2012). Such difficulties are most pronounced in those studies that concern vulnerable groups and studies in the field of public health when ethical decision making must be flexible and reasonably balanced in the safeguards provided instead of strict adherence to rules (Levine et al., 2004).

Such applied point of view underlines the idea that ethical research conduct does not merely rely on past history or academic fashion but rather on the sound governance, responsibility of the institution and accountability of the researcher. Making more emphasis on these implications applied would make the chapter more relevant to applied research ethics and bring it closer to the goals of a case-based ethics volume.

3. BIBLIOMETRIC ANALYSIS

The evolution of the studies concerning the topic of the Protection of Human Subjects during the last 5 decades is the extension and change of the biomedical, clinical and ethics researches. Bibliometric mapping of concept groups, key pattern, and topic trends, is applied to 1975-2025 VOSviewer and PubMed data. Each of these visualizations (Figure 1, Figure 2 and Figure 3) contribute to our knowledge on its own but together create an entire story.

Figure 1 presents a network visualization of terms that co-occur due to search on the topic of human subject protection, and it reflects the multi-dimensionality of the topic. The subject human beings and the word human are the most large and interconnected terms and here we can find that the subject human beings are the still core of the biomedical and clinical research. It is here that one will find a number of clusters which represent various yet related areas of study. On the left, the cluster is purple/blue and the key words are antioxidants, vitamin E, carotenoids, inflammation, lipids and cholesterol. This institute is represented by such areas as nutritional and biochemical research, in particular, the effect of dietary factors on health. Its strong associations with blood pressure, cardiotonic medication, apoptosis and anti-nervous system agents articulate the relationships between the literature of the antioxidants, cardiovascular health, cell dynamics and antitoxic process. The inclusion of both animals and rats indicates that most of these findings are supported by comparative studies that attribute the translational dissimilarity in the use of animal models and humans. On the right, a cluster of green surrounds the research subjects, clinical trials, research ethics, research informed consent and government. This is an indication of the ethical and governance aspects of the human-subject research. It is also stressed by the connections with the behavioural research, health knowledge, attitude and ethics committees that it is necessary to keep in mind that it concerns the welfare of the researchers, control over the trial and the social element of medicine. The yellow cluster, involving vaccination, antibodies, viral infections and immunoglobulin, is below. Isolated in this group is infectious diseases and immunology in which human role has been very crucial in the vaccines and antibody-based therapy. The reference to HIV, flu cases, etc. proves the topicality of such research in the real world, yet the reference to the ethics implies that there are also some questions about the safety of the individuals involved in the vaccine tests. Last but not least, the orange cluster can be applied to caries, dental care, fluoride and oral health, and saliva. It is a smaller, but an important field that deals with preventive dentistry, and health interventions on a community level.

Figure 1. Bibliometric analysis of "Protection of human subject" using VOSViewer with data from PubMed. (Data may vary slightly).

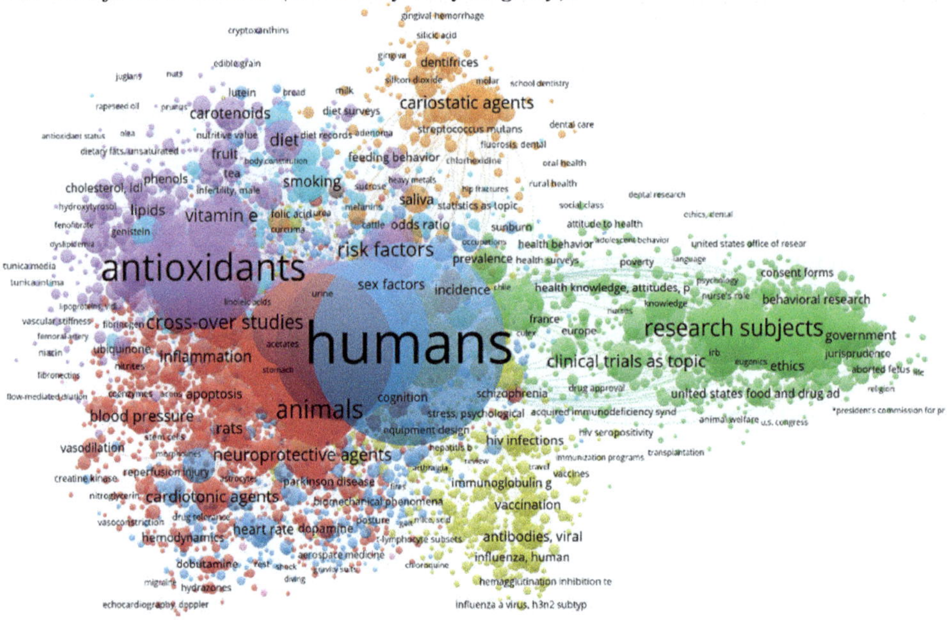

Figure 2, displays a graph of sonar records of salt pattern of the patter has been covering the protection of human subject in the study in PubMed of the time between 1975 and the year 2025. The human in the biomedical and clinical research is the most generalized and largest node that is positioned at the centre of the map. Themes and chronological sequences around this nexus have been noted to exist in groups. The problems described in blue and purple clumps are outdated literature (1975 to 1995) according to the studies topics, clinical trials, consent in studies and ethics. The latter terminologies were more popularized in the later 20 th century where a greater focus was put on the regulation of research and milestones that included things such as the Belmont Report (1979) and other ethical codes existed.

The contemporary fascination with the tenets of responsible research conduct and policymaking is captured in the language (government, jurisprudence, and even ethics committees). The green and the yellow clusters can be related to the growth of the research during the mid-1990s-2010s. Central words e.g. antioxidants, carotenoids, vitamin E, lipids and inflammation indicate an increase in nutritional and biochemical studies linking diet, oxidative stress and chronic disease such as cancer and cardiovascular disease. The reality of the greater reflection of the growth of apoptosis and blood pressure simultaneously implies the development of the evidence-based medicine and clinical nutrition, the results of experiments and it's in the activity of the human health sphere. Since the 2010s, infectious diseases and immunology entered the spotlight and the notions of vaccination, antibodies, viral infections, HIV and influenza became the focus. The clusters can be compared to the global health pandemics (e.g. HIV / AIDS pandemic, influenza pandemics, and the more recent coronavirus pandemics) that prompted the vaccine development research and led to the emergence of new ethical concerns. The newer groups like oral health, caries, fluoride and dentifrices resonate of preventive community-based interventions, which the terminology of behavioural research, health knowledge, attitudes and social class reflect the increasing significance of psychosocial and cultural determinants of the health. Over time, the network will grow more complex and interdisciplinary in nature; examples of a move towards less fragmented research focus to a more integrated biomedical and social research environment include nutrition and cardiovascular health, or vaccines and ethics.

Figure 2. Bibliometric analysis of "Protection of human subject" using VOSViewer with data from PubMed. (Data may vary slightly).

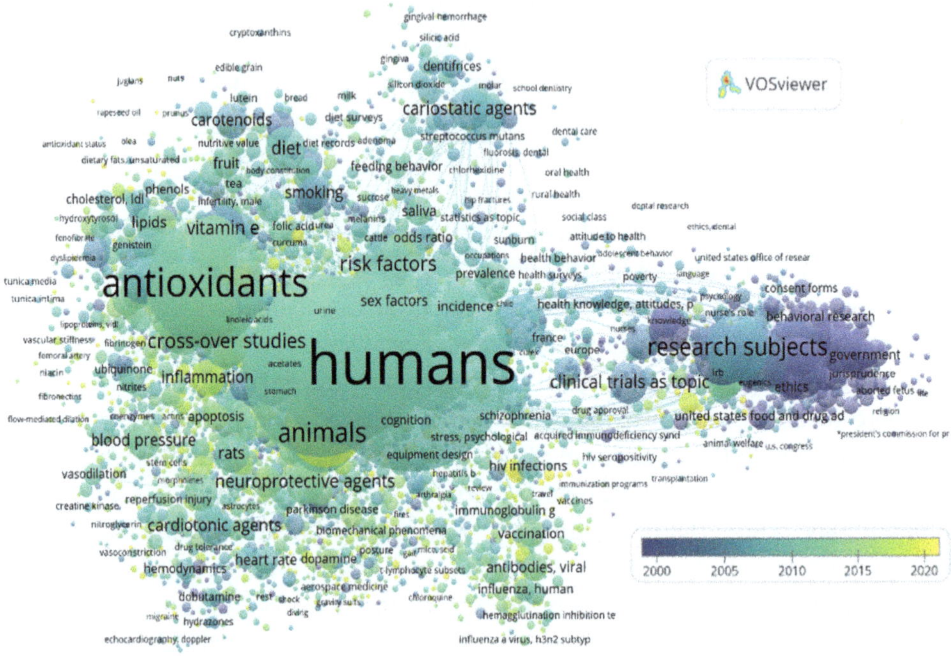

Figure 3 Visualisation of density of bibliometric data at graduation nodes impresses the intensity of the development of research. Outliers are the terms mentioned on the right hand of the map, which are thicker i.e. represent a high density of ethical and regulatory concerns such as research subjects, clinical trials as topic, ethics, consent forms and government. This is echoed in the history of the late 20 th century era whereby there was an established set of international norms that outlined the right of autonomy, beneficence, and justice. This demonstrates that ethics in research was an unquestionable aspect of a research study that was conducted on human beings. On the left-hand side, we may have nutrition and biochemistry with the following names: antioxidants, vitamin E, carotenoids, lipids and inflammation. The period between 1990s and 2010s can be described as a decade of mounting interest related to oxidative stress, dietary health and chronic disease prevention. The nomenclature of these involves apoptosis, neuroprotective agents, blood pressure and cross-over studies highlighting the possibilities of migrating laboratory work toward clinical application. The concentration at the lower right of the vaccination, antibodies, HIV infections, influenza and immunoglobulin show how much research has risen in response to global health threat, especially, epidemics like HIV/AIDS, influenza.

This also justifies why the global crisis is likely to stimulate innovation in the public health and immunology.

Lastly, there is the recent clusters (2015-2025) and it is the focus on oral and behavioural health, including oral health, dentifrices, Streptococcus mutans, health knowledge, and behavioural research. These are signs of more of an interest in preventative health and social and behavioural predispositions to wellbeing. In most cases, however, the map shows that it has over the years displayed a transition between ethics and governance and subsequently to translational research on the biochemical research and most recently to the control of infectious disease and subsequently interventionist and transdisciplinary approaches.

Figure 3. Bibliometric analysis of "Protection of human subject" using VOSViewer with data from PubMed. (Data may vary slightly).

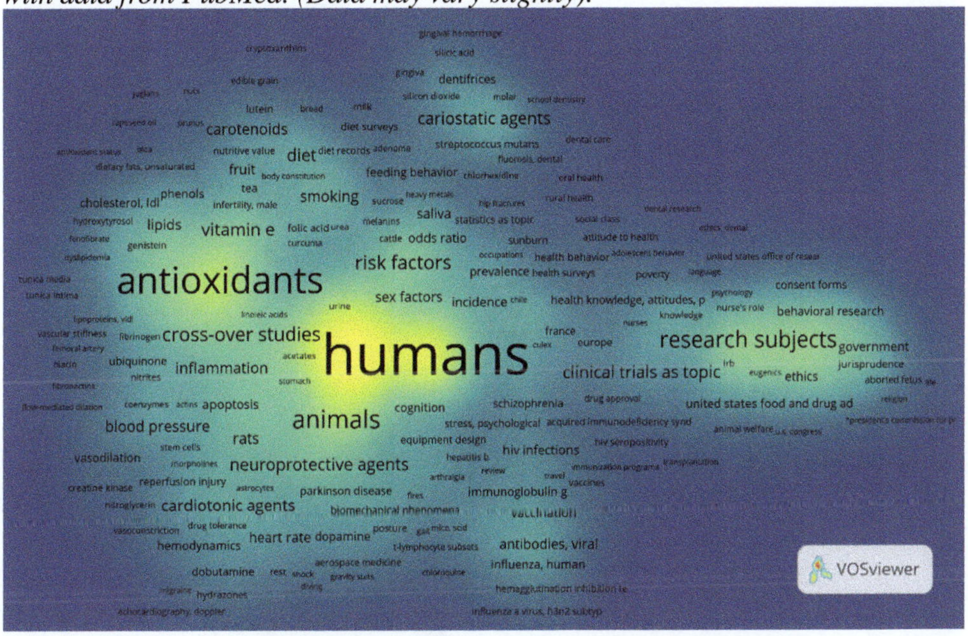

Figure 4 presents a network. It means that the Lens data have been utilized to draw the bibliometric map of the VOSviewer which shows the field of the study in the human experimentation and in the ethics. Human would be the most central and dominant node that illustrates that the purpose of biomedical and behavioural studies is to protect the subjects. This linking of the keywords and catch words that are used in a combination of biomedical and behavioural research, human testing, ethics and the United States provides the insight of how ethical, legal and policy frame work of human subject research have been instrumental in the debate. The red cluster has such ethical ideas informed consent, privacy, confidentiality and vulnerable populations, and the history factors in the Helsinki Declaration put into the spotlight. This green cluster is further narrowed to administration and control and terms such as government, risk assessment, public policy and ethical review have been labelled green and thus classified as government regulation and funding which have effects on the commencement of the research activities. The code, upon which the practice is focused on political and institutional practices, is the one that focuses on legal means, and terms related to the laws in the centre of focus govern the government. The other weak themes include the minor ones like the demographic ones (younger age, adults), researcher evaluation and staff. It implies that because research ethics are interdisciplinary, then there is an interaction between these categories in a direct sense of intersection of law, policy, medicine and human-rights in a sense that they interact to form the effect of well ethically developed scientific development that is ethically developed to meet the provisions of dignity, safety and justice.

Figure 4. Bibliometric analysis of "Protection of human subject" using VOSViewer with data from Lens. (Data may vary slightly).

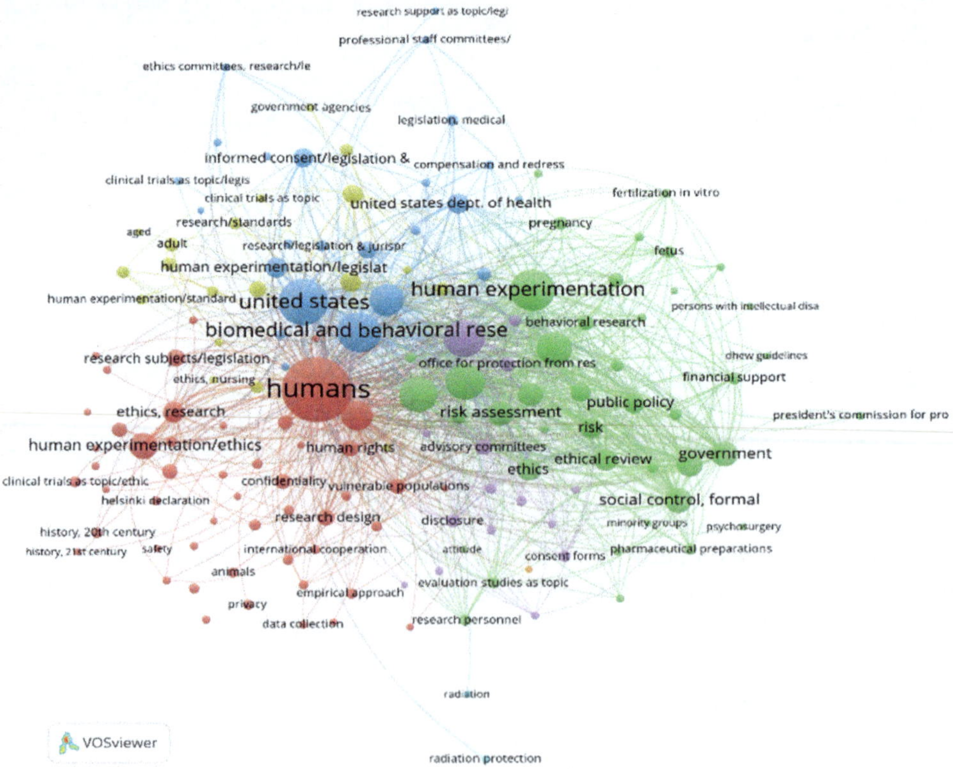

Figure 5 shows the growth and fall in the bibliometric index of the Open access (OA) and Non-Open access (Non-OA) articles due to the protection of the human subjects over the years. The following statistics was given by Lens, based on which the OA and Non-OA publications on this topic became a noticeable and force citation only in the 1980s. Since the 1990s, and since 2000 in particular, there has been an exponentially increasing number of references to Open Access. This is in the wake of the world research ethics debate, informed consent of the research subject and protection of vulnerable groups that have been developed through the international research statements and regulatory paradigms. It becomes quite clear that the application of the Open Access sources is considerable, in terms of citation, in comparison with the non-OA sources i.e., the information on ethical principles, risk assessment and human rights in research is highly diverse and it is accessible to the scholars and policymakers. Peaks in OA citations during the 2000s and 2010s coincide with policy reforms, digital repository expansion, and public demands for transparency in biomedical research. Non-OA works are comparatively less established and, thus, their presence in the international discussion of the human subject protection is less widespread due to the inaccessibility. Overall, this trend testifies to the fact that the Open Access facilitated the dissemination of the knowledge that is essential to guarantee human subjects and reach a wider access to the ethical values and principles of regulation.

Figure 5. Comparison between open access and non- open access over time analysis of "Protection of human subject" using VSCode with data from Lens. (Data may vary slightly).

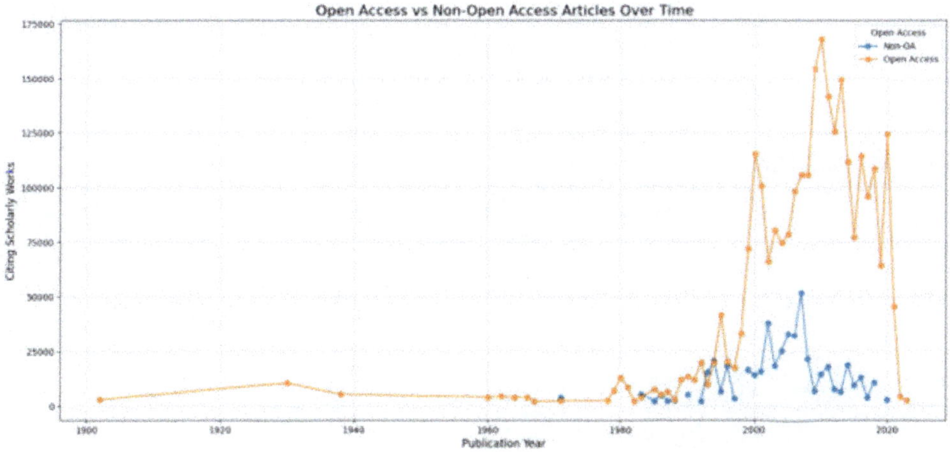

Figure 6 illustrates the tree map, which gives a summary of the most visited journals by the number of documents that have been added to Lens data and the fact that scholarship is being written on such topics as ethics, biomedical research and protection of human subjects. The largest unit of the PLOS ONE is nearly 19,000 documents, and, in the case in point, the polytheistic character of the magazine, and publication of the articles, where medical ethics, protection of human subjects, and preservation of the well-being of the population are discussed. SSCN Electronic Journal and scientific Reports are no exception and it is the testament of the contribution preprints and general scientific publishers to the communication of the discussions of ethical structures and responsible research practices. The remaining ones that are of importance are Frontiers in Immunology International Journal of Environmental Research and Public Health, Sustainability and Nature. These journals delimit the safeguarding of human subject's domains as wide as they are immunology, environmental hazards, social duty and biomedical invention. Scientific journals like Journal of Virology, Infection and Immunity and Journal of Immunology have contributed in illuminating the significance of clinical and laboratory sciences in these fields and in the case of a study which is about humanity the moral aspect has to be placed in a perspective. The fact that today Open Access journals such as PLOS ONE, Scientific Reports and Sustainability take the leading positions in the ranking can help to demonstrate that the free research will be capable of attaining the international level in terms of the question of the safety of human subjects. The latter distribution results in transparency and access based multi-disciplinary research environment that is ethics-related.

Figure 6. Tree map of journals of "Protection of human subject" using VSCode with data from Lens. (Data may vary slightly).

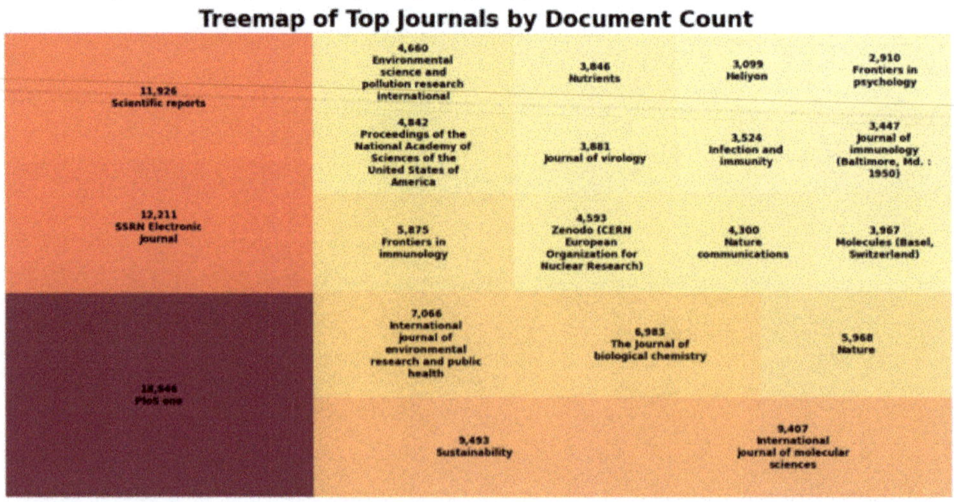

Figure 7 displays the world map as the visualization of the number of research documents created by each country, according to the Lens database. This map has been used to determine the world distribution of research, especially research using human subjects. As seen in the map, the United States records the largest number of publications as it has a long historical record of ontological and empirical research in both biomedical and social sciences. Some of the best-established ethical theories in these areas have been found to be central in the research conducted on human participants, one of them being the Belmont Report. Another significant contributor is China, which signifies that it invests the significant amount of money in research and cares more about ethical standards concerning human subject protection. A variety of European nations, including the United Kingdom, Germany, and France, have a high presence, indicating a high level of research ability and compliance with strict ethical guidelines, such as the General Data Protection Regulation (GDPR) by the European Union and existing code of medical ethics. The international character of the research in the field is also manifested in other countries, such as India, Japan, and Canada. Comparatively low research production in some areas of Africa and South America, in contrast, demonstrates unequal opportunities in research capacity and ethical infrastructure in the research field. All in all, this bibliometric review demonstrates that despite the fact that the knowledge production is still concentrated in the developed areas, the ethical issues, which concerns the human subject protection, are becoming global.

Figure 7. World map of document count by county using VSCode with data from Lens. (Data may vary slightly).

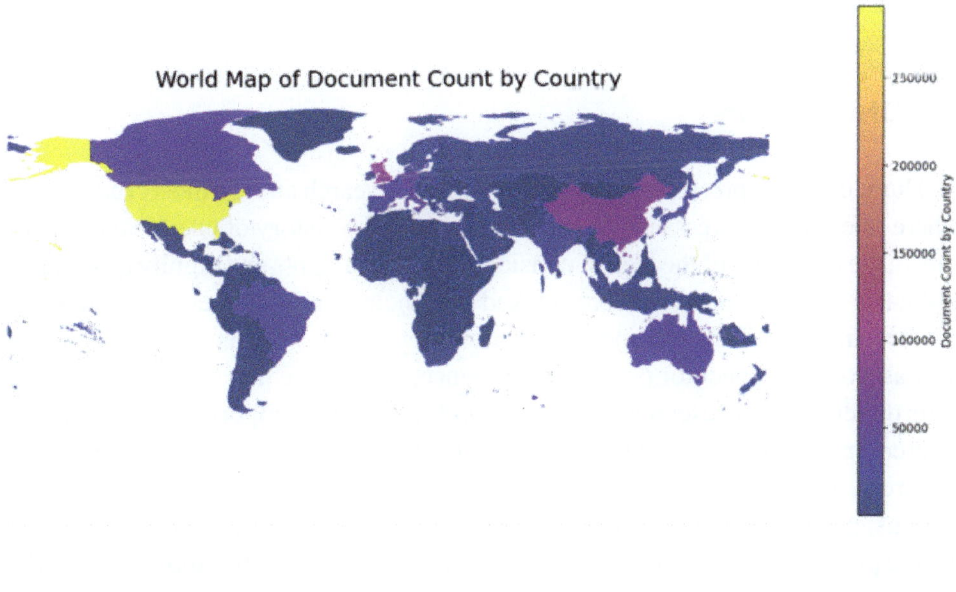

Based on a bibliometric comparison of the PubMed and Lens databases, it is observed that the two databases are similar in terms of overlaps and their differences regarding the coverage, focus, and thematic trends regarding human subject protection. PubMed being a database whose main focus is biomedical and clinical researches, focuses on concerns related to the medicine and health sciences. The bibliometric maps of it present the prevalent clusters around human participants, clinical trials, informed consent, and research ethics that are supported by such subject areas like clinical nutrition and biochemistry, infectious diseases and immunology, cardiovascular health, and preventive dentistry. Between 1975 and 1995, the literature was mainly interested in the background ethical principles, informed consent, and regulations like the Declaration of Helsinki and the Belmont Report. The period of the 1990s-2010s is characterized by the growth of research in clinical nutrition, biochemistry, and evidence-based medicine. Following 2010, health problems worldwide (HIV/AIDS, influenza, the COVID-19 pandemic, etc.) turned the focus on the ethics of vaccination, immunology, and behavioural health. The Lens database on the other hand offers a more interdisciplinary view and incorporates biomedical research with governance, legal, social, and policy aspects. Major themes are privacy, confidentiality, vulnerable populations, government regulation, risk assessment and public policy. Another point Lens makes concerns the trends in publications and in particular the development of open access literature since the beginning of the 21st century that has had a significant contribution to the spread of ethical standards around the world. PubMed and Lens provide a complementary view of the issue together, demonstrating how the protection of human subjects is formed by the collaboration of clinical practice and the more general ethical governance structures.

4. CONCLUSION

Human subject protection is the key to ethical research and the most crucial point where scientific progress and ethical duties meet. The history of unethical trials has shown the outcomes of the lack of consideration of the rights and dignity of subjects, and basic ethical principles were worked out, including the Nuremberg Code and the Declaration of Helsinki and the Belmont Report. These principles have formed the basis of the respect for person, beneficence, and justice that are still used as the main direction of the research ethics today and serve as the support of the participant-centred approach. As pointed out in this chapter, ethical protection is not a fixed requirement but a dynamic process which is influenced by the changing research designs, populations and technologies. The use of informed consent is a primary ethical protection, but there has been a growing complexity in the application of this ethical protection especially in the context of research on a vulnerable group like children, prisoners, the elderly, and individuals with disabilities. In ethical research, there must be a delicate balance to ensure that these groups are not exploited and at the same time, do not feel excluded of any fair stake in studies which could have some beneficial value. Innovations in genomics, artificial intelligence, biobanking and research driven by data also present a challenge to conventional ethical frameworks that brings with it a complex of issues surrounding privacy, data ownership, future data use and algorithmic bias. Those developments require changes in ethics governance systems, which requires more than procedural compliance, but focusing on transparency, accountability, and trust in participants. Institutional Review Boards and Research Ethics Committees are important and critical in this respect since they operationalize the principle of ethics, evaluate the ratio of risks and benefits, and advise the researcher to make complicated ethical choices in practice. The bibliometric analysis introduced in this chapter is in line with the conceptual discussion in that it shows how the ethical issues of human subject protection have been changing with the scientific and societal transformations. The theoretical, applied, and empirical viewpoints taken collectively highlight the fact that ethical safeguards must be regarded not as barriers to innovation, but as mandatory requirements of a responsible and believable research. In the end, scientific advancement cannot be achieved without considering ethical integrity, protecting the dignity, rights and well-being of the people on whom research is possible.

REFERENCES

Bertholf, R. L. (2001). Protecting human research subjects. *Annals of Clinical and Laboratory Science*, *31*(1), 119–127. PMID: 11314861

Felzmann, H., Sixsmith, J., O'Higgins, S., Chonnachtaigh, S. N., & Gabhainn, S. N. (2014). Ethical Review and Children's Research in Ireland. Centre of Bioethical Research and analysis (www.nuigalway.ie/cobra) and health Promotion Research Centre, National University of Ireland, Galway (www.nuigalway.ie/hprc)

Gostin, L. O. (1991). Ethical principles for the conduct of human subject research: Population-Based Research and Ethics. https://scholarship.law.georgetown.edu/facpub/764

Gupta, U. (2013). Informed consent in clinical research: Revisiting few concepts and areas. *Perspectives in Clinical Research*, *4*(1), 26. DOI: 10.4103/2229-3485.106373 PMID: 23533976

Hébert, J. R., Satariano, W. A., Friedman, D. B., Armstead, C. A., Greiner, A., Felder, T. M., Coggins, T. A., Tanjasiri, S., & Braun, K. L. (2015). Fulfilling Ethical Responsibility: Moving Beyond the Minimal Standards of Protecting Human Subjects from Research Harm. *Progress in Community Health Partnerships, 9*(S), 41–50. https://doi.org/DOI: 10.1353/cpr.2015.0021

Klovdahl, A. S. (2005). Social network research and human subjects' protection: Towards more effective infectious disease control. *Social Networks*, *27*(2), 119–137. DOI: 10.1016/j.socnet.2005.01.006

Kue, J., Szalacha, L. A., Happ, M. B., Crisp, A. L., & Menon, U. (2017). Culturally Relevant Human Subjects Protection Training: A case study in Community-Engaged Research in the United States. *Journal of Immigrant and Minority Health*, *20*(1), 107–114. DOI: 10.1007/s10903-017-0548-x PMID: 28239756

Largent, E. A. (2016). Recently proposed changes to legal and ethical guidelines governing human subjects research. *Journal of Law and the Biosciences*, *3*(1), 206–216. DOI: 10.1093/jlb/lsw001 PMID: 27774244

Lens. (2025). Retrieved, 2025, from https://www.lens.org/

Levine, C., Faden, R., Grady, C., Hammerschmidt, D., Eckenwiler, L., & Sugarman, J. (2004). A response to commentators on "The limitations of 'Vulnerability' as a protection for human research participants.". *The American Journal of Bioethics*, *4*(3), W32. DOI: 10.1080/15265160490508954 PMID: 16192138

Lynn, J. (2004). When does quality improvement count as research? Human subject protection and theories of knowledge. *BMJ Quality & Safety*, *13*(1), 67–70. DOI: 10.1136/qshc.2002.002436 PMID: 14757803

Manti, S., & Licari, A. (2018). How to obtain informed consent for research. *Breathe (Sheffield, England)*, *14*(2), 145–152. DOI: 10.1183/20734735.001918 PMID: 29875834

Manton, A., Wolf, L. A., Baker, K. M., Carman, M. J., Clark, P. R., Henderson, D., & Zavotsky, K. E. (2014). Ethical Considerations in Human Subjects research. *Journal of Emergency Nursing: JEN*, *40*(1), 92–94. DOI: 10.1016/j.jen.2013.11.001 PMID: 24384483

Parija, S., & Mandal, J. (2014). Informed consent and research. *Tropical Parasitology*, *4*(2), 78. DOI: 10.4103/2229-5070.138533 PMID: 25250226

Pisani, A. R., Wyman, P. A., Mohr, D. C., Perrino, T., Gallo, C., Villamar, J., Kendziora, K., Howe, G. W., Sloboda, Z., & Brown, C. H. (2016). Human Subjects Protection and Technology in Prevention Science: Selected Opportunities and challenges. *Prevention Science*, *17*(6), 765–778. DOI: 10.1007/s11121-016-0664-1 PMID: 27220838

PubMed. (2025). Retrieved, 2025, from https://pubmed.ncbi.nlm.nih.gov/

Roets, L. (2017). Protection of the human research participant: A structured review. *South African Medical Journal*, *107*(10), 847–853. DOI: 10.7196/SAMJ.2017.v107i10.12533 PMID: 29022527

Singer, E., & Levine, F. J. (2003). PROTECTION OF HUMAN SUBJECTS OF RESEARCH: RECENT DEVELOPMENTS AND FUTURE PROSPECTS FOR THE SOCIAL SCIENCES. *Public Opinion Quarterly*, *67*(1), 148–164. DOI: 10.1086/346013

Sng, B., Yip, C., & Han, N. (2016). Legal and ethical issues in research. *Indian Journal of Anaesthesia*, *60*(9), 684. DOI: 10.4103/0019-5049.190627 PMID: 27729698

Stineman, M. G., & Musick, D. W. (2001). Protection of human subjects with disability: Guidelines for research. *Archives of Physical Medicine and Rehabilitation*, *82*(12), S9–S14. DOI: 10.1016/S0003-9993(01)65984-7 PMID: 11805914

Tsan, M. F. (2025). Impact of the revised Common Rule on enhancing human research subject protections and reducing researcher burdens [Research Article]. *Qeios*, *7*(3), TU6VRN.2. Advance online publication. DOI: 10.32388/TU6VRN.2

Vanclay, F., Baines, J. T., & Taylor, C. N. (2013). Principles for ethical research involving humans: Ethical professional practice in impact assessment Part I. *Impact Assessment and Project Appraisal, 31*(4), 243–253. DOI: 10.1080/14615517.2013.850307

VOSviewer. https://www.vosviewer.com/

Wendler, R. (2012). Human Subjects Protection: a source for Ethical Service-Learning practice. In *Michigan Journal of Community Service Learning* (pp. 29–39).

White, M. G. (2020). Why human subjects research protection is important. In *Ochsner Journal* (Vol. 20, pp. 16–33). https://doi.org/DOI: 10.31486/toj.20.5012

Xu, A., Baysari, M. T., Stocker, S. L., Leow, L. J., Day, R. O., & Carland, J. E. (2020). Researchers' views on, and experiences with, the requirement to obtain informed consent in research involving human participants: A qualitative study. *BMC Medical Ethics, 21*(1), 93. Advance online publication. DOI: 10.1186/s12910-020-00538-7 PMID: 33008387

ADDITIONAL READING

Capron, A. (1991). Protection of research subjects: Do special rules apply in epidemiology? *Journal of Clinical Epidemiology, 44*, 81–89. DOI: 10.1016/0895-4356(91)90180-H PMID: 2030401

Davis, A. M., Hull, S. C., Grady, C., Wilfond, B. S., & Henderson, G. E. (2002). The Invisible Hand in Clinical Research: The Study Coordinator's Critical role in Human Subjects Protection. *The Journal of Law, Medicine & Ethics, 30*(3), 411–419. DOI: 10.1111/j.1748-720X.2002.tb00410.x PMID: 12497701

Islam, M. R. (2014). Informed consent of human subjects: A review. *Bangladesh Journal of Bioethics, 5*(1), 20–35. DOI: 10.3329/bioethics.v5i1.18444

Jonsen, A. R. (1978). Research involving Children: Recommendations of the National Commission for the Protection of Human Subjects of Biomedical and Behavioral Research. *Pediatrics, 62*(2), 131–136. DOI: 10.1542/peds.62.2.131 PMID: 693150

Loe, J. D., Winkelman, D. A., & Robertson, C. T. (2016). An assessment of the Human Subjects Protection review Process for exempt research. *The Journal of Law, Medicine & Ethics, 44*(3), 481–491. DOI: 10.1177/1073110516667944 PMID: 27587452

Oquendo, M. A., Stanley, B., Ellis, S. P., & Mann, J. J. (2004). Protection of human subjects in Intervention Research for Suicidal Behavior. *The American Journal of Psychiatry, 161*(9), 1558–1563. DOI: 10.1176/appi.ajp.161.9.1558 PMID: 15337642

Ross, L. F., Loup, A., Nelson, R. M., Botkin, J. R., Kost, R., Smith, G. R.Jr, & Gehlert, S. (2010). Nine Key functions for a Human Subjects Protection Program for Community-Engaged Research: Points to consider. *Journal of Empirical Research on Human Research Ethics; JERHRE, 5*(1), 33–47. DOI: 10.1525/jer.2010.5.1.33 PMID: 20235862

Wolf, L. E., Patel, M. J., Tarver, B. W., Austin, J. L., Dame, L. A., & Beskow, L. M. (2015). Certificates of Confidentiality: Protecting human subject research data in law and practice. *The Journal of Law, Medicine & Ethics, 43*(3), 594–609. DOI: 10.1111/jlme.12302 PMID: 26479569

KEY TERMS AND DEFINITIONS

Artificial intelligence: Machines that have been taught to think and learn like people are capable of simulating human intelligence by seeing, reasoning, solving problems, and acting in ways that accomplish particular objectives.

Autonomy: In the context of safeguarding human research, autonomy refers to the moral precept of upholding a study participant's right to self-governance and self-determination.

Human dignity: Human dignity is the idea that all people are valuable and valuable just by virtue of being human, regardless of their accomplishments, social standing, or other traits.

Human subject research: A systematic study that employs interaction or intervention with real individuals to produce knowledge that can be applied to other contexts is known as research with human subjects. It involves either directly contacting or coercing individuals to obtain information or biospecimens, or studying their personally identifiable information or biospecimens in order to further knowledge.

Justice: Justice research is a branch of study that aims to understand and advance justice in a variety of circumstances, especially in social policy and research ethics, by guaranteeing an equitable distribution of advantages and disadvantages.

Research ethics: Research ethics are the moral principles and guidelines that govern the planning, execution, and reporting of research to ensure its ethical conduct, protect participant welfare, and maintain the integrity of scientific findings.

Trust in research: People's or communities' readiness to expose themselves to researchers and organisations with the hope that the researchers will behave competently, honourably, and in the participants' best interests

Chapter 4
Borrowed Words, Broken Ethics: Addressing Plagiarism in Academia

Jashanpreet Kaur Sarao
http://orcid.org/0009-0003-9920-3990
Chandigarh Group of Colleges Jhanjeri, Mohali, India

EXECUTIVE SUMMARY

Plagiarism in academia is a notorious and emerging phenomenon that violates the essence of values in the academic fraternity, such as intellectual honesty, trust and originality. It is a theoretically, empirically, and technologically informed chapter that provides a deep analysis and starts by providing a contextual picture of the historical and philosophical basis of plagiarism and intellectual property. It uses ethical theories, including ethical deontology, consequentialism, and virtue ethical theories, to understand moral reasoning, and discusses cultural and educational factors that affect how people perceive plagiarism. There are challenges in establishing intent and accountability that are reflected in global case studies. Academic pressure, cultural misperceptions, and lack of policy are some root causes explored in the chapter, as well as their consequences to individuals and institutions. Finally, it states that fighting plagiarism not only needs enforcement but also needs a culture of academic integrity that is durable to absorb the transformation of technology.

INTRODUCTION

Plagiarism, which is traditionally described as a representation of another person's words, thoughts, or intellectual activities without due credit, has become one

DOI: 10.4018/979-8-3693-8302-5.ch004

of the key ethical problems in science and research circles. It touches the core of academic honesty and corrupts the concept of the knowledge creation and spread on which it is premised. The practice of plagiarism has been on the rise because it has become more visible and more complex, starting at the student level with young people in undergraduate and postgraduate courses, moving up to faculty and researchers and gaining a lot of concern among society and institutions take a lot of specific interest. The circle of academics survives on originality, openness, and proper source accreditation. Plagiarism is an irregularity that challenges not just the morals of individuals but the institutional futility of academia in publishing, managing and rating scholarly contributions. Even though tradition has regarded plagiarism as an issue of personal malpractices, modern-day rhetoric is establishing a wider socio-academic context of the sins, such as performance pressures, technological advancement, education disparities, and cultural relativism.

Caused by the various forces and having numerous forms, plagiarism is a matter that demands a delicate interpretation. It is not bothersome to just copy the material, but could be through small misattributions, improper paraphrasing, or even repurposing of the already published material without any disclosure. Velmurugan (2024) and Akbar (2018) note that the very definition of plagiarism is not unanimous either in terms of disciplines or cultures. The divergence between permissible reutilization and the misappropriation of ideas has been progressively obscured; however, the emergence of generative AI tools that generate text at an intense level with comparatively little human input is another reason (Eacersall et al., 2024; Ka & Chan, 2023).

The honesty during the research and learning process is significant. It is through integrity that scholarship maintains credibility, the students come out with the skills they are assumed to have, and research provides truth to the body of knowledge. The breaches to such integrity undermine such trust, not only trust in particular people but also institutions, or even academic fields. As it is stressed by Macfarlane, Zhang, and Pun (2014), academic integrity underlies the legitimacy of higher education itself.

Moreover, the digitalization of the process of teaching and researching has also brought about opportunities and weaknesses. On the one hand, online databases and tools for managing citations have simplified the work of scholarship in general; on the other, an online space can simplify copying, pasting, spinning, or creating with minimal moderation. Indeed, as Eshet (2025) shows in his analysis of plagiarism trends in the context of the COVID-19 pandemic, academic dishonesty has been pronounced due to the sudden transition to online learning and evaluations, and it is even more critical that institutions change their ethical standards.

This chapter seeks to unpack the complex nature of academic plagiarism regarding the history surrounding the phenomenon, the philosophies behind this malpractice, its categorization, and its case studies in the real world. It also explores how plagiarism changes in an age of artificial intelligence and digital tools, coupled with the provision of effective rules on prevention, detection, and ethics capacity building. The chapter uses the existing international literature, empirical research findings and the recent trends in research ethics as its resource to give a thorough understanding of the occurrence of plagiarism in academic documents and the solution to deal with it. After all, the aim is not simply to chastise indiscipline but to develop an academic culture of honesty, accountability and regard towards intellectual effort. Resting on the old ethical standards that guide the work of scholars and modifying them to suit the new challenges is how academia will be more able to guide students, teachers, and researchers to be in a position to cope with the ethical complexities within the knowledge economy of the 21st century.

After defining the increasing importance and changing nature of plagiarism in academic institutions, it is also important to learn how the phenomenon has been evolving over the years. The next part of the paper will follow historical backgrounds and philosophical views, which have minimalized the current demands of originality, authorship, and ethical scholarship.

HISTORICAL AND PHILOSOPHICAL CONTEXT

1. Evolution of Intellectual Property and Originality

Although plagiarism is regarded as an academic offense that is relevant today, the concept is rather complex and has an extensive history. During ancient times, imitation had, in many cases, been praised and not punished. The presupposition of learning by imitating the master rhetoricians was a feature of ancient Greece and Rome, and originality was not the major value as it is nowadays. The idea of intellectual property, that is, ideas and expressions as belonging to individuals, was developed substantially later, being especially dominant in the Enlightenment and the emergence of print culture. The technical legal protection of authorship began to pick up speed in the 18th century, with statutes such as the Statute of Anne (1710) in Britain being thought to be the first copyright law. This change was a process that indicated the increasing view of intellectual works as an individual, proprietary product. By the 19th century, as the professional scholar and author had come into their own, originality was not merely something to be appreciated but required. The concept of originality has come to play with academic excellence in the contemporary world of academia, especially in the 20th century, leading to the present structure of citation, attribution, and evaluation by peers in the academic world.

The academic publishing ecosystem is based on the integrity of citation practice to ensure the maintenance of the integrity of knowledge.

This historical and ethical basis shows that academic honesty is not only technically necessary, but also a very deep moral and cultural expectation. Having these bases, the second part focuses on the appearance of plagiarism in the modern academia and throws light on both the traditional and new forms of misconduct.

2. Ethical Theories Supporting Academic Honesty

As a philosophical condemnation, plagiarism is defined by accepted ethical thought as a concept lying behind the moral arguments in academia. Deontological Ethics is based on the writings of Immanuel Kant, who suggests that people must do right by the moral rules, no matter what the consequences might be. Deontologically, plagiarism is wrong in itself; one is acting against what is right regarding the aspects of truthfulness, respecting the work of other people, and being just. It is kind of a fraud, deceiving audiences of the origin of ideas, and thus contravenes the basic responsibility of telling the truth.

Consequentialism (most frequently, in the guise of utilitarianism) considers actions in terms of results. Plagiarism may be viewed as an unethical practice because the implications and effects are negative: it compromises academic integrity, disorients the readers, distorts the measures that define academic achievements, and may result in unreliable research being disseminated. In addition, it denies original writers fame and prevents real scholarship.

The school of Virtue Ethics is based on the Aristotelian school of thought, and its focus is on the training of character and moral growth. An honest scholar appreciates humility, hard work and integrity-qualities that cannot be combined with plagiarism. Based on this perspective, the necessity to avoid plagiarism cannot be reduced to the matter of rules but concerns the need to become an ethically responsible scholar. Schools can, thus, contribute to the nurture of virtues via mentoring, moral education, and example offered by schools (Macfarlane et al., 2014).

Together, these schools of thought add weight to the value of academic honesty as a moral requirement that has a principled basis behind both institutional and personal responsibility.

3. Cultural Differences in Understanding Plagiarism

Although plagiarism is usually taboo in international academic policy, customs interpret it differently, and the concept differs in different conditions. Originality, authorship, and referencing are some of the non-negotiable academic standards in most institutions in the West. Though in other parts of the world, especially collectivistic ones, e.g., parts of Asia, Africa or the Middle East, knowledge has long been held as a communal property, and memorisation of authoritative works is considered a sign of honor (as opposed to intellectual plagiarism). Research findings like the ones stated by Sowden (2005) and Pennycook (1996) have been able to demonstrate that international students might not see replication of existing ideas without credit as having unethical issues, or at least putting it as a major priority, considering that the learning systems in their respective countries have demanded rote learning instead of critical participation. In addition to that, the concept of voice in writing, which is such an essential Western concept in academic style, is not always given much stress. This cultural conflict usually results in unintended plagiarism by the non-with English speakers and foreign students who have not been initiated into the Western citation rules. The ensuing difficulties are both pedagogical and ethical: not only do the students have to be trained to adopt a new form of referencing technique, but also adjusted to a new world outlook in which original synthesis is more important than imitation (Introna, Hayes, Blair & Wood, 2003).

The existing cultural tensions thus have to be met sensitively, with institutions finding a balance between propagating academic standards and teaching international norms. Violations only punished can isolate further and only continue misunderstanding. The historic and philosophical picture of plagiarism demonstrates that the problem has wide moral, cultural, and legal aspects and is not only a policy violation. Knowing how it has become an acceptable mimicry to a punishable offense and the moral values that guide strict academic integrity allows educators, policymakers and learners a more informed, understanding framework of prevention and correction. With further globalization and digitization of academia, recognition of this complexity will be a significant step in developing a just and inclusive environment in the scholarly world.

FORMS AND TYPES OF PLAGIARISM

Plagiarism as a crime is not a single character - there are many versions, both consciously malicious, and unwittingly, as a fault of a novice. Categorizing plagiarism facilitates the explanation of its degree, motivation, and educational or disciplinary actions to deal with plagiarism. This topic discusses the continuum of plagiarism, ranging from the traditional to new online issues.

1. Intentional Plagiarism

Intentional plagiarism can be defined as the conscious practice of presenting the work or ideas, or the expressions of another person and portraying them as one's own, as a remarkable violation of academic ethics and usually accompanied by severe consequences (Bretag et al., 2019). Perhaps the most obvious and widely accepted one is copy-pasting without references, where a quoted phrase, sentence or paragraph is taken verbatim, without any reference to the original writer. This kind of plagiarism is often identified with the assistance of such tools as Turnitin, which reveals similarities in a text. As an example, the student may copy direct paragraphs of a journal article into their term paper, omitting quotation marks or references thereof. The next variation is ghostwriting, wherein a person hands in a piece of work written by another person, usually as payment, in circumstances generally referred to as essay mills. Academic integrity is at risk from such services, and the essay mill market is growing fast, occupying the institutional gaps that are difficult to discover (Lancaster & Clarke, 2016). Tightly connected is the concept of contract cheating, which is similar to outsourcing assignments to third parties, such as freelancers and online platforms, but there is an obvious transactional relationship between the student and the recipient of the service. The said practices serve to compromise fair evaluation and destroy authentic student learning.

2. Unintentional Plagiarism

Accidental plagiarism is the product of ignorance, misunderstanding and uneventful scholarly habits, and although the offspring has no evil intention, it is still textbook infringement (Pecorari, 2003). Poor paraphrasing is one of the common ways in which plagiarism is done, where a student copies material with words too close to those of the source or uses synonyms, but in the same format. With these changes in place, the content is still not out of the problem, barring that it has been properly cited (Sutherland-Smith, 2008). As an example, instead of giving a source, modifying a phrase such as the following, "Climate change impacts on agricultural yields", to read as "Agricultural outputs are impacted by climatic changes" would be plagiarism. Inaccurate or incomplete citation is another common problem, and in this case, students only address some of the borrowed material or fail to include vital information like the name of the author, date of publication, or page numbers, thus cluttering the intellectual property and bewilderment of the readers (Ashworth et al., 1997). Further, ignorance is also an important factor, especially among first-year students or other international academics who fail in their first year to get used to the art of referencing and academic writing formatting (Gullifer & Tyson, 2010). This is a challenge that necessitates the adoption of formal training in research ethics and correct referencing in ensuring integrity in academic work (Devlin & Gray, 2007).

3. Self-Plagiarism

Self-plagiarism is described as the repetition of already submitted or published work of an individual with the lack of due cognizance. This becomes unethical when the author deceives the readers that the work is purely original, even though he or she remains the same. The most common of them are the attempts to submit the same paper to a variety of courses, or present a substantial part of previous research as new work without any notification. These actions, as analyzed by the Committee on Publication Ethics (COPE, 2019), create a distortion of academic history and an unreal increase in the number of works by an author, which in the end will weaken the level of scholarly research appraisal itself.

4. Digital Plagiarism

As the number of digital tools and online platforms increased, plagiarism has taken new forms, which are usually hard to identify and morally problematic (Jereb et al., 2018). Among them is the implementation of AI-produced work, which allows essays or research summaries to be written using such instruments as ChatGPT and Writesonic. Although this text might not be copied in the conventional meaning, the absence of crediting the AI assistance reduces the clarity of transparency and authorship and leaves question marks behind whether the AI should be referenced as a co-author or contract cheating (Floridi & Chiriatti, 2020). Despite the lack of firm policies in most universities, academic publishers have started demanding the disclosure of AI use. Text spinners and paraphrasing tools are also an emerging practice; these are pieces of software that automatically reframe writing to avoid detection by plagiarism software. Although such tools can contribute to the enhancement of languages, they are usually used to hide plagiarized work, thus crossing the ethical limits. Likewise, patchwriting, the strategic stitching of sentences assembled by several sources without references, forms the tapestry of uncited ideas, which is often found with poorly skilled academic writers and those who have insecurities about what and how to reference work (Howard, 1995).

5. Visual and Non-Textual Plagiarism

Plagiarism is not only verbal, as written text plagiarism, but one can plagiarize images, charts, tables of data, PowerPoint presentations, and source code. As much as in STEM and design fields, copying of graphs or visualizations and not giving credit or authorization are contraventions of the copyright laws and also against the ethics codes (Ercegovac & Richardson, 2004). In the same way, unethical academic practices are the infringement of presenting academic presentations with the inclusion of infographics of another person without attribution, since it falsifies the contribution of work and ideas.

The variety of different types of plagiarism, like a deliberate copying or the creation of a text by AI, helps to understand how overwhelming the problem can be. To further put the forms into context, the following section resorts to real-life case studies, which explain how plagiarism is played out in academic practice and what ethical, institutional, and personal dilemmas it involves.

CASE STUDIES ON ACADEMIC PLAGIARISM

Case studies can be seen as an effective method of determining the complexities of the real-life issue of plagiarism as compared to theoretical definitions. They put ethical breaches within the context of real-world realities, pointing at the motivation behind such breaches, their outcome, as well as the moral issues that emanate in academic situations.

Case 1: A Master's Thesis Under Scrutiny

One of the postgraduate students of sociology at a prestigious university, who subscribed to a master thesis, included significant blocks of passages that resembled published journal articles at the word-by-word level without quotation marks or in-text citations, yet the sources were listed in the bibliography, deceiving readers to think that they are original (Bretag et al., 2019). It created the question of whether this was a voluntary act of plagiarism or it arose due to a lack of formal training on how to write as an academic, as researchers observe that one of the common causes of poor citation skills is not out of nefarious intent, but rather a lack of professional training. The stakeholders were the student, the student's thesis supervisor, the academic committee and the original authors. After an inquiry, the university ascertained that the student had no formal citation training and, beyond implementing any punitive actions, the thesis needed to be revised under supervision and that there were mandatory research writing workshops that would be introduced to postgraduates (Bretag et al., 2019; Gullifer & Tyson, 2010).

Case 2: The Publish-or-Perish Trap

One example of plagiarism is by a mid-career member of faculty in a public university who was discovered copying paragraphs of another article conducted by a researcher, even after confessing that he/she was under intense pressure to achieve promotion requirements and departmental research statistics. In this case, there was an ethical question as to whether the systemic pressure, including the publish-or-perish culture, has a rationale to condone academic dishonesty since the literature recommends that such atmospheres could potentially encourage unethical behavior without the actual intention. The stakeholders were the faculty member who had written the scholarship, the editors and peer reviewers of the journals, the original author and the academic board of the university. The compromise was the retraction of the submission by the journal, a formal warning by the ethics committee of that university, and a tightening up of the policies of the evaluation department in the university to stress research quality instead of quantity.

Case 3: Plagiarism in Collaborative Research

A cross-institutional absoluteness of a paper on climate imperfection, presented to a high-impact journal, was discovered (after its publication) to have a duplication in the literature review, added to the earlier composition without citation that had been by a colleague at another college institute (Steneck, 2006). This case has created the ethical question of whether the blame for the instance of plagiarism on shared works should lie on the individual offender or all the writers included in shared works, since literature states that although accountability shared by all the authors is necessary, it is not always followed (Bretag et al., 2019; Steneck, 2006). The stakeholders were farmers and co-writers, the editorial board of the journal where the study was to be published, the funding institutions of the research, and the institutions with which they were associated. The solution was that the journal put a correction notice, the guilty co-author was suspended by his university, and the main author established the protocols of cross-checking to avoid this in the future (Resnik & Shamoo, 2017).

Case 4: The AI Conundrum

One undergraduate advanced entry in a final-year submitted an essay that was perfectly grammatical, and very analytical, in a way that was highly unpredicted by earlier works. With the exploration, the student confessed to utilizing ChatGPT to write the whole paper and has made a few corrections. This brought into question the issue of plagiarism of unacknowledged AI-generated materials and lately, the question of whether AI and the associated ritual strategy interframe-present a threat to pre-existing notions of authorship (Floridi & Chiriatti, 2020). The stakeholders were a student, a course instructor, university administration and the developers of AI tools. The resolution consisted of the student being given a zero mark on the assignment and not facing any consequences in regards to discipline because there was no defined policy used to deal with AI usage in the institution. The university then established the guidelines that require disclosure of AI assistants and introduced training of the faculty on AI literacy (Floridi & Chiriatti, 2020).

These examples have made it clear that plagiarism is not an isolated or ad hoc instance, but it intersects with institutional pressures, academic cultures and the changing technologies. In order to resolve this issue in a comprehensive way, the analysis of the underlying factors that contribute to the inclinations of individuals towards unethical behavior was to be performed, and the following section discusses them in detail.

LESSONS FROM THE CASES

In these scenarios, four themes keep reemerging. One, the issue of intent versus impact reveals that although intentional plagiarism should be punished harshly, inadvertent ones tend to create an awareness that there may be a lack of education in the field of academia, that shouldn points to the aspect of immorality (Devlin, 2006). Second, pressures of the system, namely the widespread culture of publish or perish, become visible factors in the induction of misconduct and thus require some institutional changes to strike a balance between productivity and ethical research conduct. Third, the principle of shared responsibility in collaborative projects denotes that every author should take the precautionary measures because mutual accountability is necessary in ensuring the integrity of scholarship (Steneck, 2006; Bretag et al., 2019). Lastly, technological disruption by the use of AI technology like ChatGPT is helping to topple the traditional ideas of authorship and originality, resulting in an urgent sense of policy and ethical guidelines that deal with the new frontier and its challenges.

ROOT CAUSES OF PLAGIARISM

It is important to understand the reason behind plagiarism to control and prevent it. Although plagiarism is generally viewed as a conscious attempt at deceit, the reasons behind it are not simple and one-dimensional. These factors include flaws with a structural approach to education to psychological pressure, poor teaching, and the changing online standards (Abbasi et al., 2021; Awasthi, 2019; Neupane, 2023).

1. Academic Pressure and Competition

The high levels of performance pressure by students and faculty are cited as one of the most common reasons attributable to plagiarism. Having achieved high grades associated with scholarships, internships, or postgraduate opportunities, the students are usually motivated to cheat when exam day comes (Comas-Forgas & Sureda-Negre, 2010; Macfarlane et al., 2014). Faculty members are likely to indulge in unethical endeavors to achieve institutional research goals or to be able to get promotions, particularly in the cases of publish-or-perish cultures (Aguinis & Henle, 2004; Cavaliere et al., 2020). In this kind of competitive undertaking in academia, an emphasis on the end product instead of the procedure creates environments whereby integrity is often sacrificed to achieve an end outcome (Taylor, 2024).

2. Lack of Training in Research Ethics

Lack of research ethics and academic writing education is the cause of many cases of plagiarism, especially among students (Akbar, 2018; Mulenga & Shilongo, 2024). It is possible that students are not fully aware of what makes up plagiarism and even subtle behaviors that can be regarded as plagiarism, like paraphrasing without referencing to original material or creating and repurposing their own previously written material without giving credit (Velmurugan, 2024; Roka, 2017). In academic courses, achieving proficiency with the content rather than creating methodological and ethical competencies is the focus (Reijers et al., 2018). Moreover, foreign scholars who do not know about Western rules of citation can violate academic norms without intentionality because of cultural variances of the terms concerning intellectual property and authorship perception (Vlad & Vlad, 2021).

3. Poor Time Management and Academic Burnout

The cost of mismanaging time contributes significantly to plagiarism because students who fail to manage their time challenge themselves and may be tempted to use shortcuts like selecting and pasting text from online sources, resubmission of previously completed assignments, or even using generative AI without knowing much about its consequences (Ka & Chan, 2023; Pudasaini et al., 2024). Additionally, motivation and judgment can be lost due to the academic burnout generated by excessive working loads, which leads people to make ethically corrupted decisions just to get it done (Abbasi et al., 2021).

4. Misunderstanding of Citation and Referencing Standards

It is not only the technical requirement because it reflects academic honesty; however, several students have issues with distinguishing between direct quotation and paraphrasing, knowing when they should cite, what style to use (APA, MLA, Chicago, etc.), and figuring out the purpose of secondary citations. This miscommunication, especially in the fields that imply the use of textual proofs, leads to unintended theft most of the time (Akbar, 2018; Velmurugan, 2024).

5. Overreliance on Digital Tools and Online Resources

The digital era found poor lines between the creation and access due to the ease of access to large quantities of information, but helped to blur the edges between authorship and accessibility (Foltynk et al., 2019; Khaled & Al-Tamimi, 2021). Copy-paste culture actively invites students to copy the work verbatim and reduce integration into their voices (Awasthi, 2019), and conversely, students believe that the copypaste culture belief that they might think of stealing the content without any citation (Taylor, 2024). Further, the increased prevalence of generative AI applications such as ChatGPT devices in the absence of an effective set of ethical principles threatens to promote hidden plagiarism (Eacersall et al., 2024; Pudasaini et al., 2024).

6. Language Barriers and Communication Skills

The non-English speakers may find it difficult to explain complex ideas thus tend to use a lot of source material to represent the arguments. The dependence can be expressed in the form of a high frequency of copying/pasting direct text rather than putting it in their own words because of linguistic vulnerability, or in the belief that a couple of words swapped will produce a work of their own (Neupane, 2023; Vlad & Vlad, 2021). Although such practices might not necessarily bear any malaise intent, most academic policies consider such to be plagiarizing (Mulenga & Shilongo, 2024).

7. Weak Institutional Policies and Enforcement

Plagiarism policies in certain academic contexts are unclear or inconsistently applied, and there is uncertainty as to what is acceptable (Loban et al., 2023; Cavaliere et al., 2020). In cases when lessening codes are not reported, they de facto precede more significant transgressions, and faculty might be disinterested in submitting cases because the procedure takes too much time, and they might face adverse effects (Taylor, 2024).

8. Cultural Differences in the Concept of Authorship

Different cultures give different interpretations to plagiarism (Roka, 2017; Neupane, 2023). Some Eastern traditions look at copying texts authoritatively as a symbol of respect, whereas in oral knowledge or indigenous knowledge traditions, the flow of ideas is considered to be a community and not a personal property (Vlad & Vlad, 2021). Reliance on direct replication can also be promoted by educational systems that put less weight on critical analysis and more on rote learning and memorization (Macfarlane et al., 2014).

9. Inadequate Supervision and Mentorship

The insufficiency of meaningful mentoring on academic writings and research may leave the students feeling abandoned (Mulenga & Shilongo, 2024). Even though graduate students have very low levels of feedback, they might resort to dubious means to accomplish their task (Miranda-Rodriguez et al., 2024), and faculty mentors might accidentally condone such practices by ignoring initial indicators of a low-quality research approach (Aguinis & Henle, 2004).

10. Normalization of Minor Offenses

Minor cases of dishonesty, like copying the sources of the term on Wikipedia, receive no punishment or, at best, are simply dismissed, thus creating a culture where plagiarism comes quite naturally to students, they tend to feel that everyone does it and ethical standards become slowly downgraded (Comas-Forgas & Sureda-Negre, 2010; Awasthi, 2019).

The identification of underlying plagiarism causes as complex (structural, psychological, cultural, and technological) builds a platform of formulating some of the most specific and effective interventions. Having this background, the chapter now moves on to the institutional responsibilities, which is how the universities and the academic systems can be responsive and ethically responsible.

CONSEQUENCES OF PLAGIARISM

Plagiarism is not only an act that goes against the rules of the particular institution, but also a misconduct of ethical behaviors with serious academic, legal, and professional consequences. It is done both purposefully and due to the lack of intent, but either way, it damages the credibility of scholars, as well as the confidence of the general population and distorts the core principles of education and research.

1. Academic Penalties

Depending on the severity, intent, and frequency of the offense, they are also proportional to the consequences in academic establishments, where minor or first-time plagiarism may result in a warning, or even a failing grade, and repeat or serious violations may be cause to fail the course or be suspended or put on academic probation to undergo some process of rehabilitation or risk expulsion, especially in thesis or dissertation work (Bretag et al., 2019), and even the loss of degrees when discovered, sometimes years later, which can lead permanently to reputational damage.

2. Legal and Reputational Damage

In some situations, plagiarism encompasses copyright infringement thus creating possible legal implications like unauthorized reproduction of information under copyright infringement and in violation to the intellectual property acts, expensive suits and settlements involving high profile cases and even high claims of libel and defamation depending on allegations, made which are proven to be false; beyond legal prosecution, plagiarism may also cause irreparable reputational problems that may jeopardize job security, publication rights, and academic credibility.

3. Impact on Institutions and Research Credibility

Failure to deal with plagiarism adequately may result in the deterioration of institutional reputations because of declining international rating as the perceived slackness in academic standards, possible loss of accreditation or funding because misconduct becomes too persistent to be tolerated by the granting agencies and foundations, collapse of international co-research due to a breach of trust, and retractions of already published works, which negatively affect not only the reputation of the individual authors but also the institution itself (Bretag et al., 2019).

4. Career and Professional Setbacks

The consequences of plagiarism can permeate into the workplace, having included loss of job opportunities leaving some employers to avert the job application of a candidate with a documented track record of dishonesty, watering down of careers by resigning the services of faculty members or researchers especially under publicly funded projects where misconduct is detected, loss of professional licenses in professions like law or medicine after established academic malpractices, and blanket publication ban lasting indefinitely on career and publications of persons by academic institutions and professional associations.

5. Psychological and Emotional Consequences

Individual costs may be as serious as the academic, legal, and career costs made by plagiarism, and as a sanction may lead to emotional discomfort potentially including anxiety, depression, or even shame, undermines self-confidence through the labelling of an individual as a plagiarist, prompts social isolation among peers through its effect on reputation, and instills a permeating fear of engaging in further academic or career opportunities due to the long-term presence of the stain of an earlier plagiarism (Pecorari, 2003).

6. Consequences for Whistleblowers and Faculty

Reporters face some major problems, such as the fear of retaliation or marginalization, especially in a case of a high-profile offender, the bureaucracy of extended proceedings that use faculty time and energy away from research and teaching, and ethical dilemmas in serving the principal mission of teaching, research, and mentoring the offender to behave ethically.

7. Long-Term Erosion of Academic Values

When not controlled, plagiarism undermines the pillars of scholarship since it undermines academic degrees and discredits them. It also results in a loss of trust between the scholars and their partners as the integrity of the published literature is undermined. Moreover, when plagiarism becomes a repeated event, it has the capability of inducing a kind of moral numbness that gets people to accept such behavior as the norm, and contributes to a lack of dedication to the social theme of academic integrity.

INSTITUTIONAL RESPONSIBILITIES AND RESPONSES

Plagiarism is not just a personal weakness but may be a sign of breakdowns in the education system, monitoring, and academic culture (Bretag et al., 2019; Sutherland-Smith, 2020). The institutions are essential to creating the conditions that restrain or encourage unethical conduct. An active, organized, open policy on academic integrity does not just decrease plagiarism incidences, it also builds upon the certainty of the institution itself (Macfarlane et al., 2014).

1. Developing a Clear and Consistent Policy Framework

Clear and easier to access plagiarism policies have to be formulated in every school (Macfarlane, 2021). Different forms of plagiarism, such as intentional, unintentional, self-, and digital plagiarism, should be clearly defined in these policies, and the process of reporting and investigating misconduct should be illustrated as well. They have to set punishments commensurate with the nature of actions and means of appeal or grievance. Incorporating these policies into academic handbooks, course syllabi, orientation materials, and institutional websites will help make these policies uniformly applied throughout the institution and is particularly necessary among students with diverse educational or cultural backgrounds.

2. Strengthening Ethics Committees and Review Boards

An academic or research ethics committee is usually consulted to evaluate the alleged misconduct in the institutions. These bodies should be objective and have the skills and authority to explore honestly and effectively. Inclusion of different types of membership provides fairness, e.g., faculty, legal advisors, etc. There are also cases where student representation issues are taken up by the members (Steneck, 2006). Besides issuing resolutions to problems, ethics committees act as a safeguard to emerging issues by advising changes on policies when confronted with new challenges, like AI-generation text and contract cheating (Lancaster & Clarke, 2016).

3. Promoting a Culture of Research Integrity

Policies cannot be used on their own because they need a culture of appreciating originality, hard work, and intellectual integrity (Bretag et al., 2019). Ethical conduct should start with the leadership, and an example should be set by senior faculty by adopting good citation and other transparency practices in research. Institutional values can be reinforced when ethical actions are celebrated by such means as awards, newsletters, or inclusion in performance evaluation. Safe spaces to discuss issues, including the creation of ethics forums, will reduce stigma and are likely to bring about the active discussion of plagiarism (Pecorari, 2003). Honor codes, integrity pledges, and community-based responsibility are other community efforts that reinforce accountability.

4. Utilizing Plagiarism Detection Tools Effectively

The use of technological tools, including Turnitin, iThenticate, and Unicheck, is the best companion in confronting plagiarism. There ought to be granting of access, training of both faculty and students, and integration of the technology into learning management systems such as Moodle or Blackboard so that monitoring can be performed more efficiently (Sutherland-Smith, 2020). However, similarity reports cannot be interpreted by computers; rather, people should do the work; false positives are possible to get using structural common phrases or properly cited material (Introna & Hayes, 2003). The dangers of excessive dependence on software-based interventions are that one may forget the curative nature of plagiarism detection.

5. Supporting Faculty and Supervisors

Instructors have to be at the forefront of identifying and dealing with plagiarism, but usually, they do not receive sufficient institutional support. Through the mandatory training on academic integrity of all members of the teaching staff, including adjuncts, there is less uncertainty and inconsistency in enforcement (Bretag et al., 2019). Proactive engagements can be fostered through the recognition of faculty who mentor students in ethical scholarship (Resnik, 2017). Workloads also need to be levelled in ways that will discourage unethical efforts in both research and instruction.

6. Addressing Plagiarism in Research and Publishing

The study required by a graduate and postgraduate research requires strict originality. Institutions ought to impose plagiarism scanning of theses and dissertations, have plagiarism declarations signed and train students on predatory publishing activities. The productive rules of responsible authorship, peer review, and data management should be discussed in special workshops that doctoral candidates and early-career researchers will need (Macfarlane et al., 2014).

7. Procedural Fairness and Restorative Approaches

Although sanctions are also at times requisite, restoration techniques, including learning sessions and reflection tasks, for the first-time criminal, can instill everlasting integrity. Restorative conversations reconstruct faith between the students and the faculty. The appeal processes and confidentiality measures ensure that the students are not subjected to undue stigma as well as injustice (Pecorari, 2003). These moderate reactions are compatible with the studies that support rehabilitative as opposed to a punitive approach.

8. Institutional Accountability and Transparency

Regularly, institutions ought to conduct integrity reviews, perform audits annually and use peer review to check the effectiveness of the plagiarism policies (Steneck, 2006). Relevance is provided through benchmarking the documents with international best practices-COPE, ICMJE, APA, and UGC guidelines (Bretag et al., 2019). The result of the feedback of the stakeholders in terms of students, faculty, and administrators is a feedback loop that leads to continuous improvement. Use of transparent procedures gives credibility to institutions and increases the faith of the population in academic degrees (Resnik, 2017).

The solution to plagiarism in the long term, although not as important as institutional policies and systems, is also the empowerment of the people through education, mentorship, and ethical growth. To this end, the following section has provided viable prevention and learning tools that are meant to promote authenticity and academic honesty at all levels of scholarship.

PREVENTION AND EDUCATION STRATEGIES

The problem of plagiarism can be best resolved not only by penalties, but also by the long-term process of education, the positive support network, and the transition to an integrity-based learning process. The basis of prevention should consist of awareness, accessibility, and empowerment- the students and the faculty should be trained, mentored, and stimulated to understand the importance of originality and academic honesty throughout the entire process of creating scholarly works (Bretag et al., 2019; Macfarlane et al., 2014).

1. Curriculum Integration of Research Ethics

Integrating ethics of research into the academic curricula at every level is a policy that would be of interest in preventing plagiarism (Macfarlane et al., 2014). Instead of having general ethical education in modules, institutions ought to introduce ethics courses in all undergraduate and postgraduate studies as a compulsory course (Bretag et al., 2019). The inclusion of case-based learning would motivate students to discuss the real-life ethical dilemmas, focusing on a better understanding and developing critical thinking. There is also the necessity of subject-specific training that is sensitive to disciplinary differences that are found in unique citation of certain subjects, be they law, medicine, and the humanities, among others. These values are also supported by the assessment-driven ethics projects, such as peer-reviewed essays or integrity reflective assignment work. A holistic element of ethics in the curriculum would convey a clear message that academic integrity is central to the learning process and is a non-negotiable aspect.

2. Orientation and Induction Programs

It is through orientation of the new students that ethical expectations should be given a clear and explicit introduction to the new students. Plagiarism can be explained with these sessions, where its consequences are also described so that students realize the severity of academic dishonesty (Pecorari, 2003). They may also add different citation styles like APA, MLA, and Chicago and provide students with the instruments that will allow them to use references properly (American Psychological Association, 2020). Plagiarism detection software demonstration, including live instances, will make students aware of how plagiarism detection and prevention occur. Also, the orientation needs to emphasize institutional policies and resources of support to advise the students on how to conduct their studies with academic integrity (Bretag et al., 2019). Another thing that could be done to anticipate the significance of ethical scholarship is to share real-life examples of successful and unsuccessful experiences of academic integrity cases. The international and first-generation students will especially benefit through such early interventions, as they help students to overcome cultural and educational differences in the best way (Sowden, 2005).

3. Workshops and Skill-Building Sessions

Plagiarism is usually a misdeed that was the result of a lack of writing or citation prowess as opposed to malicious cheating (Pecorari, 2003). The answer to this is that learning institutions must conduct lessons regularly that would be aimed at the subject of academic writing and how to paraphrase content well. Also, librarians or writing centers may hold citation and referencing clinics since they can offer specific help to students. Interactive classes that introduce students to plagiarism-detecting applications increase their understanding of what those programs do and how to use and utilize them responsibly. Peer editing clubs also establish a learning environment that encourages students to learn to criticize the ideas of peers constructively and ethically. Along with this is the fact that all these initiatives empower the students to speak out brilliantly and yet maintain academic integrity.

4. Mentorship and Academic Supervision

A healthy supervisory relationship forms an important initial barrier to academic misconduct (Bretag et al., 2019). Good mentors have clear expectation criteria in this area of originality, which include teaching correct citation techniques and suggestions on how to complete a comprehensive literature review. The process of reviewing the drafts repeatedly allows the supervisors to be familiar with unintentional problems very early in the research procedures (Pecorari, 2003). Further, engaging the students in open communication on aspects related to ethics promotes clarity and confidence among the students and supervisors (Sowden, 2005). To improve this support, colleges ought to invest with the view of training supervisors as ethics coaches, hence maximizing the integrity of postgraduate and doctoral research (Macfarlane et al., 2014).

5. Assessment Design that Encourages Originality

Conventional evaluation methods have the potential to encourage shortcut solutions, which leads to an increase in plagiarism. Faculty is encouraged to set up individualized tasks that decrease the chance of copying and involve process-based assessment, like annotated bibliographies and literature maps, to reinforce the idea of continuous work with research (Bretag et al., 2019). Oral presentation and viva components of assessment provide another element of responsibility and an individual level of checking knowledge, where group assignments offer peer responsibility and the moral worth of accountability. Also, the assessments that use open books with a focus on interpretation and critical thinking rather than rote memorization are even further discouraging the dishonest behavior (Macfarlane et al., 2014). These authentic forms of assessments are creative and diverse and can cause genuine learning and greatly diminish the inclination to plagiarize.

6. Providing Access to Resources

Avoidance of plagiarism also needs to eliminate logistical impediments that can compel students to make unethical decisions. The institutions ought to make citation management tools easily accessible to simplify the proper referencing and have writing support centers both online and offline to help students develop their skills (Pecorari, 2003). Intuitive how-to guides to different referencing styles also help to attribute work properly (American Psychological Association, 2020), and AI-sensitive writing courses may help students make sense of what they can do with emerging technologies, negating any ethical issues concerning AI assistance (Bretag et al., 2019). This way, institutions help students in understanding how to uphold academic integrity and to minimize the use of dishonest shortcuts by equipping them with these tools and knowledge.

7. Encouraging a Growth Mindset Toward Writing

This aspect is part of the process of becoming an academic writer: it is a skill honed by evaluation and practice, and a culture of failure, revision and ongoing learning is the best way to encourage ethical practice (Sowden, 2005). This is possible through permitting successive submission of drafts before the final assessment to ensure refinement and improvement, setting up feedback loops to offer constructive feedback, rewarding progress and effort instead of grades alone, and making revision as a fundamental aspect of academic development the norm (Bretag et al., 2019). When the art of writing is treated as a growth process and not as a task performed once in a lifetime, the urge to plagiarize is reduced massively.

8. Faculty Development and Ethical Pedagogy

The core training needs by instructors should include principles and best practices of assignment designs where the least chances of plagiarism occur (Macfarlane et al., 2014), responsible use of similarity report results to differentiate between potentially acceptable overlaps and unethical behaviors, assist the learner in ethical dilemmas to make sense and promote integrity, and cultural sensitivity regarding their enforcement to treat all student populations fairly and respectfully (Sowden, 2005). Moreover, working seminars and meetings of faculty also contribute to the creation of a supportive community that advocates academic honesty and ethical scholarship as a whole.

9. Transparency in Detection and Remediation

Misinformation and misunderstanding concerning the use of plagiarism detection software by students can sometimes cause a lot of fear towards the plagiarism detection software (Pecorari, 2003). In response to this, detection measures must be transparent, and people should know how reports are created and consumed by the institutions (Bretag et al., 2019). Engaging students in the process of checking out their similarity reports may take the mystery out of the procedure and encourage them to learn. Furthermore, great clarity in explaining the differences between normal citation or rather the permissible citation and patchwriting serves better comprehension of the students about academic integrity. Correcting misdeeds and assisting in the case of the first mistakes also promotes ethical conduct and a culture of trust and honesty.

10. Promoting Integrity Through Recognition

The use of positive reinforcement is a good measure to support academic integrity. Ethical research practices must be identified and rewarded by the institutions to facilitate the culture of honesty (Bretag et al., 2019). Quality student and faculty work should be published so that it may become a source of inspiration to others. Moreover, when students are encouraged and supported to conduct their projects regarding integrity, this will instill a sense of peer influence and control of ethical principles (Sowden, 2005). This recognition causes a culture that does not revolve around fear of going to punishment but the aspirational values, making integrity a good and desired thing to achieve.

Despite the fact that the strong education and prevention practices are the background to moral scholarship, the environment of academic honesty is still shifting rather quickly. The new challenges to plagiarism are emerging technologies, international partnerships and changing learning contexts, which the next section will discuss as the paper explores the future of plagiarism in higher education.

FUTURE CHALLENGES AND EMERGING ISSUES

Due to the changes that are taking place in academic landscapes regarding technological advancements and globalization, the profile of plagiarism is altering exponentially (Eshet, 2025; Mulenga & Shilongo, 2024). Although the traditional ones are still present (Macfarlane et al., 2014), newer and more complicated problems darken the borders between primal uniqueness, authorship, and ethical subjection (Taylor, 2024). Institutions need to be proactive, focusing on the changes and implementing dynamic and prospective practices to maintain the integrity of academic matters (Loban et al., 2023; Vlad & Vlad, 2021).

1. Plagiarism in the Age of Generative AI

Writing assistance has evolved with the emergence of large language models (LLMs) in the form of ChatGPT, Bard, and Claude (Eacersall et al., 2024; Pudasaini et al., 2024). They can produce complete essays, paraphrase and mimic the tone of scholars, leaving people to wonder whether or not they are being cheated and how far it is ethical or unethical (Ka & Chan, 2023). The majority of the detection tools are trailing to detect the text generated by AI, which establishes a gap in the regulations (Khaled & Al-Tamimi, 2021). To protect this, policies that involve the use of AI, transparency, and AI literacy must be covered in plagiarism education (Reijers et al., 2018).

2. Paraphrasing Tools and Text Spinners

The ethically hazy areas that are promoted through on-demand platforms such as Quill bot and Spin bot involve only mechanical rewriting and anti-critical thinking (Velmurugan, 2024; Akbar, 2018). The teachers need to correct the wrong notion that students might be developing that the automation or paraphrasing they can do does not require citation (Neupane, 2023). It is very necessary to prevent the improper use of such tools and lightened learning effects (Comas-Forgas & Sureda-Negre, 2010). Of these, solutions can be the teaching of how not to confuse paraphrasing as a skill with algorithmic shortcuts (Awasthi, 2019).

3. AI-Generated Citations and Fabrication

Generative AI can fabricate yet potentially plausible-looking references, called AI hallucination (Pudasaini et al., 2024). This type of error compromises academic reliability and tracking of the sources (Abbasi et al., 2021). To avoid misinformation about academic work, it is imperative to train students to check the results of AI (Miranda-Rodriguez et al., 2024).

4. The Rise of Contract Cheating Platforms

Essay mills and ghostwriting services allow contract cheating, which is spreading around the world (Cavaliere et al., 2020; Roka, 2017). Such platforms increasingly employ the use of AI to go unnoticed (Folt The best means of fighting this is with face-to-face evaluations, authorship verification, and international implementation (Mulenga & Shilongo, 2024).

5. Globalization and Cultural Diversity in Academic Norms

The definitions of plagiarism are different in different cultures (Macfarlane et al., 2014). Some learning cultures appreciate replication and memorisation (Neupane, 2023). The confusion concerning citation standards may also cause unaware contravention (Akbar, 2018). To solve this, culturally responsive plagiarism education must take place (Awasthi, 2019; Loban et al., 2023).

6. Interdisciplinary and Collaborative Research Ethics

The co-authorship and authorship credit as well as ownership and rights over ideas are raised by cross-functional, interdisciplinary projects (Reijers et al., 2018; Vlad & Vlad, 2021). The fact of ambiguity may lead to the duplication of literature reviews or frameworks (Taylor, 2024). There should be a clear practice of agreement and conflict resolution between the contributors (Abbasi et al., 2021).

7. Inadequacies in Detection and Institutional Response

Preventive tools tend to generate false positives or fail to identify (and prevent) contract cheating and artificial-intelligence text (Foltynk et al., 2019; Khaled and Al-Tamimi, 2021). Policy credibility can also be weakened by institutional responses that are inconsistent (Mulenga & Shilongo, 2024). Timestamped reporting procedures and frequent updates of tools should be provided (Eacersall et al., 2024).

8. Legal and Copyright Complexities

Copyright issues become complex due to the spread of open-access repositories and AI-generated works (Vlad & Vlad, 2021). Pictures, sounds, and multimedia provoke plagiarism paradigms (Roka, 2017). There is a need to clarify the applicability of the copyright law to the current AI-generated and digital content by the institutions (Reijers et al., 2018).

9. Ethics Beyond the Classroom: Industry and Publishing Pressures

The faculty, researchers, and publishers are involved with plagiarism (Cavaliere et al., 2020; Macfarlane et al., 2014). Self-plagiarism and dubious authorship practices may be brought about by publication pressures (Taylor, 2024). There is a need to have ethical modeling by senior academics and also training in ethics, not just at the student level (Aguinis & Henle, 2004).

10. Toward Global Standards for Academic Integrity

Integrity has underscored a convergence of integrity policies in the context of global academic mobility and online education (Mulenga & Shilongo, 2024). The existing inequality leaves international students ethically confused (Neupane, 2023). COPE and ENAI are international organizations that might lead the standardization (Loban et al., 2023; Vlad & Vlad, 2021).

Tradition, innovation and constant change are the three elements that need to be balanced to ensure academic integrity as these future-oriented challenges depict. It is in this light that the final section draws a conclusion on the main arguments and emphasizes on the pathways towards the creation of an ethical, resilient, and a future-ready academic culture.

CONCLUSION

Still, plagiarism, which could be understood as simple stealing of words, has become a sophisticated and multifaceted problem that penetrates all areas of the academic environment. This chapter not only emphasizes plagiarism as a rule breach, but it is an in-depth ethical dilemma that depicts structural conflict in learning systems, research settings, and the dawn of technology. The scholarly literature indicates that plagiarism in most cases is a result of missing parts of the ethics course, specialized training, or an institutional tradition, not necessarily an incident of personal laxity (Bretag et al., 2019).

Taking the ethical nature of plagiarism as a crisis changes the basis of plagiarism from an isolated event to the whole picture of educational direction, culture, and pressure affecting scholars. Regardless of whether they are done through ignorance, stress or even intentional conduct, plagiaristic acts hurt the credibility of scholarly work, raising the distrust of academic systems (Sutherland-Smith, 2008). Therefore, to deal with plagiarism, it is necessary to abandon the punitive response in favor of educative, preventive, and restorative measures by incorporating ethics not only as a code of conduct but as a value experienced. The ability to detect the different types of plagiarism- intentional System 1 cheat, self-plagiarism and AI-assisted writing requires a high level of nuance and realization of the context behind such plagiarism, which can include academic pressure, shoddy training and cultural misunderstanding.

Institutions have a vital role as gatekeepers of academic integrity, ensuring that clear policies are in place, that there is training, that there is fair enforcement of standards, and resources to serve ethical scholarship. These values are supported by faculty and administrator leadership through action. Technology has both a positive and negative influence, introducing novel dilemmas via AI and plagiarism detection software, but also providing a chance to detect better and get ethically diverse through using the combination of digital literacy and technology. The inclusion of case study-based learning in the classroom process enhances ethical knowledge and awareness in students, and in an age of international cooperation, there must be synchronized flexible policies that can be adaptable to local, cultural and other emerging problems, including AI authorship. In the end, it is an effort shared by all to create a culture that appreciates honesty, responsibility, and respect when it comes to intellectual work; it will only be important to facilitate the preservation of the pursuit of the truth in academic and research institutions, as we continue to explore the future of complexities (Bretag et al., 2019).

REFERENCES

Abbasi, P., Yoosefi-Lebni, J., Jalali, A., Ziapour, A., & Nouri, P. (2021). Causes of the plagiarism: A grounded theory study. *Nursing Ethics*, *28*(2), 282–296. DOI: 10.1177/0969733020945753 PMID: 32909912

Aguinis, H., & Henle, C. A. (2004). Ethics in research. *Handbook of research methods in industrial and organizational psychology*, 34-56.

Akbar, A. (2018). Defining plagiarism: A literature review. *Ethical Lingua: Journal of Language Teaching and Literature*, *5*(1), 31–38. DOI: 10.30605/ethicallingua.v5i1.750

Ashworth, P., Bannister, P., & Thorne, P.Students on the Qualitative Researc. (1997). Guilty in whose eyes? University students' perceptions of cheating and plagiarism in academic work and assessment. *Studies in Higher Education*, *22*(2), 187–203. DOI: 10.1080/03075079712331381034

Awasthi, S. (2019). Plagiarism and academic misconduct: A systematic review. *DESIDOC Journal of Library and Information Technology*, *39*(2), 94–100. DOI: 10.14429/djlit.39.2.13622

Bretag, T., Harper, R., Burton, M., Ellis, C., Newton, P., Rozenberg, P., Saddiqui, S., & Van Haeringen, K. (2019). Contract cheating: A survey of Australian university students. *Studies in Higher Education*, *44*(11), 1837–1856. DOI: 10.1080/03075079.2018.1462788

Cavaliere, P., De Souza, D., Fenton, A. L., Giridharan, B., Gralla, C., Inshakova, N., & Zaharuk, G. (2020). *Academic misconduct and plagiarism: Case studies from universities around the world.* Rowman & Littlefield.

Comas-Forgas, R., & Sureda-Negre, J. (2010). Academic plagiarism: Explanatory factors from students' perspective. *Journal of Academic Ethics*, *8*(3), 217–232. DOI: 10.1007/s10805-010-9121-0

. Devlin, M. (2006). Challenging accepted wisdom about conceptions of teaching in academic development.

Devlin, M., & Gray, K. (2007). In their own words: A qualitative study of the reasons Australian university students plagiarize. *Higher Education Research & Development*, *26*(2), 181–198. DOI: 10.1080/07294360701310805

Eacersall, D., Pretorius, L., Smirnov, I., Spray, E., Illingworth, S., Chugh, R., . . . Howell, K. A. (2024). Navigating ethical challenges in generative AI-enhanced research: The ethical framework for responsible generative AI use. *arXiv preprint arXiv:2501.09021.*

Ercegovac, Z., & Richardson, J. V. (2004). Academic dishonesty, plagiarism included, in the digital age: A literature review. *College & Research Libraries*, *65*(4), 301–318. DOI: 10.5860/crl.65.4.301

Eshet, Y. (2025). Examining the dynamics of plagiarism: A comparative analysis before, during, and after the COVID-19 pandemic. *International Journal for Educational Integrity*, *21*(1), 3. DOI: 10.1007/s40979-024-00178-z

Floridi, L., & Chiriatti, M. (2020). GPT-3: Its nature, scope, limits, and consequences. *Minds and Machines*, *30*(4), 681–694. DOI: 10.1007/s11023-020-09548-1

Gullifer, J., & Tyson, G. A. (2010). Exploring university students' perceptions of plagiarism: A focus group study. *Studies in Higher Education*, *35*(4), 463–481. DOI: 10.1080/03075070903096508

Howard, R. M. (1995). Plagiarisms, authorships, and the academic death penalty. *College English*, *57*(7), 788–806. DOI: 10.58680/ce19959094

Introna, L., Hayes, N., Blair, L., & Wood, E. (2003). *Cultural attitudes towards plagiarism*. University of Lancaster.

Jereb, E., Perc, M., Lämmlein, B., Jerebic, J., Urh, M., Podbregar, I., & Šprajc, P. (2018). Factors influencing plagiarism in higher education: A comparison of German and Slovene students. *PLoS One*, *13*(8), e0202252. DOI: 10.1371/journal.pone.0202252 PMID: 30096189

. Ka, C., & Chan, Y. (2023). Is AI Changing the Rules of Academic Misconduct. *An In-depth Look at Students' Perceptions of "AI-giarism.*

Khaled, F., & Al-Tamimi, M. S. H. (2021). Plagiarism detection methods and tools: An overview. *Iraqi Journal of Science*, **•••**, 2771–2783.

Lancaster, T., & Clarke, R. (2016). Contract cheating: The outsourcing of assessed student work. Handbook of academic integrity, 639-654.

. Loban, G. A., Faustova, M. O., & Chumak, Y. V. (2023). ACADEMIC INTEGRITY AS AN IMPORTANT COMPONENT IN ENHANCING EDUCATION QUALITY. *Актуальні проблеми сучасної медицини: Вісник Української медичної стоматологічної академії, 23*(2.2), 91-95.

Macfarlane, B., Zhang, J., & Pun, A. (2014). Academic integrity: A review of the literature. *Studies in Higher Education*, *39*(2), 339–358. DOI: 10.1080/03075079.2012.709495

Miranda-Rodríguez, R. A., Sánchez-Nieto, J. M., & Ruiz-Rodríguez, A. K. (2024, May). Effectiveness of intervention programs in reducing plagiarism by university students: A systematic review. [). Frontiers Media SA.]. *Frontiers in Education*, *9*, 1357853. DOI: 10.3389/feduc.2024.1357853

Mulenga, R., & Shilongo, H. (2024). Academic integrity in higher education: Understanding and addressing plagiarism. *Acta Pedagogia Asiana*, *3*(1), 30–43. DOI: 10.53623/apga.v3i1.337

Neupane, N. (2023). Ins and outs of plagiarism in academia: Reconnoitring concepts, causes, and consequences. *Janapriya Journal of Interdisciplinary Studies*, *12*(1), 95–105. DOI: 10.3126/jjis.v12i1.62244

Pecorari, D. (2003). Good and original: Plagiarism and patchwriting in academic second-language writing. *Journal of Second Language Writing*, *12*(4), 317–345. DOI: 10.1016/j.jslw.2003.08.004

Pennycook, A. (1996). Borrowing others' words: Text, ownership, memory, and plagiarism. *TESOL Quarterly*, *30*(2), 201–230. DOI: 10.2307/3588141

Pudasaini, S., Miralles-Pechuán, L., Lillis, D., & Salvador, M. L. (2024). Survey on plagiarism detection in large language models: The impact of chatgpt and gemini on academic integrity. *arXiv preprint arXiv:2407.13105*.

Reijers, W., Wright, D., Brey, P., Weber, K., Rodrigues, R., O'Sullivan, D., & Gordijn, B. (2018). Methods for practising ethics in research and innovation: A literature review, critical analysis and recommendations. *Science and Engineering Ethics*, *24*(5), 1437–1481. DOI: 10.1007/s11948-017-9961-8 PMID: 28900898

Resnik, D. B., & Shamoo, A. E. (2017). Reproducibility and research integrity. *Accountability in Research*, *24*(2), 116–123. DOI: 10.1080/08989621.2016.1257387 PMID: 27820655

Roka, Y. B. (2017). Plagiarism: Types, causes and how to avoid this worldwide problem. *Nepal Journal of Neuroscience*, *14*(3), 2–6. DOI: 10.3126/njn.v14i3.20517

Sowden, C. (2005). Plagiarism and the culture of multilingual students in higher education abroad. *ELT Journal*, *59*(3), 226–233. DOI: 10.1093/elt/cci042

Steneck, N. H. (2006). Fostering integrity in research: Definitions, current knowledge, and future directions. *Science and Engineering Ethics, 12*(1), 53–74. DOI: 10.1007/PL00022268 PMID: 16501647

Sutherland-Smith, W. (2008). *Plagiarism, the Internet, and student learning: Improving academic integrity*. Routledge. DOI: 10.4324/9780203928370

Taylor, J. S. (2024). Reassessing academic plagiarism. *Journal of Academic Ethics, 22*(2), orari211–230. DOI: 10.1007/s10805-023-09478-4

Velmurugan, V. S. (2024). Types and Definitions of Plagiarism: An Overview. *Global Research Journal of Social Sciences and Management, 2*(1), 13–18. DOI: 10.55306/GRJSSM.2024.2102

Vlad, M. D., & Vlad, S. (2021). The Evolution of Intellectual Property Concepts at National and International Level. *SCIENTIA MORALITAS-International Journal of Multidisciplinary Research, 6*(1), 31–39.

Chapter 5
Untangling Conflicts:
Personal, Financial, and Institutional Dilemmas in Research Ethics

Amandeep Kaur
http://orcid.org/0009-0004-5716-7298
Dhandigarh School of Buisness, Chandigarh Group of Colleges Jhanjeri, Mohali, India

Gaganpreet Kaur
http://orcid.org/0009-0006-2957-5352
Chandigarh Group of Colleges Jhanjeri, Mohali, India

Ramandeep Sandhu
http://orcid.org/0000-0003-2595-4030
Lovely Professional University, India

EXECUTIVE SUMMARY

In the dynamic world of research, conflicts of interest whether personal, financial, or institutional can quietly undermine credibility, objectivity, and public trust. This chapter explores the complex ethical terrain of competing interests in academic and applied research environments. Through real-world case studies and theoretical insights, it sheds light on how subtle biases, hidden incentives, and institutional pressures can compromise research integrity. Personal conflicts often stem from relationships or individual ambitions, subtly influencing research decisions. Financial conflicts, including funding sources, stock ownership, and consultancy roles, pose significant risks of data manipulation or selective reporting. Meanwhile, institutional conflicts arise when universities or organizations prioritize prestige, funding, or

DOI: 10.4018/979-8-3693-8302-5.ch005

political interests over academic rigor. This chapter presents a comparative analysis of each conflict type, supported by landmark cases and recent controversies from diverse disciplines.

1. INTRODUCTION

Research, as both a scientific endeavor and a societal imperative, is anchored in principles of honesty, transparency, objectivity, and accountability. These foundational values uphold the trust that society places in the academic and scientific community. However, the expanding interface between research and external interests be they personal ambitions, financial gains, or institutional objectives has given rise to a complex web of ethical dilemmas collectively known as *conflicts of interest* (COIs). These conflicts do not necessarily imply wrongdoing or unethical behavior per se, but they signal a divergence between a researcher's obligations to unbiased inquiry and secondary influences that may consciously or unconsciously bias their actions (Lo & Field, 2009).

The prevalence of COIs in research is not a new phenomenon, but their implications have become increasingly consequential in a globalized, high-stakes knowledge economy. Universities, industries, and governments are now more interconnected than ever through funding collaborations, policy-influenced research agendas, and technology transfer mechanisms. While such relationships may spur innovation, they also carry inherent risks to research integrity if not transparently managed. As Steneck (2006) points out, the line between collaboration and coercion becomes blurred when institutional pressures for productivity and visibility override methodological rigor and ethical considerations.

One of the most insidious aspects of COIs is their ability to remain undetected. Unlike overt misconduct such as fabrication or plagiarism conflicts of interest are often invisible, operating through subtle shifts in priorities, judgment, and reporting (Resnik, 2018). These invisible biases can shape the selection of research topics, influence peer review outcomes, skew data interpretation, or determine whether certain findings are published at all. Importantly, COIs are not confined to one level of the research enterprise; they manifest across individual, financial, and institutional domains, each with its own challenges and consequences.

Personal COIs may emerge from interpersonal relationships, professional rivalries, or aspirations for career advancement. Financial COIs are often linked to external funding, consulting roles, or equity holdings, which can exert direct or indirect pressure on researchers. Institutional COIs occur when universities or research organizations prioritize funding acquisition, industry alignment, or political influence over academic independence (Krimsky, 2003). These conflicts are not mutually exclusive they often overlap, reinforcing each other and creating an ecosystem vulnerable to ethical compromise.

The urgency of addressing these concerns has been amplified by high-profile research scandals that have eroded public trust in science. Cases such as the Merck-Vioxx controversy, where financial incentives led to the suppression of adverse clinical findings, or the tobacco industry's long-term manipulation of health research, highlight the far-reaching consequences of unresolved COIs (Topol, 2004; Proctor, 2012). Beyond the biomedical domain, similar issues have been reported in climate science, psychology, economics, and engineering demonstrating that no discipline is immune.

To better understand the significance of these conflicts and their implications, Table 1.1 presents a cross-disciplinary overview of how COIs typically manifest in various research domains:

Table 1. Typical Manifestations of COIs Across Research Disciplines

Discipline	Common Type of COI	Example Scenario	Impact on Integrity
Biomedical Sciences	Financial	Drug trial sponsored by pharma company leading to positive bias	Patient safety risks; suppressed adverse data
Social Sciences	Personal	Researchers studying their own community or identity group	Subjectivity in data interpretation
Engineering & Tech	Institutional & Financial	Sponsored project promoting proprietary tech solutions	Favoritism; poor reproducibility
Climate Science	Political/Institutional	Pressure to underreport anthropogenic factors to appease policymakers	Misinformation; policy misdirection
Psychology	Personal & Institutional	Bias toward replicating mentor's findings to sustain academic lineage	Reproducibility crisis; publication bias
Economics	Financial	Consulting for industry while researching regulatory impacts	Biased policy recommendations

As research continues to evolve in complexity and scale, so too must our ethical frameworks. Tackling COIs requires a dual focus: preventing harm through transparency and fostering a proactive culture of ethical literacy. Leading bodies such as the International Committee of Medical Journal Editors (ICMJE), the Committee on Publication Ethics (COPE), and the National Academies of Sciences, Engineering, and Medicine (NASEM) have proposed comprehensive guidelines for managing, disclosing, and mitigating COIs. Yet, as noted by Bero (2017), policies alone are insufficient without institutional will and individual commitment to integrity.

This chapter seeks to provide a detailed exploration of the three dominant categories of conflict personal, financial, and institutional through theoretical analysis, landmark case studies, and ethical recommendations. By untangling these dilemmas, we aim to equip researchers, educators, and policymakers with the tools to foster a research environment that upholds credibility, objectivity, and public trust.

2. UNDERSTANDING CONFLICTS OF INTEREST IN RESEARCH

Conflicts of interest (COIs) are widely recognized as a critical challenge to the credibility and objectivity of scientific research. The essence of a COI lies in the presence of competing interests often personal, financial, or institutional that can interfere with or appear to interfere with a researcher's professional responsibilities or judgments. While the presence of a COI does not inherently imply unethical behavior or research misconduct, the failure to identify, disclose, and manage such conflicts can have far-reaching consequences for the reliability of scientific output and the trustworthiness of the research enterprise (Lo & Field, 2009).

At its core, COI represents a tension between two sets of obligations: the *primary interest* of producing objective and credible scientific knowledge, and a *secondary interest*, such as personal gain, financial benefit, or institutional advancement. These secondary interests, even if subconscious or unintentional, can subtly bias research design, data interpretation, publication practices, and even peer review processes (Thompson, 1993). As research becomes more interdisciplinary and externally funded, these risks multiply, requiring a clearer understanding of their typologies and implications.

A particularly important aspect of COIs is the distinction between actual, perceived, and potential conflicts:

- Actual conflict refers to a situation where a secondary interest has directly influenced research behavior.
- Perceived conflict arises when others believe that a conflict exists, even if the researcher maintains objectivity.
- Potential conflict refers to circumstances that could develop into actual conflicts if not appropriately managed (Resnik, 2018).

The seriousness of COIs lies not only in the actual bias they may introduce but in their capacity to undermine public trust in science. For example, when a government agency relies on academic studies to inform public health guidelines, any suspicion that those studies were influenced by industry funding can lead to skepticism, public backlash, or policy inertia. This is particularly true in areas of high social impact, such as pharmaceuticals, environmental science, food safety, or artificial intelligence (Krimsky & Rothenberg, 2001).

To further conceptualize the landscape of COIs, Table 2.1 provides an overview of the key dimensions along which conflicts can be analyzed:

Table 2. Key Dimensions of Conflicts of Interest in Research

Dimension	Description	Example
Source of Conflict	The origin of the competing interest (e.g., financial, relational, institutional)	Receiving funding from a company whose product is under study
Level of Influence	The degree to which the secondary interest can affect research judgment	A faculty member sitting on a thesis committee of a family member
Timing	When conflict arises in the research lifecycle	Conflict during peer review vs. after publication
Disclosure Status	Whether the conflict is disclosed, undisclosed, or undiscovered	Disclosing financial ties in a published article vs. hiding them
Control Mechanism	Strategies to manage or mitigate the conflict	Use of third-party data auditing; peer-review anonymity; ethics board oversight

In addition to the typology and dimensions of COIs, understanding their consequences is vital. According to a meta-analysis by Bekelman et al. (2003), industry-funded studies are more likely to yield favorable results for sponsors, indicating that financial COIs do indeed skew research findings. In another landmark study, Lundh et al. (2017) confirmed that industry-sponsored trials reported more positive outcomes compared to non-industry-funded research, underscoring the risk of systematic bias introduced by COIs.

The consequences of unmanaged COIs extend across the research ecosystem:

- For individual researchers, COIs can damage reputations, jeopardize careers, and lead to retractions or sanctions.
- For institutions, they can undermine credibility, reduce public trust, and threaten funding eligibility.
- For society, COIs can distort evidence-based decision-making, leading to misinformed policies and adverse outcomes in health, safety, or justice.

However, it is also important to note that not all COIs are avoidable. In many fields, collaboration with industry or policy institutions is essential to resource mobilization and innovation. The key, therefore, lies not in eliminating all conflicts, but in recognizing, disclosing, and managing them ethically. Mechanisms such as pre-registration of studies, data transparency, independent review boards, and robust ethics education serve as safeguards to mitigate these risks (Steneck, 2006).

Recent regulatory developments have emphasized institutional responsibility in managing COIs. The U.S. National Institutes of Health (NIH) and the International Committee of Medical Journal Editors (ICMJE) have introduced mandatory disclosure policies, while journals increasingly require authors, reviewers, and editors to publicly report any potential conflicts. Furthermore, organizations like the Committee on Publication Ethics (COPE) provide detailed flowcharts and guidance on handling undisclosed COIs during peer review or post-publication (COPE, 2021).

In conclusion, understanding nature, sources, and impact of conflicts of interest is foundational to preserving research integrity. A nuanced and proactive approach one that balances transparency with practicality is essential for navigating the ethical terrain of modern research. The subsequent sections will unpack the specific dimensions of personal, financial, and institutional COIs, providing case studies and management frameworks for each.

3. PERSONAL CONFLICTS: BIAS IN DISGUISE

Among the spectrum of conflicts of interest (COIs) in research, personal conflicts are the most subtle and insidious. Unlike financial or institutional COIs, which may involve formalized relationships and tangible transactions, personal conflicts are deeply rooted in psychological and relational dimensions. These may include emotional attachments, career ambitions, intellectual rivalries, or loyalties to colleagues, students, or mentors. The danger of personal COIs lies in their invisibility. They rarely leave a paper trail, are seldom declared, and are often mischaracterized as merely subjective preferences or disciplinary norms (Fanelli, 2009).

Personal conflicts often manifest in ways that challenge the researcher's impartiality. For instance, a scholar reviewing the grant application of a former student, a colleague evaluating a rival's manuscript, or an editor favoring authors within their professional network all these are examples where personal relationships or sentiments may interfere with objective judgment. Even when no malicious intent is present, such conflicts can subtly influence decisions, leading to bias in peer review, data interpretation, or hiring and promotion processes (Lo & Field, 2009).

Moreover, the aspirational nature of academic work itself contributes to personal COIs. Researchers strive for publication, recognition, tenure, and funding goals that can unconsciously pressure them to present results in a more favorable light. This phenomenon, often referred to as *motivated reasoning* or *confirmation bias*, occurs when individuals selectively interpret evidence to support their pre-existing beliefs or desired outcomes (Nickerson, 1998). When researchers become invested in a hypothesis or theory especially one they have published or built their career on they may resist contradictory evidence or design studies that inadvertently favor the expected result.

3.1 Case in Point: The Replication Crisis

A notable illustration of personal COIs can be found in the ongoing *replication crisis* across the social sciences, particularly in psychology and behavioral economics. In several instances, foundational studies once highly cited and celebrated failed to replicate under rigorous scrutiny. While multiple factors contributed to the crisis, including publication bias and methodological flaws, personal COIs were also implicated. Researchers, facing pressure to publish novel and significant results, often omitted null findings or engaged in *p-hacking* (manipulating data until statistically significant results emerge) to maintain their academic reputation (Simmons et al., 2011).

Such biases may not stem from explicit dishonesty but from a confluence of personal motivations and structural incentives. When a researcher's identity or career becomes entwined with a particular finding or methodology, the ability to remain objective can be compromised even if unconsciously.

Table 3. Common Sources and Manifestations of Personal Conflicts in Research

Source of Conflict	Manifestation	Potential Consequences
Relationship with authors/ students	Favorable peer reviews, undue authorship credit	Erosion of meritocracy, publication bias
Intellectual ownership of ideas	Suppression of contradictory evidence	Distorted scientific record, replication failure
Desire for professional advancement	Selective reporting, inflated claims	Misleading conclusions, loss of public trust
Competitive rivalries	Unfair peer reviews, gatekeeping	Reputational harm to others, bias in editorial decisions
Disciplinary or theoretical loyalty	Preferential treatment of like-minded scholars	In-group bias, narrow scholarly discourse

While journal editors and funding agencies increasingly require disclosure of financial conflicts, personal COIs are often left unaddressed due to their nebulous and subjective nature. Yet their effects can be equally corrosive. The case of Jan Hendrik Schön, a physicist at Bell Labs whose high-profile research was later exposed as fraudulent, raised questions about the oversight of personal and reputational motivations in elite scientific circles (Service, 2002). His rapid rise to prominence may have led colleagues and reviewers to overlook inconsistencies a form of *halo effect* linked to prestige bias.

It is important to emphasize that personal COIs are not inherently unethical. Personal relationships and ambitions are normal aspects of human interaction and academic life. What distinguishes ethical from unethical conduct is how these relationships and interests are managed. Transparency, critical self-awareness, and institutional safeguards such as double-blind review processes, ethics training, and recusal policies are key to minimizing bias. According to the Committee on Publication Ethics (COPE), editors and reviewers should recuse themselves from evaluating work where personal relationships may impair impartiality (COPE, 2021).

Educational institutions also have a role in equipping early-career researchers with tools to recognize and address personal conflicts. Ethical training programs, mentorship on academic integrity, and case-based learning can foster a culture where transparency and fairness are valued as much as scientific rigor. Moreover, the cultivation of *epistemic humility* the willingness to acknowledge uncertainty and the possibility of error can counteract the overconfidence that often underpins biased decision-making (Whitcomb et al., 2017).

In sum, personal conflicts are perhaps the most overlooked yet pervasive ethical challenge in research. They operate at the intersection of individual psychology and institutional culture, often disguised as scholarly judgment or academic freedom. Recognizing their presence, understanding their mechanisms, and instituting safeguards to minimize their impact are essential steps in fostering a more ethical and accountable research environment.

4. FINANCIAL CONFLICTS: INFLUENCE OF CAPITAL

Financial conflicts of interest (FCOIs) are perhaps the most visible and systematically documented category of ethical dilemmas in research. These conflicts arise when monetary relationships or financial incentives have the potential to influence, directly or indirectly, the objectivity of scientific inquiry. While financial support is often critical for advancing research particularly in fields like pharmaceuticals, engineering, and biotechnology it introduces a competing interest that, if left unmanaged, can compromise the integrity, transparency, and reproducibility of findings (Bekelman, Li, & Gross, 2003).

A financial conflict exists when a researcher or institution receives financial benefits such as grants, consultancy fees, honoraria, equity shares, or patent royalties that may be affected by the outcome of the research. Such relationships are not inherently unethical, but they become problematic when they bias the research agenda, suppress negative results, or distort the communication of findings to stakeholders or the public (Lo & Field, 2009). Studies have consistently shown that industry-funded research is more likely to produce favorable results for sponsors compared to independently funded studies, suggesting a troubling trend of outcome reporting bias (Lundh et al., 2017).

Forms of Financial Conflicts in Research

FCOIs can emerge at various points across the research lifecycle, from conceptualization to commercialization. The following are common sources of FCOIs:

- Sponsorship Bias: Occurs when industry funders such as pharmaceutical companies finance studies and influence research design, data interpretation, or publication to favor their products.
- Equity Interests: Researchers or their institutions may hold stock or intellectual property rights in companies whose products they are studying, creating direct financial stakes in research outcomes.
- Consulting and Advisory Roles: When scientists serve as paid consultants for companies, their independence may be compromised, especially if they are simultaneously conducting evaluations of those companies' products.
- Ghostwriting and Ghost-Authorship: Pharmaceutical companies have been known to hire professional writers to produce favorable studies under the names of respected academics, a practice that has come under sharp ethical scrutiny (Ross et al., 2008).

4.1 Case Study: The Vioxx Scandal

One of the most notorious examples of FCOIs distorting scientific integrity is the case of rofecoxib (Vioxx), a painkiller developed by Merck & Co. While initial studies many of them funded and authored with Merck's involvement suggested that Vioxx was safe and effective, later independent analyses revealed a significantly increased risk of heart attacks among users. Investigations showed that adverse data were deliberately omitted or downplayed in industry-sponsored publications, and that prominent researchers receiving funding from Merck had failed to disclose these relationships (Topol, 2004). The drug was eventually withdrawn in 2004, but not before it was linked to tens of thousands of cardiovascular-related deaths, highlighting the catastrophic consequences of unchecked FCOIs.

Table 4. Typology of Financial Conflicts of Interest in Research

Form of Conflict	Description	Examples	Potential Impact
Sponsorship Bias	Favorable research outcomes due to funding influence	Industry-sponsored trials showing higher success rates	Misleading evidence; erosion of public trust
Equity Ownership	Holding stocks or shares in a company under study	Researcher owns shares in a biotech firm whose drug is under trial	Biased interpretation; investment over patient safety
Consulting & Honoraria	Payments for advisory services to companies	Academic serving as consultant while publishing favorable results	Conflict of loyalty; dual obligations
Intellectual Property Rights	Holding patents that may benefit from positive research findings	Inventor publishes on own patented device	Pressure to validate the product; self-citation bias
Ghostwriting & Ghost Authorship	Writing research for publication under another's name	Pharma ghostwriting drug efficacy studies for key opinion leaders	Fabricated credibility; unethical publication practices

4.2 The Risk of Non-Disclosure

The core issue with FCOIs lies in the failure to disclose them transparently. Non-disclosure creates an environment of suspicion, and once uncovered, can delegitimize otherwise valid scientific work. Disclosure allows journals, funders, and readers to assess the potential influence of financial relationships on the research. However, studies show that even when disclosure is mandated, compliance is inconsistent. For example, Cooper et al. (2006) found that more than 20% of U.S.-based researchers had failed to disclose relevant financial ties in studies submitted for publication.

Global Policy and Ethical Guidelines

To address the growing concern surrounding FCOIs, several regulatory bodies and professional organizations have developed ethical frameworks and mandatory disclosure policies:

- The International Committee of Medical Journal Editors (ICMJE) requires all authors to complete a standardized disclosure form detailing financial relationships that could bias their submissions (ICMJE, 2019).
- The National Institutes of Health (NIH) mandates that researchers disclose significant financial interests and that institutions implement conflict management plans for federally funded projects (NIH, 2020).
- The Committee on Publication Ethics (COPE) offers practical guidance for editors in handling undisclosed conflicts during and after peer review (COPE, 2021).

Furthermore, some journals now include Conflict of Interest Statements as part of each published article, and funders are encouraging greater transparency through open-access policies, clinical trial registries, and data sharing requirements.

4.3 Balancing Innovation and Integrity

It is important to acknowledge that financial support from industry is not inherently detrimental to research. In fact, industry-academic collaborations have led to major breakthroughs in medicine, engineering, and environmental technologies. The ethical issue arises not from the funding itself, but from the lack of transparency and accountability in managing the resulting relationships. A balanced approach one that encourages partnerships but ensures that conflicts are disclosed and independently evaluated is critical for maintaining both scientific progress and public trust.

Best practices include:

- Establishing independent conflict-of-interest review boards within institutions.
- Requiring third-party data analysis for industry-sponsored studies.
- Enforcing firewalls between researchers and funders during data collection.
- Promoting open science practices such as pre-registration and open peer review.

5. INSTITUTIONAL CONFLICTS: PRESTIGE AND POWER DYNAMICS

While personal and financial conflicts of interest (COIs) often center on individuals, institutional conflicts arise at the organizational level, where universities, research centers, hospitals, and corporate-affiliated labs face competing pressures between ethical integrity and strategic interests. These conflicts occur when an institution's priorities such as prestige, funding acquisition, donor satisfaction, political alignment, or public image interfere with its responsibility to uphold academic independence and scientific rigor (Krimsky, 2003).

Institutional COIs are particularly complex because they are often systemic, embedded in governance structures, and culturally normalized within academic organizations. Unlike personal or financial conflicts, institutional COIs rarely appear in disclosure forms. Instead, they operate behind institutional policies, funding decisions, and internal power dynamics making them harder to detect, manage, and rectify (Campbell et al., 2007).

5.1 Nature and Sources of Institutional COIs

Institutions often depend on external funding to sustain research, infrastructure, and innovation goals. This creates an inherent tension between preserving research autonomy and aligning with funders' expectations. In such environments, ethical decision-making can be compromised by the desire to maintain long-term collaborations, attract prestigious partnerships, or avoid reputational harm.

Common sources of institutional COIs include:

- Sponsored Research Pressures: Universities or research hospitals may encourage researchers to tailor findings to suit sponsor interests, especially in lucrative fields like pharmaceuticals, energy, or defense.
- Donor Influence: Wealthy donors may exert pressure on academic programs to support particular ideologies, restrict certain lines of inquiry, or shape curriculum in their interest.
- Reputation Management: Institutions may suppress or delay the publication of negative results, especially when such findings threaten ongoing collaborations or public image.
- Intellectual Property and Commercialization Interests: When institutions hold patents or licensing rights, they may prioritize product development over open science or scientific accuracy.

5.2 Case Study: Tobacco Industry and University Partnerships

A well-documented example of institutional COIs is the historical partnership between universities and the tobacco industry. Throughout the 20th century, tobacco companies funded research at major institutions to obscure the health risks of smoking. In many instances, universities accepted this funding without transparency or oversight, providing scientific legitimacy to industry propaganda. Researchers who challenged the safety of tobacco were often marginalized or defunded, while institutions prioritized the maintenance of financial ties (Proctor, 2012). This manipulation of science not only endangered public health but damaged the credibility of the academic institutions involved.

Table 5. Examples of Institutional COIs and Their Impact

Institutional Context	Conflict Source	Example Scenario	Potential Ethical Breach
Research Universities	Industry-sponsored funding	Pressuring researchers to produce favorable data for pharma or energy companies	Data distortion; publication bias
Academic Hospitals	Clinical trial partnerships	Suppressing negative trial outcomes to retain funding	Patient harm; loss of public trust
Endowed Chairs and Programs	Influential donors with political agendas	Donors requiring control over hiring or curriculum decisions	Erosion of academic freedom
Technology Transfer Offices	Commercialization interests	Withholding negative results on a patented drug to boost licensing deals	Inhibited transparency; misinformed investors
Institutional Review Boards (IRBs)	Internal conflicts in approving studies	Favoring projects with high institutional benefit over ethically sound design	Inadequate ethical oversight

5.3 Challenges in Identifying and Managing Institutional COIs

One of the main challenges with institutional COIs is their diffuse accountability. Unlike individuals who can be held directly responsible, institutions often function through collective decision-making, complex hierarchies, and layers of administrative discretion. As a result, the ethical breach may not be attributable to a single actor but to an organizational culture that incentivizes questionable practices.

Furthermore, institutional self-interest can lead to *selective enforcement* of COI policies. For example, a university may rigorously enforce financial COI disclosures among junior faculty but ignore potential conflicts involving senior researchers who bring in significant funding. Such disparities can erode internal ethics and discourage whistleblowing (Thacker, 2005).

Another issue is the opacity of institutional agreements. Memorandums of Understanding (MoUs) with private corporations, classified research deals, or donor contracts are often not publicly accessible, preventing transparency and external scrutiny. In the absence of full disclosure, it becomes nearly impossible to assess whether the institution is compromising its ethical responsibilities.

5.4 Regulatory Approaches and Best Practices

Several organizations have emphasized the importance of identifying and managing institutional conflicts:

- The Association of American Medical Colleges (AAMC) recommends that academic health centers establish standing committees to review institutional COIs and ensure that key decisions are made independently of financial considerations (AAMC, 2008).
- The National Academies of Sciences, Engineering, and Medicine (NASEM) advise the separation of research funding from research oversight functions to prevent ethical compromise (NASEM, 2017).
- Internationally, research integrity frameworks stress the necessity of transparent governance, audit systems, and open access to contracts that involve research funding or intellectual property.

Some universities have implemented firewall structures, in which units responsible for commercialization (e.g., tech transfer offices) are administratively separated from those involved in research ethics. Others have adopted public COI registries, where institutional relationships with funders and corporations are disclosed online.

Still, implementation remains uneven, and the lack of standardized global policies leaves room for selective compliance. A study by Boyd and Bero (2006) found that many U.S. academic medical centers lacked adequate policies for disclosing or managing institutional COIs, even when receiving substantial industry funding.

5.5 Toward Ethical Institutional Governance

To ensure ethical integrity at the institutional level, a cultural shift is needed one that prioritizes transparency and accountability over prestige and profitability. Institutions must go beyond formal policies and foster a research environment that values truth-seeking and independence. This includes:

- Encouraging open debate and dissent.
- Protecting whistleblowers and critics of institutional decisions.
- Publicly disclosing donor agreements and commercial partnerships.
- Establishing independent ethics governance structures free from financial influence.

Institutional ethics must also be included in research training curricula. Educators should not only teach students about personal and financial COIs but also help them critically examine how institutional priorities shape knowledge production.

6. COMPARATIVE ANALYSIS OF CONFLICT TYPES

To fully grasp the scope and severity of ethical dilemmas in research, it is essential to compare the three major categories of conflicts of interest personal, financial, and institutional not only in terms of their definitions but also in how they differ in visibility, prevalence, impact, and mitigation. Each category carries distinct ethical risks, yet they are interrelated and often co-occur, creating complex and layered ethical landscapes. Understanding these distinctions enables stakeholders researchers, institutions, funders, and policymakers to devise tailored strategies for identifying, disclosing, and managing such conflicts (Lo & Field, 2009).

Although academic discourse often emphasizes financial conflicts due to their traceable nature and regulatory frameworks, recent scholarship emphasizes that personal and institutional COIs may be even more pervasive and insidious due to their invisibility and systemic embedding in organizational culture (Krimsky, 2003; Fanelli, 2009). This comparative analysis highlights both the uniqueness and the overlap between conflict types, drawing attention to the need for holistic ethics policies in research governance.

6.1 Dimensions of Comparison

6.1.1. Nature of Influence

- Personal COIs typically stem from subjective human relationships, academic loyalties, or professional rivalries. They often manifest in subtle, emotional, or cognitive biases such as favoritism, selective reporting, or confirmation bias (Nickerson, 1998).
- Financial COIs involve tangible monetary relationships that can create direct incentives for manipulating data or outcomes. These are more likely to be externally scrutinized and have clearer thresholds for disclosure (Bekelman et al., 2003).
- Institutional COIs are embedded within the organizational framework and concern the alignment of research practices with institutional priorities like profitability, reputation, or political influence. They are systemic and harder to localize to individual misconduct (Campbell et al., 2007).

6.1.2. Level of Transparency

- Financial COIs are often subject to formal disclosure requirements, particularly by journals and funding agencies.

- Personal COIs, due to their subjective and relational nature, are rarely declared unless mandated by strict peer-review protocols or university guidelines.
- Institutional COIs are the least transparent, frequently concealed under non-disclosure agreements, private MoUs, or implicit political alignments.

6.1.3. Consequences for Research Integrity

Each conflict type has distinct ethical repercussions:

- Personal COIs may compromise peer review, hiring decisions, and authorship, leading to reputational harm and skewed academic evaluation metrics.
- Financial COIs can directly affect public health, environmental policy, or technological safety, especially when biased results inform high-stakes decision-making.
- Institutional COIs can lead to long-term systemic damage by suppressing dissent, narrowing research agendas, and undermining public trust in science.

Table 6. Comparative Matrix of Conflict Types in Research

Conflict Type	Primary Source	Visibility	Regulatory Oversight	Impact Severity	Mitigation Strategies
Personal	Relationships, biases, ambition	Low	Weak, often informal	Medium to high	Blind peer review, recusal policies, ethical training
Financial	Funding, equity, consulting	High	Strong (journals, funders, IRBs)	High	Mandatory disclosure, independent auditing, firewalls
Institutional	Reputation, funding, politics	Very low	Moderate to weak	Very high	Transparent contracts, independent governance, public oversight

6.2 Intersections and Overlaps

In practice, these conflict types often intersection. A senior professor may simultaneously have:

- Personal conflict: reviewing a student's paper,
- Financial conflict: holding shares in a startup linked to the research,
- Institutional conflict: working at a university seeking funding from that startup.

Such overlaps compound the ethical risk, especially when conflicts are undisclosed. Research ethics scholars refer to this as multi-dimensional conflict layering, which requires an integrated ethical response rather than isolated interventions (Resnik, 2018).

Moreover, power asymmetries exacerbate these risks. For instance, junior researchers may feel pressured to comply with institutional COIs due to job insecurity or fear of professional retaliation. In such environments, even when personal integrity exists, systemic forces may push individuals toward ethically questionable decisions (Steneck, 2006).

6.3 Toward Integrated Conflict Management

Recognizing that no single type of COI exists in isolation, institutions must embrace multi-layered ethics policies that combine disclosure, education, governance, and cultural change. Best practices emerging from leading institutions and journals suggest that:

- Hybrid Ethics Committees composed of internal and external reviewers can offer unbiased assessments of overlapping conflicts.
- Conflict of Interest Registries for both individuals and institutions can promote transparency.
- Pre-emptive Training Modules in graduate education can equip researchers to identify subtle conflicts before they escalate.

Importantly, ethics management must move beyond compliance checklists toward values-based frameworks rooted in transparency, fairness, and public accountability. The complexity of contemporary research demands that ethical reasoning be adaptive, evidence-informed, and sensitive to context.

7. ETHICAL GUIDELINES AND REGULATORY FRAMEWORKS

The growing complexity of research ecosystems marked by interdependencies among academia, industry, government, and civil society has necessitated the development of comprehensive ethical guidelines and regulatory frameworks to manage conflicts of interest (COIs). These frameworks aim not only to prevent unethical conduct but to enhance public trust in science, promote transparency, and create institutional cultures of accountability (Resnik, 2018).

While the nature of COIs may vary personal, financial, or institutional their common thread is the potential to compromise objectivity and integrity in research.

Therefore, managing COIs must go beyond ad hoc policies or voluntary codes. Instead, it requires systematic, enforceable, and transparent mechanisms, embedded in both organizational practice and professional norms. Over the past two decades, multiple national and international organizations have established such frameworks to guide researchers, editors, reviewers, and institutions.

7.1 Key Global Regulatory Bodies and Their Guidelines

A number of prominent regulatory and advisory organizations have formulated guidelines specifically addressing conflicts of interest in research. These frameworks provide a valuable reference point for institutions and individuals navigating ethical dilemmas:

7.1.1. International Committee of Medical Journal Editors (ICMJE)

- The ICMJE provides one of the most widely accepted standards for COI disclosure in scholarly publishing.
- All authors submitting to ICMJE-compliant journals must complete a standardized disclosure form that details financial relationships within the past 36 months.
- ICMJE emphasizes the reader's right to know about any interests that may influence the credibility of published research.

"Transparency is essential to maintain public trust in scientific findings and ensure the integrity of the research enterprise." – ICMJE, 2019

7.1.2. Committee on Publication Ethics (COPE)

- COPE offers a comprehensive set of guidelines, flowcharts, and decision tools to assist journal editors and publishers in managing ethical issues including COIs.
- COPE's position is that both perceived and actual COIs must be managed with equal seriousness.
- It recommends that journals create policies requiring disclosure from all participants in the publication process authors, reviewers, and editors and provides structured actions to take when COIs are undisclosed.

7.1.3. National Institutes of Health (NIH), USA

- The NIH defines Significant Financial Interest (SFI) and requires federally funded investigators to disclose any such interests.
- Institutions must develop and enforce Conflict of Interest Management Plans (CMPs).
- The NIH requires annual reporting and public accessibility of information regarding COIs linked to federally funded research.

7.1.4. World Health Organization (WHO)

- WHO mandates comprehensive COI disclosures from external experts participating in guideline development or research committees.
- Emphasizes conflict mitigation strategies, such as exclusion from decision-making, third-party evaluations, and independent peer reviews.

7.1.5. National Academies of Sciences, Engineering, and Medicine (NASEM), USA

- NASEM's 2017 report, *Fostering Integrity in Research*, underscores that institutions must embed COI policies within broader research integrity programs.
- The report stresses the need for ethics education, whistleblower protection, and institutional independence in COI evaluation.

7.2 Summary Table

Table 7. Comparative Snapshot of COI Guidelines

Organization	Key Focus	Disclosure Scope	Enforcement Mechanisms	Tools Provided
ICMJE	Publishing Ethics	Authors in academic journals	Journal rejection or retraction	Standard forms, publication statements
COPE	Editorial Integrity	Authors, reviewers, editors	Editorial decisions and COI investigations	Flowcharts, case studies
NIH	Research Funding (US)	Investigators on federal grants	Institutional compliance; audits	CMP templates, reporting tools

continued on following page

Table 7. Continued

Organization	Key Focus	Disclosure Scope	Enforcement Mechanisms	Tools Provided
WHO	Global Health Research	External advisors and committee members	Expert exclusion, public disclosure	COI forms, ethics panels
NASEM	Research Culture Reform	Institutions and individuals	Policy guidance, not legal enforcement	Reports, workshops, training resources

7.3 Common Elements of Effective COI Frameworks

Despite differing contexts, these global frameworks share several core components that collectively strengthen research integrity:

1. Disclosure: Open declaration of any secondary interest financial, relational, or institutional that may influence research judgment.
2. Documentation: Standardized forms, conflict registries, and digital repositories that track disclosures over time.
3. Mitigation: Strategies to minimize influence, including recusal from decisions, independent reviews, and third-party data analysis.
4. Monitoring: Ongoing evaluation through audits, ethics committees, or compliance offices to ensure effectiveness and accountability.
5. Education: Ethics training, scenario-based learning, and mentorship for early-career researchers to identify and navigate COIs.

7.4 Challenges in Implementation

Despite the existence of comprehensive guidelines, implementation gaps persist, especially across lower-resourced institutions or in countries with weak regulatory infrastructure. Key challenges include:

- Inconsistent compliance: Many researchers misunderstand or underreport their conflicts, particularly when disclosure is seen as optional or symbolic.
- Fear of career repercussions: Young researchers may avoid disclosure if they fear losing funding or advancement opportunities.
- Lack of enforcement: Institutions may adopt policies but fail to enforce them rigorously, especially when the implicated researchers are high-performers or fund-raisers.
- Cultural differences: Ethical norms regarding COIs vary globally, leading to disparities in what constitutes an "unethical" conflict.

To address these gaps, scholars suggest international harmonization of disclosure standards and the creation of transnational ethics databases that track recurring ethical breaches (Steneck, 2006; Resnik, 2018).

7.5 Role of Institutions and Journals

Institutions and journals serve as gatekeepers of ethical standards. Their commitment to enforcing COI policies can determine the effectiveness of national and global frameworks. Best practices include:

- Creating dedicated COI oversight bodies within research organizations.
- Publishing COI statements with every article, including both author and editor disclosures.
- Auditing compliance and periodically reviewing policies for relevance and effectiveness.
- Rewarding transparency, rather than penalizing honest disclosures.

By moving beyond punitive approaches and toward a culture of trust, institutions can promote a climate where ethical reflection is integral to the research process, not an afterthought.

8. BUILDING A CULTURE OF ETHICAL RESEARCH

Managing conflicts of interest (COIs) is not solely a matter of enforcing policies or regulatory compliance it requires nurturing an institutional and intellectual climate that values ethics as a foundational pillar of research practice. Building a culture of ethical research means embedding values of integrity, transparency, accountability, and fairness into the everyday decisions of researchers, educators, administrators, and policymakers. This cultural shift is especially vital in an era where the line between academic inquiry and external influence is increasingly blurred (NASEM, 2017).

Despite robust frameworks and disclosure systems, many COIs remain undeclared or poorly managed. One reason is the lack of normative commitment to ethical self-regulation within research institutions. Policies, while essential, cannot substitute for a shared moral orientation that promotes responsible conduct as a collective duty rather than an individual burden (Steneck, 2006).

8.1 Ethics Beyond Compliance

Ethical research culture involves moving from a compliance-based model (where rules are followed to avoid penalties) to a values-based model, where researchers internalize ethical principles as part of their professional identity. This transformation involves three levels of intervention:

- Individual level: Cultivating ethical awareness, critical thinking, and self-reflection among researchers through mentorship and training.
- Organizational level: Creating institutional structures and norms that reward transparency, protect whistleblowers, and ensure independence of ethics review.
- Systemic level: Aligning funding mechanisms, publication practices, and career progression criteria with ethical integrity rather than just productivity or impact metrics.

According to Macfarlane (2009), research ethics education must evolve beyond didactic instruction to include case-based learning, dilemma simulations, and reflective dialogues that explore real-life scenarios of ethical conflict.

8.2 Role of Education and Mentorship

Early intervention is crucial. Graduate programs must treat research ethics not as a peripheral component but as a core element of research training. Exposure to real-world case studies, participatory ethics workshops, and continuous dialogue with mentors helps young scholars recognize, question, and responsibly navigate COIs (Kalichman & Plemmons, 2007).

Mentorship also plays a pivotal role in shaping ethical conduct. Experienced researchers serve as role models for research integrity or, conversely, as facilitators of unethical normalization. Therefore, institutions should not only support formal mentorship programs but also assess mentors based on their ethical leadership and transparency in practice (Shamoo & Resnik, 2015).

8.3 Institutional Strategies to Promote Ethical Culture

To reinforce a culture of ethics institutionally, the following proactive strategies can be adopted:

Table 8. Proactive strategies

Strategy	Description	Expected Impact
Institutional Ethics Committees	Independent panels to oversee COI disclosures and handle ethical dilemmas	Ensures objectivity and fairness in COI management
Open Science Initiatives	Promoting pre-registration, open data, and reproducibility	Enhances transparency and reduces result manipulation
Recognition and Rewards for Integrity	Awards, career points, or funding linked to ethical behavior	Incentivizes good conduct beyond publication metrics
Whistleblower Protection Policies	Legal and institutional safeguards for reporting unethical practices	Encourages accountability and early detection of misconduct
Ethics Climate Surveys	Periodic anonymous assessments of ethical environment	Provides feedback and benchmarks for institutional improvement

These strategies should not be implemented in isolation. Rather, they must be part of an ethics ecosystem that includes faculty, administration, journal editors, and external stakeholders such as funders and professional bodies.

8.4 Changing Incentive Structures

A significant challenge to ethical research is the existing reward system, which tends to prioritize publication quantity, impact factor, and grant income over research integrity. This can create perverse incentives that encourage researchers to cut corners, exaggerate findings, or avoid disclosing conflicts.

To address this, new metrics of success should be introduced, such as:

- Ethical Impact Statements in CVs and promotion portfolios,
- Evaluation of data transparency and reproducibility,
- Institutional audits that consider ethical compliance as a KPI.

The San Francisco Declaration on Research Assessment (DORA) and the Leiden Manifesto are leading initiatives pushing for reform in how research success is measured, emphasizing quality over quantity and ethical contribution over mere productivity (Hicks et al., 2015).

8.5 Fostering International Solidarity in Ethics

Given the global nature of research, especially in collaborative studies across countries with different ethical standards, there is a need for cross-border alignment on COI policies. Ethical culture must transcend national boundaries and create shared norms of professional conduct.

International collaborations can promote:

- Joint ethics training programs for multinational research teams.
- Cross-institutional COI registries for shared transparency.
- Global codes of conduct in specific fields (e.g., AI, public health, climate science).

UNESCO's 2021 *Recommendation on Open Science* and the OECD's *Best Practices for Research Integrity* emphasize the collective responsibility of the global research community to maintain high ethical standards and build public trust.

A strong ethical culture is not a hindrance to research excellence it is the foundation of it. When researchers are free from undue influence, protected from retaliation, and encouraged to act transparently, science flourishes in both credibility and social relevance. By reimagining ethics not as an obstacle but as a value-driven compass, institutions can lead a transformative shift toward sustainable, inclusive, and responsible research.

9. CONCLUSION

In an era where scientific research profoundly shapes public policy, technology, healthcare, and social progress, the ethical integrity of research practices is more critical than ever. This chapter examines the nuanced and multifaceted landscape of conflicts of interest (COIs) focusing on personal, financial, and institutional dimensions and illustrated how these conflicts, if not properly managed, can compromise the reliability of knowledge, erode public trust, and perpetuate systemic inequities in academic environments.

While each category of COI is distinct in origin and manifestation, they are bound by a common thread: the risk of subordinating truth and objectivity to secondary interests. Personal COIs often emerge subtly, through emotional attachments or intellectual ego, undermining peer review and authorship credibility. Financial COIs, though more visible and legally regulated, present potent challenges due to the tangible influence of industry funding, consultancy roles, or intellectual property stakes. Institutional COIs are the most insidious operating systemically through organizational priorities that favor prestige, revenue, or donor loyalty over academic independence.

As the chapter demonstrated through historical cases (e.g., Vioxx and tobacco research), COIs are not abstract ethical dilemmas but real-world threats with material consequences, ranging from flawed clinical guidelines to misinformed environmental policy. Therefore, the ethical response cannot be passive or reactive. It must be deliberate, dynamic, and embedded across all levels of the research ecosystem.

Global regulatory bodies such as the NIH, ICMJE, COPE, and WHO have laid essential groundwork in requiring disclosure, standardizing ethical review, and enforcing accountability. However, regulatory compliance alone is not sufficient. A meaningful response to COIs must prioritize the cultivation of an ethical research culture, where researchers are not only aware of their obligations but actively engaged in ethical reflection, peer accountability, and transparent communication.

To achieve this, institutions must:

- Design incentive structures that reward integrity as much as impact.
- Ensure that ethical training is central to research education.
- Support whistleblowers and dissenters as guardians of truth.
- Mandate transparency in all research-related agreements, including donor contracts and corporate sponsorships.

Moreover, ethics must evolve in tandem with science. Emerging fields such as artificial intelligence, synthetic biology, and climate engineering introduce new ethical frontiers, where the boundaries of COIs are still being drawn. In these areas, anticipatory ethics a forward-looking framework that assesses potential risks before they materialize will be essential in guiding responsible innovation.

In conclusion, untangling conflicts of interest is not merely a regulatory task, but a philosophical and cultural undertaking. It demands that we, as scholars, practitioners, and institutions, continuously interrogate our motivations, challenge our assumptions, and recommit to the ideals of academic honesty and public good. Only then can we ensure that research serves not just markets or careers, but humanity itself.

REFERENCES

AAMC. (2008). *Protecting Patients, Preserving Integrity, Advancing Health: Accelerating Implementation of COI Policies in Human Subjects Research*. https://www.aamc.org

Bekelman, J. E., Li, Y., & Gross, C. P. (2003). Scope and impact of financial conflicts of interest in biomedical research: A systematic review. *Journal of the American Medical Association*, *289*(4), 454–465. DOI: 10.1001/jama.289.4.454 PMID: 12533125

Bero, L. (2017). Addressing bias and conflict of interest among biomedical researchers. *Journal of the American Medical Association*, *317*(17), 1723–1724. DOI: 10.1001/jama.2017.3854 PMID: 28464166

Boyd, E. A., & Bero, L. A. (2006). Institutional policies on faculty conflicts of interest: A survey of medical school deans. *Journal of the American Medical Association*, *296*(9), 1105–1112. DOI: 10.1001/jama.296.9.1105

Campbell, E. G., Weissman, J. S., Ehringhaus, S., Rao, S. R., Moy, B., Feibelmann, S., & Goold, S. D. (2007). Institutional academic-industry relationships. *Journal of the American Medical Association*, *298*(15), 1779–1786. DOI: 10.1001/jama.298.15.1779 PMID: 17940234

Campbell, E. G., Weissman, J. S., Vogeli, C., Clarridge, B., Abraham, M., Marder, J. E., & Koski, G. (2007). Financial relationships between institutional review board members and industry. *The New England Journal of Medicine*, *356*(18), 1811–1820. DOI: 10.1056/NEJMsa066498 PMID: 17135585

Cooper, R. J., Gupta, M., Wilkes, M. S., & Hoffman, J. R. (2006). Conflict of interest disclosure policies and practices in peer-reviewed biomedical journals. *Journal of General Internal Medicine*, *21*(12), 1248–1252. DOI: 10.1111/j.1525-1497.2006.00598.x PMID: 17105524

COPE. (2021). *COPE Guidelines on Conflicts of Interest in Peer Review*. https://publicationethics.org

Fanelli, D. (2009). How many scientists fabricate and falsify research? A systematic review and meta-analysis of survey data. *PLoS One*, *4*(5), e5738. DOI: 10.1371/journal.pone.0005738 PMID: 19478950

Hicks, D., Wouters, P., Waltman, L., de Rijcke, S., & Rafols, I. (2015). The Leiden Manifesto for research metrics. *Nature*, *520*(7548), 429–431. DOI: 10.1038/520429a PMID: 25903611

ICMJE. (2019). *Recommendations for the Conduct, Reporting, Editing, and Publication of Scholarly Work in Medical Journals*. https://www.icmje.org/recommendations/

Kalichman, M., & Plemmons, D. (2007). Reported goals for responsible conduct of research courses. *Academic Medicine, 82*(9), 846–852. DOI: 10.1097/ACM.0b013e31812f78bf PMID: 17726389

Krimsky, S. (2003). *Science in the Private Interest: Has the Lure of Profits Corrupted Biomedical Research?* Rowman & Littlefield.

Krimsky, S., & Rothenberg, L. S. (2001). Conflict of interest policies in science and medical journals: Editorial practices and author disclosures. *Science and Engineering Ethics, 7*(2), 205–218. DOI: 10.1007/s11948-001-0041-7 PMID: 11349360

Lo, B., & Field, M. J. (2009). *Conflict of Interest in Medical Research, Education, and Practice*. National Academies Press.

Lundh, A., Lexchin, J., Mintzes, B., Schroll, J. B., & Bero, L. (2017). Industry sponsorship and research outcome. *Cochrane Database of Systematic Reviews, 2017*(2). Advance online publication. DOI: 10.1002/14651858.MR000033.pub3 PMID: 28207928

Macfarlane, B. (2009). *Researching with Integrity: The Ethics of Academic Enquiry*. Routledge.

NASEM. (2017). *Fostering Integrity in Research*. National Academies Press., DOI: 10.17226/21896

Nickerson, R. S. (1998). Confirmation bias: A ubiquitous phenomenon in many guises. *Review of General Psychology, 2*(2), 175–220. DOI: 10.1037/1089-2680.2.2.175

NIH. (2020). *Financial Conflict of Interest*. https://grants.nih.gov/grants/policy/coi/

Proctor, R. N. (2012). *Golden Holocaust: Origins of the Cigarette Catastrophe and the Case for Abolition*. University of California Press. DOI: 10.1525/9780520950436

Resnik, D. B. (2018). *The Ethics of Research with Human Subjects: Protecting People, Advancing Science, Promoting Trust*. Springer. DOI: 10.1007/978-3-319-68756-8

Ross, J. S., Hill, K. P., Egilman, D. S., & Krumholz, H. M. (2008). Guest authorship and ghostwriting in publications related to rofecoxib: A case study of industry documents from rofecoxib litigation. *Journal of the American Medical Association, 299*(15), 1800–1812. DOI: 10.1001/jama.299.15.1800 PMID: 18413874

Service, R. F. (2002). Bell Labs fires star physicist found guilty of falsifying data. *Science, 298*(5595), 30–31. DOI: 10.1126/science.298.5595.30 PMID: 12364753

Shamoo, A. E., & Resnik, D. B. (2015). *Responsible Conduct of Research* (3rd ed.). Oxford University Press.

Simmons, J. P., Nelson, L. D., & Simonsohn, U. (2011). False-positive psychology: Undisclosed flexibility in data collection and analysis allows presenting anything as significant. *Psychological Science, 22*(11), 1359–1366. DOI: 10.1177/0956797611417632 PMID: 22006061

Steneck, N. H. (2006). Fostering integrity in research: Definitions, current knowledge, and future directions. *Science and Engineering Ethics, 12*(1), 53–74. DOI: 10.1007/PL00022268 PMID: 16501647

Thacker, P. D. (2005). Searching for transparency: Conflicts of interest cloud medical research. *Journal of the American Medical Association, 294*(6), 689–692. DOI: 10.1001/jama.294.6.689

Thompson, D. F. (1993). Understanding financial conflicts of interest. *The New England Journal of Medicine, 329*(8), 573–576. DOI: 10.1056/NEJM199308193290812 PMID: 8336759

Topol, E. J. (2004). Failing the public health—Rofecoxib, Merck, and the FDA. *The New England Journal of Medicine, 351*(17), 1707–1709. DOI: 10.1056/NEJMp048286 PMID: 15470193

Whitcomb, D., Battaly, H., Baehr, J., & Howard-Snyder, D. (2017). Intellectual humility: Owning our limitations. *Philosophy Compass, 12*(8), e12415. DOI: 10.1111/phc3.12415

World Health Organization (WHO). (2021). *WHO Handbook for Guideline Development.* https://www.who.int/publications

Chapter 6
Navigating Ethical Challenges in Sports and Exercise Science Research

Debajit Karmakar
http://orcid.org/0000-0001-9272-0627
Lakshmibai National Institute of Physical Education, India

Anindita Das
http://orcid.org/0000-0002-6502-9619
Lakshmibai National Institute of Physical Education, India

Karuppasamy Govindasamy
http://orcid.org/0000-0002-3019-5545
Symbiosis International University, India

Tashi Ongmu Bhutia
http://orcid.org/0009-0000-0127-5709
Government B.Ed. College, Soreng, India

Atisha Mandal
http://orcid.org/0009-0003-2731-5985
Lakshmibai National Institute of Physical Education, India

EXECUTIVE SUMMARY

Sports and exercise science research presents unique ethical challenges extending beyond traditional biomedical frameworks. This chapter examines moral considerations in studying human performance within competitive athletic contexts. Fundamental principles of respect for persons, beneficence, justice, and integrity require careful adaptation for athletic populations facing distinctive pressures and

DOI: 10.4018/979-8-3693-8302-5.ch006

vulnerabilities. Critical issues include informed consent procedures for diverse populations, risk assessment in high intensity protocols, confidentiality protection in elite populations, and data privacy with emerging technologies. Research design challenges encompass participant selection, randomization ethics, and emergency preparedness. Intervention studies raise concerns regarding ergogenic aids, fatigue boundaries, and placebo controls. The chapter emphasizes research integrity through proper data handling, conflict management, and transparent reporting.

INTRODUCTION

Sports and exercise science research occupies a distinctive position within the broader scientific landscape, combining rigorous empirical investigation with the practical demands of athletic performance enhancement and health promotion. This field presents unique ethical challenges that stem from its focus on human performance, its intersection with competitive athletics, and its potential for both significant benefits and substantial risks to participants. The ethical landscape of sports science research has evolved considerably over recent decades, driven by increasing recognition of the complex moral considerations inherent in studying human movement, performance, and adaptation (Harriss et al., 2022).

The ethical imperative in sports science research extends beyond traditional biomedical research ethics to encompass considerations specific to athletic populations, performance optimization, and the competitive environment. Researchers in this field must navigate competing interests between advancing scientific knowledge, optimizing athletic performance, protecting participant welfare, and maintaining research integrity. The stakes are particularly high given that research findings often directly influence training practices, policy decisions, and the health and careers of athletes ranging from recreational participants to elite professionals (Harriss & Atkinson, 2009).

Contemporary sports science research encompasses diverse methodological approaches, from controlled laboratory investigations of physiological responses to field-based studies of training interventions and observational research on athletic populations. Each approach presents distinct ethical challenges related to participant selection, risk assessment, informed consent, data privacy, and result interpretation (Rhodes, 2005). The increasing sophistication of technology in sports science, including wearable devices, motion capture systems, genetic testing, and advanced imaging techniques, has introduced new dimensions of ethical complexity regarding data ownership, privacy protection, and the appropriate use of personal information. The globalization of sports and the commercial interests surrounding athletic performance have further complicated the ethical landscape. Researchers often work with international athletes, navigate different cultural contexts, and face pressure from sponsors, governing bodies, and commercial partners. These factors can create conflicts of interest and ethical dilemmas that require careful consideration and transparent management (Harriss et al., 2022).

FOUNDATIONS OF RESEARCH ETHICS IN SPORTS SCIENCE

- **Core Ethical Principles**

The ethical foundation of sports science research rests upon four fundamental principles that guide decision-making and establish the moral framework for research conduct. These principles, originally articulated in biomedical research ethics, have been adapted and refined for application in sports and exercise science contexts.

Respect for Persons encompasses two primary components: autonomy and protection of individuals with diminished autonomy. In sports science research, respecting autonomy means recognizing participants' capacity to make informed decisions about their involvement in research. This principle requires researchers to provide comprehensive information about study procedures, risks, benefits, and alternatives, enabling participants to make voluntary and informed choices. The principle becomes particularly complex in sports contexts where athletes may face external pressures from coaches, sponsors, or team management that could compromise their autonomous decision-making (Stewart & Reider, 2016).

The protection aspect of this principle is especially relevant when working with vulnerable populations common in sports science research, including minors, athletes with cognitive impairments, or individuals in hierarchical sporting structures where power imbalances may exist. Researchers must implement additional safeguards to ensure these participants' rights and interests are protected, even when they cannot fully advocate for themselves.

Beneficence obligates researchers to maximize benefits while minimizing harms. In sports science research, benefits may include improved performance, enhanced health outcomes, advancement of scientific knowledge, and contributions to evidence-based practice in sport and exercise. However, determining and weighing these benefits against potential risks requires sophisticated analysis, particularly given that many sports science interventions involve physical stress, fatigue, or discomfort that would be considered harmful in other research contexts but may be acceptable or even beneficial within athletic training (Waddington, 2013).

The principle of beneficence extends beyond individual participants to consider broader societal benefits. Sports science research often aims to improve training methods, prevent injuries, or enhance performance across athletic populations. Researchers must balance individual participant welfare against these broader benefits while ensuring that the research design is scientifically sound and likely to produce meaningful results.

Justice demands fair distribution of research benefits and burdens. This principle has particular relevance in sports science where research participants often come from specific athletic populations that may not directly benefit from research findings. Elite athletes, for example, frequently participate in studies that may primarily benefit recreational athletes or general populations. Similarly, research conducted with male athletes has historically been generalized to female populations without adequate consideration of sex-specific differences (Rhodes, 2005).

Justice also encompasses issues of access and inclusion. Sports science research must consider whether study populations adequately represent the diversity of individuals who may be affected by research findings. This includes considerations of sex, age, race, ethnicity, socioeconomic status, and disability status. The principle of justice challenges researchers to ensure that the benefits of research are accessible to all relevant populations and that no group bears a disproportionate burden of research participation (Stewart & Reider, 2016).

Integrity encompasses honesty, accuracy, efficiency, and objectivity in research conduct. This principle requires researchers to design studies that can answer the proposed research questions, collect and analyze data appropriately, report findings accurately and completely, and acknowledge limitations and uncertainties. In sports science, integrity is particularly challenged by commercial interests, competitive pressures, and the temptation to overstate findings that may provide performance advantages (Milanović et al., 2021).

Research integrity also encompasses responsible stewardship of resources, including participant time, funding, and institutional support. Given the physical and time demands often placed on sports science research participants, researchers have an obligation to ensure that studies are well-designed, adequately powered, and likely to contribute meaningful knowledge to the field.

- **International Guidelines and Declarations**

The ethical conduct of sports science research is governed by multiple international guidelines and declarations that provide frameworks for ethical decision-making and establish minimum standards for research conduct. These documents reflect international consensus on research ethics while allowing for adaptation to specific cultural and disciplinary contexts.

The **Declaration of Helsinki**, developed by the World Medical Association, serves as a cornerstone document for research ethics globally. While originally focused on biomedical research, its principles are directly applicable to sports science research involving human participants. The Declaration emphasizes the primacy of participant welfare over scientific and societal interests, requires independent ethical review of research protocols, and mandates informed consent procedures. Recent revisions have strengthened provisions for vulnerable populations, post-trial access to interventions, and registration of clinical trials, all of which have implications for sports science research (M. J. McNamee et al., 2006).

The Declaration's emphasis on risk-benefit assessment is particularly relevant for sports science research, where interventions may involve significant physical stress or discomfort. The document requires that risks be minimized through careful study design and that potential benefits justify any remaining risks. This framework provides guidance for evaluating high-intensity training protocols, fatigue studies, and other physically demanding research procedures common in sports science (Stewart & Reider, 2016).

The **Belmont Report**, commissioned by the United States National Commission for the Protection of Human Subjects of Biomedical and Behavioural Research, articulated the three core principles of research ethics: respect for persons, beneficence, and justice. While developed in response to abuses in biomedical research, the Report's principles have been widely adopted across disciplines, including sports science. The Report's emphasis on informed consent, risk-benefit assessment, and fair participant selection provides practical guidance for sports science researchers (M. McNamee et al., 2025).

The Belmont Report's discussion of vulnerable populations is particularly relevant for sports science research, which often involves minors, individuals in institutional settings, and participants who may have impaired decision-making capacity due to competitive pressures or hierarchical relationships within sporting organizations.

The **World Anti-Doping Agency (WADA) Code** establishes the framework for anti-doping policies and procedures in sport. While not primarily an ethics document, the Code has significant implications for sports science research, particularly studies involving nutritional supplements, ergogenic aids, or other substances that could enhance performance. The Code's prohibited substances list and criteria for inclusion create ethical obligations for researchers to ensure that research interventions do not inadvertently violate anti-doping rules or create situations where participants might be sanctioned for research participation (Rhodes, 2005).

The Code's emphasis on athlete health protection aligns with research ethics principles while introducing sport-specific considerations around fair play and competition integrity. Researchers must navigate the intersection between scientific inquiry and anti-doping regulations, ensuring that research contributes to knowledge advancement while respecting the values and rules of competitive sport.

UNESCO's Universal Declaration on Bioethics and Human Rights provides a global framework for bioethics that extends beyond medical research to encompass all research involving humans. The Declaration's principles of human dignity, human rights, and non-discrimination are directly relevant to sports science research. Its emphasis on cultural diversity and pluralism is particularly important for researchers working with international athletes or conducting cross-cultural studies (M. McNamee et al., 2025).

The Declaration's articles on consent, privacy and confidentiality, and non-discrimination provide guidance for sports science researchers working in diverse cultural contexts. Its recognition of the importance of traditional knowledge systems and local values offers a framework for respectful engagement with different sporting cultures and traditions.

ETHICAL CHALLENGES IN RESEARCH DESIGN

Research design decisions in sports and exercise science carry profound ethical implications that extend beyond methodological considerations to encompass fundamental questions of fairness, safety, and scientific validity. The unique characteristics of athletic populations, the physical nature of many interventions, and the competitive context of sport create distinctive challenges that require careful ethical analysis and innovative solutions.

- **Participant Selection and Inclusion Criteria**

The process of participant selection in sports science research involves complex ethical considerations that intersect with scientific validity, practical feasibility, and justice concerns. Traditional approaches to participant selection, often driven by convenience and homogeneity considerations, have resulted in research populations that inadequately represent the diversity of individuals who may be affected by research findings. Gender representation presents one of the most significant challenges in sports science research (Rhodes, 2005). Historically, research has overrepresented male participants, leading to knowledge gaps regarding female physiology, performance, and response to training interventions. This gender bias reflects both practical considerations, such as controlling for menstrual cycle effects, and institutional factors, including greater availability of male research participants in some sporting contexts. However, this bias violates principles of justice by excluding women from research that may directly benefit them while potentially exposing them to interventions that have not been adequately tested in female populations (M. J. McNamee et al., 2006).

Contemporary ethical frameworks demand more inclusive approaches to participant selection that account for sex differences in physiology, metabolism, and response to exercise interventions. This requires researchers to either include adequate representation of both sexes with appropriate statistical power to detect sex-specific effects or to provide clear justification for single-sex studies based on scientific rationale rather than convenience. Age-related considerations present additional complexity in participant selection (Sachs & Siegler, 1993). Much sports science research focuses on young adult populations, typically university-aged individuals who are readily accessible to researchers. However, this age bias limits the generalizability of findings to older athletes, masters athletes, and youth populations. Each age group presents unique physiological characteristics and ethical considerations that must be addressed through appropriate study design and participant selection strategies. The inclusion of youth participants introduces heightened ethical obligations related to assent procedures, parental consent, and protection from harm. Researchers must carefully balance the scientific value of including young athletes against the additional risks and ethical complexities inherent in research with minors. This includes consideration of developmental appropriateness of research procedures, long-term impacts on growth and development, and the potential for research participation to interfere with normal social and educational activities (Macrina, 2007).

Elite athlete participation presents unique ethical challenges related to coercion, confidentiality, and conflicts of interest. Elite athletes may feel pressured to participate in research due to relationships with coaches, sponsors, or sporting organizations that support the research. Their small population size creates challenges for maintaining anonymity and confidentiality, particularly in sports where individual performance data is closely monitored and widely reported.

- **Randomization and Control Groups in Performance Studies**

Randomization serves as a cornerstone of rigorous experimental design, but its implementation in sports science research presents distinctive ethical challenges related to fairness, athlete development, and competitive integrity. The tension between scientific rigor and ethical obligations to participants becomes particularly acute when randomization may result in some athletes receiving potentially beneficial interventions while others receive standard care or placebo treatments. The ethical justification for randomization rests on the principle of equipoise, which requires genuine uncertainty about the relative merits of different interventions (Sachs & Siegler, 1993). In sports science research, achieving and maintaining equipoise can be challenging due to strong theoretical rationales for certain interventions, pilot data suggesting effectiveness, or pressure from coaches and athletes who believe they know which intervention is superior. Randomization becomes ethically problematic when it potentially disadvantages certain participants in ways that could affect their athletic development, competition performance, or long-term career prospects. For example, randomizing athletes to different training interventions during important preparation periods could create unfair advantages or disadvantages that extend beyond the research context. This is particularly concerning when working with young athletes whose development trajectories may be significantly influenced by training interventions received during critical periods (Macrina, 2007).

The use of control groups in sports science research presents additional ethical complexities. No-treatment control groups may be ethically problematic when effective interventions exist and withholding them could disadvantage participants. Wait-list control designs can address some ethical concerns by ensuring all participants eventually receive the intervention, but they may not be feasible in contexts where timing is critical for athletic development or competition preparation (van Stekelenburg, 2024).

- **Use of Deception and Its Moral Limits**

Deception in research involves deliberately misleading participants about study procedures, purposes, or other relevant information. While generally discouraged in research ethics, deception is sometimes considered necessary to obtain valid results, particularly in studies examining psychological factors, placebo effects, or behaviours that might be altered by participant awareness. In sports science research, deception may be employed in various contexts, including studies of pacing strategies where participants are given false information about their performance or competitors, investigations of placebo effects where participants receive inert treatments described as active interventions, or examinations of psychological factors where revealing the true study purpose might influence participant behaviour (Milanović et al., 2021).

The ethical justification for deception requires satisfaction of several stringent criteria. First, the research must address important questions that cannot be answered without deception. Alternative methods that do not involve deception must be infeasible or inadequate for addressing the research questions. Second, the deception must be minimal and must not involve withholding information about significant risks or aspects of the research that might affect participants' willingness to participate. Third, participants must be debriefed about the deception as soon as feasible, typically immediately after their participation concludes. This debriefing must include an explanation of why deception was necessary, an opportunity for participants to ask questions, and information about how to contact researchers if concerns arise later. Participants should also be given the opportunity to withdraw their data if they object to the deception after learning about it (Sachs & Siegler, 1993).

- **Risk-Benefit Assessment in High-Intensity Protocols**

Risk-benefit assessment represents one of the most critical ethical challenges in sports science research, particularly for studies involving high-intensity exercise protocols, fatigue studies, or interventions that push participants to their physiological limits. The assessment requires careful consideration of potential physical, psychological, and social risks while balancing these against scientific and societal benefits.

Physical risks in sports science research can range from minor discomfort and temporary fatigue to serious injury or medical complications. High-intensity exercise protocols may cause muscle damage, cardiovascular stress, metabolic disturbances, or exacerbation of underlying health conditions. The challenge for researchers is distinguishing between acceptable risks that are inherent to athletic training and research-specific risks that require additional justification and safeguards. The concept of minimal risk, commonly used in research ethics, is particularly complex in sports science contexts. Minimal risk is typically defined as risks no greater than those encountered in daily life or routine medical examinations. However, for athletes, daily life often involves significant physical stress and risk of injury. This creates ambiguity about what constitutes minimal risk for athletic populations and whether more intensive research procedures might be justified based on participants' baseline risk exposure (Milanović et al., 2021).

INFORMED CONSENT IN ATHLETE POPULATIONS

Informed consent represents a fundamental ethical requirement that takes on unique complexity within sporting contexts. The process of obtaining meaningful consent from athletes involves navigating power dynamics, cultural considerations, and practical challenges that distinguish it from consent procedures in other research domains.

- **Challenges with Minors and Adolescent Athletes**

Research involving minor athletes presents multifaceted ethical challenges that extend beyond standard pediatric research considerations to encompass the unique developmental, social, and competitive aspects of youth sport participation. The involvement of minors in sports science research requires sophisticated approaches to consent and assent that respect young people's developing autonomy while protecting their interests and wellbeing (Steneck & Bulger, 2007).

The legal framework for research with minors typically requires parental or guardian consent along with participant assent, but this binary approach fails to capture the complex reality of adolescent athlete decision-making. Young athletes, particularly those competing at high levels, often demonstrate considerable maturity and independence in making sport-related decisions. They may have strong opinions about research participation based on their understanding of training, competition, and performance optimization that exceed their parents' knowledge of these areas (Rhodes, 2005). However, adolescents' cognitive and emotional development continues throughout the teenage years, affecting their ability to fully comprehend long-term consequences, assess risks accurately, and resist peer or authority pressure. The competitive environment of youth sport can amplify these developmental limitations, as young athletes may be particularly motivated to pursue any opportunity that might enhance their performance or standing within their sport. The assent process for minor athletes must be developmentally appropriate while addressing the sophisticated understanding many young athletes have of their sport. This requires researchers to develop age-appropriate explanations that neither oversimplify complex concepts nor overwhelm young participants with technical information. Visual aids, interactive presentations, and opportunities for questions and discussion can enhance understanding and engagement in the assent process (Lategan, 2012).

- **Language, Comprehension, and Cultural Considerations**

The globalization of sport has created increasingly diverse research populations that challenge traditional approaches to informed consent. Athletes from different linguistic, cultural, and educational backgrounds bring varying expectations, values, and communication styles that must be considered in developing consent procedures that are both respectful and effective.

Language barriers represent an obvious but complex challenge in obtaining meaningful consent from international athletes. Simple translation of consent documents is insufficient, as direct translations may not convey intended meanings, cultural concepts may not have equivalent expressions across languages, and technical or scientific terminology may be particularly difficult to translate accurately. Beyond linguistic translation, researchers must consider cultural translation of consent concepts and procedures. Different cultures have varying expectations about authority relationships, individual versus collective decision-making, and the appropriate level of detail in information sharing. Some cultures emphasize collective decision-making processes that may conflict with individualistic consent procedures common in Western research traditions (Steneck & Bulger, 2007).

- **Consent in Team-Based Versus Individual Studies**

The distinction between team-based and individual sports research creates fundamentally different contexts for informed consent that reflect the social dynamics, competitive structures, and decision-making processes characteristic of each setting. These differences require tailored approaches to consent procedures that respect the unique aspects of each research environment while maintaining ethical standards. Team-based research often involves complex social dynamics that can influence individual consent decisions. Team members may feel pressure to participate in research to maintain team cohesion, avoid disappointing teammates or coaches, or prevent being perceived as uncooperative. The collective nature of team sports can create situations where individual athletes' preferences are subordinated to team interests or where peer pressure influences participation decisions (Bird, 1996).

- **Ongoing Consent in Longitudinal and Intervention Studies**

Longitudinal research and extended intervention studies present unique challenges for maintaining meaningful consent over time. Initial consent, no matter how comprehensive, cannot fully address all issues that may arise during extended research participation. Changes in participants' circumstances, research procedures, understanding, or preferences require ongoing attention to consent and opportunities for participants to reconsider their involvement (Edwards, 2020). The concept of ongoing consent recognizes that consent is not a single event but a continuous process that must be maintained throughout research participation. This is particularly important in sports science research, where training interventions, competitive schedules, and athlete circumstances may change significantly over time, affecting the risks, benefits, and practical implications of continued research participation (Bird, 1996).

CONFIDENTIALITY AND DATA PRIVACY ISSUES

The protection of participant confidentiality and data privacy represents a cornerstone of ethical research practice that takes on distinctive characteristics and challenges within sports science research contexts. The nature of athletic performance data, the competitive environment of sport, and the increasing sophistication of data collection technologies create complex privacy considerations that extend beyond traditional research confidentiality concerns.

- **Protection of Sensitive Personal and Health Information**

Sports science research generates vast amounts of sensitive personal and health information that requires sophisticated protection strategies. This information often includes detailed physiological data, performance metrics, training histories, injury records, psychological assessments, and lifestyle information that could have significant implications for athletes' careers, relationships, and future opportunities (Steneck & Bulger, 2007). The sensitivity of health information in sports contexts is amplified by the competitive implications of such data. Information about injuries, fitness levels, physiological capabilities, or training responses could provide competitive advantages if accessed by rivals, coaches of opposing teams, or other unauthorized parties. Similarly, psychological or behavioural data could be used inappropriately to influence selection decisions, contract negotiations, or public perceptions of athletes (Rao et al., 2024).

Medical information collected during sports science research often extends beyond what would be typical in clinical settings, including detailed assessments of cardiovascular function, metabolic profiles, biomechanical analyses, and specialized testing procedures. This information may reveal health conditions or predispositions that athletes themselves are unaware of, creating ethical obligations for researchers regarding disclosure and follow-up care (Horn, 2013).

- **Anonymity in Small Athlete Populations (Elite/Professional Levels)**

Elite athlete populations present unique challenges for maintaining anonymity due to their small size, high visibility, and distinctive performance characteristics. Traditional approaches to anonymity that rely on aggregating data across large populations may be insufficient when dealing with small numbers of highly visible individuals whose performance data may be publicly available or easily identifiable. The challenge of maintaining anonymity is particularly acute in sports with small elite populations, such as certain Olympic sports, where individual athletes may be easily identifiable even when traditional demographic identifiers are removed. Performance data, training histories, injury patterns, or even participation in specific research studies may provide sufficient information for identification by knowledgeable observers (Rao et al., 2024).

- **Digital Data Storage, Wearable Technology, and Privacy Concerns**

The proliferation of digital technologies in sports science research has created new opportunities for data collection while introducing novel privacy and security challenges. Wearable devices, smartphone applications, online platforms, and cloud storage systems generate continuous streams of personal data that require sophisticated protection strategies. Wearable technology devices commonly used in sports science research, including heart rate monitors, GPS trackers, accelerometers, and specialized sensors, collect detailed information about athletes' movements, physiological responses, and behavioural patterns. This data may be transmitted wirelessly, stored on devices, and synchronized with online platforms, creating multiple points of potential vulnerability (Horn, 2013).

- **Handling Video, Motion Capture, and Biometric Data**

Video recordings, motion capture data, and biometric information present unique privacy and confidentiality challenges in sports science research. These data types often contain highly identifiable information that cannot be easily anonymized while maintaining scientific value, requiring specialized approaches to protection and management. Video data used for biomechanical analysis, technique assessment, or behavioural studies typically captures athletes' faces, distinctive physical characteristics, and movement patterns that make them easily identifiable. Unlike other forms of research data that can be de-identified through removal of personal identifiers, video data inherently contains identifying information that cannot be separated from the scientific content (Rao et al., 2024).

Motion capture systems generate detailed three-dimensional representations of human movement that may be used to identify individuals based on their unique movement patterns or gait characteristics. Research has demonstrated that individuals can be identified from biomechanical data with high accuracy, even when traditional identifiers are removed, creating challenges for maintaining anonymity in studies using these technologies (Horn, 2013).

ETHICAL ISSUES IN INTERVENTION STUDIES

Intervention studies in sports and exercise science present some of the most complex ethical challenges in the field, as they involve deliberate manipulation of athletes' training, nutrition, supplementation, or other factors that may directly impact performance, health, and wellbeing. These studies require careful consideration of the balance between scientific advancement and participant welfare, while navigating the competitive pressures and performance expectations inherent in athletic contexts.

- **Use of Ergogenic Aids and Supplementation Research**

Research involving ergogenic aids and nutritional supplements presents unique ethical challenges that intersect with anti-doping regulations, health risks, and performance enhancement considerations. The potential for supplements to provide performance benefits creates strong incentives for athletes to participate in such research, but it also raises questions about fairness, safety, and the appropriate boundaries of performance enhancement research (Ketefian, 2015). The World Anti-

Doping Agency's prohibited substances list creates a complex regulatory environment for supplementation research. Researchers must ensure that any substances used in studies do not violate anti-doping rules or place participants at risk of sanctions. This requires ongoing monitoring of prohibited substances lists, careful sourcing and testing of supplements, and clear communication with participants about potential anti-doping implications (Watson II & Etzel, 2019).

Many dietary supplements and ergogenic aids lack rigorous safety testing, particularly in combination with intensive training or when used by elite athletes with unique physiological demands. The supplement industry's limited regulation means that products may contain undisclosed ingredients, contaminants, or adulterants that could pose health risks or result in positive doping tests. The placebo effect is particularly strong in performance contexts, creating challenges for designing appropriate control conditions in supplementation research. Athletes may experience performance improvements from inert substances simply due to their expectations and beliefs about the intervention. This necessitates careful attention to blinding procedures and placebo design to ensure valid results (Laryeafio & Ogbewe, 2023).

- **Testing Boundaries of Fatigue, Injury Risk, and Recovery**

Sports science research often involves pushing athletes to their physiological limits to understand fatigue mechanisms, performance capacity, and recovery processes. These studies present ethical challenges related to the acceptable level of physical stress, risk of injury, and potential for lasting negative effects on participants (Zhang et al., 2024).

Fatigue studies typically require participants to exercise to exhaustion or maintain high-intensity efforts until they can no longer continue. While such protocols are necessary for understanding fatigue mechanisms and performance limits, they involve significant physical and psychological stress that may extend beyond what athletes normally experience in training or competition. The assessment of injury risk in sports science research requires careful consideration of both acute injury potential during research procedures and longer-term risks associated with research interventions. High-intensity protocols, novel training methods, or biomechanical assessments may place participants at risk for musculoskeletal injuries that could affect their athletic careers (Rao et al., 2024).

Recovery research often involves manipulating athletes' normal recovery processes through sleep deprivation, altered nutrition, modified training schedules, or other interventions that may compromise their wellbeing or performance capacity. Such studies must carefully balance scientific objectives against potential negative impacts on participants' health and athletic development.

- **Sham/Placebo Interventions in Performance Trials**

The use of sham or placebo interventions in sports performance research presents unique ethical challenges related to deception, equipoise, and the potential for participants to be disadvantaged during critical training or competition periods. Unlike pharmaceutical research where inert placebos can be easily masked, sports interventions often involve behavioural or training components that are difficult to simulate convincingly (Zhang et al., 2024).

Creating effective sham interventions for training studies may require elaborate deception about the nature of the intervention, the expected mechanisms of action, or the scientific rationale for the protocol. This deception must be justified by the scientific importance of the research question and the inability to obtain valid results through alternative methods (Rao et al., 2024). The timing of placebo-controlled studies becomes particularly important in sports contexts where training periodization and competitive schedules create critical periods for athlete development. Randomizing athletes to receive ineffective interventions during important preparation phases may disadvantage them in ways that extend beyond the research context (Kang & Hwang, 2020).

- **Emergency Preparedness in High-Risk Physical Testing**

High-intensity exercise testing and performance assessments in sports science research create potential for medical emergencies that require comprehensive emergency preparedness protocols. The combination of maximal physical exertion, potential underlying health conditions, and the competitive drive of athletes creates elevated risks that must be carefully managed (Zhang et al., 2024).

Emergency action plans must address the full range of potential medical complications associated with high-intensity exercise, including cardiac events, heat illness, exercise-induced bronchospasm, metabolic disturbances, and musculoskeletal injuries. These plans should specify personnel responsibilities, equipment requirements, communication procedures, and protocols for emergency medical care. The qualifications and training of research personnel conducting high-risk testing must be appropriate for managing potential emergencies. This may require medical supervision, certified emergency response training, and regular updates on emergency procedures and equipment maintenance (Rao et al., 2024).

Participant screening procedures become critical for identifying individuals who may be at elevated risk for exercise-related complications. This includes cardiovascular screening, assessment of previous injuries or medical conditions, and evaluation of current health status and medication use.

INTEGRITY IN DATA COLLECTION, ANALYSIS, AND REPORTING

Research integrity forms the foundation of credible scientific inquiry and is particularly crucial in sports science research, where findings may directly influence training practices, policy decisions, and athlete welfare. The competitive pressures, commercial interests, and performance implications inherent in sports contexts create unique challenges for maintaining the highest standards of scientific integrity throughout the research process.

- **Fabrication, Falsification, and Selective Reporting**

Research misconduct in the form of fabrication, falsification, or selective reporting represents the most serious violations of scientific integrity. In sports science research, these practices are particularly problematic due to their potential to misinform training practices and endanger athlete health and performance. Fabrication involves the creation of false data or results that were never obtained through legitimate research procedures. In sports science contexts, fabrication might involve creating performance data, physiological measurements, or survey responses that were never actually collected. The pressure to produce positive results in performance research, combined with the difficulty of replicating some sports science studies, may create incentives for fabrication (Connelly, 2014).

Falsification encompasses the manipulation or alteration of research data, procedures, or results to misrepresent findings. This might involve selectively modifying performance times, altering physiological measurements, or changing statistical analyses to produce desired outcomes. The competitive nature of sports and the potential commercial value of positive results create particular temptations for falsification. Selective reporting involves the biased presentation of research findings through the omission of negative results, cherry-picking of favourable outcomes, or post-hoc modification of study objectives to match obtained results. In sports science research, selective reporting might manifest as presenting only positive performance outcomes while omitting adverse effects, focusing on subgroup analyses that show benefits while ignoring overall null results, or modifying primary outcome measures based on the pattern of results obtained (Resnik, 2014).

The prevention of research misconduct requires comprehensive training, clear policies and procedures, appropriate supervision and mentorship, and systems for detecting and addressing potential problems. Research institutions must foster cultures that prioritize integrity over productivity and provide support for researchers facing pressures that might compromise ethical conduct.

- **Conflicts of Interest in Sponsored Research**

The commercial interests surrounding sports performance create numerous opportunities for conflicts of interest that can compromise research integrity. Sponsorship from equipment manufacturers, supplement companies, sports organizations, or other commercial entities may create financial, professional, or personal interests that conflict with researchers' obligations to conduct objective, unbiased research (Milanović et al., 2021).

Financial conflicts of interest are perhaps the most obvious concern, occurring when researchers have financial relationships with entities that have interests in research outcomes. This might include research funding from companies whose products are being tested, consulting relationships with organizations that could benefit from positive results, or equity interests in companies that might be affected by research findings (Horn, 2013).

Professional conflicts of interest may arise when researchers have ongoing relationships with organizations or individuals that could influence their objectivity. This might include employment relationships, collaborative partnerships, or professional obligations that create pressures to produce favourable results or avoid negative findings. Personal conflicts of interest encompass relationships or commitments that might affect researchers' judgment, even in the absence of financial considerations. This could include personal friendships with athletes or coaches involved in research, commitments to specific training philosophies or approaches, or personal beliefs about the effectiveness of certain interventions (Resnik, 2014).

The management of conflicts of interest requires transparent disclosure, appropriate oversight, and sometimes recusal from certain research activities. Researchers must identify potential conflicts early in the research process and implement strategies to minimize their impact on study conduct and result interpretation.

1. Ethical Use of Statistics and Interpretation of Results

The appropriate use of statistical methods and interpretation of results is fundamental to research integrity in sports science. The complexity of athletic performance data, the multiple factors that influence outcomes, and the pressure to find significant results create numerous opportunities for statistical misconduct or misinterpretation (Milanović et al., 2021).

Statistical misconduct may include inappropriate selection of analytical methods, multiple testing without appropriate corrections, post-hoc modification of analytical plans based on results, or selective reporting of statistical outcomes. These practices can lead to false positive findings, overestimation of effect sizes, or inappropriate conclusions about intervention effectiveness. The interpretation of statistical results requires careful consideration of effect sizes, confidence intervals, practical significance, and clinical relevance, not just statistical significance. In sports science research, statistically significant findings may not translate to meaningful performance improvements or may come with unacceptable risks or costs (Resnik, 2014).

The presentation of uncertainty and limitations is crucial for honest reporting of research findings. This includes acknowledging potential confounding factors, discussing alternative explanations for findings, recognizing the limitations of study designs, and avoiding overstatement of conclusions based on available data (Rossouw et al., 2014).

- **Authorship, Collaboration, and Publication Ethics**

Ethical issues in authorship and publication are particularly complex in sports science research due to the multidisciplinary nature of the field, the involvement of multiple stakeholders including athletes and coaches, and the various forms of contribution that may occur in applied research settings. Authorship criteria should be based on substantial contributions to research conception and design, data acquisition, analysis and interpretation, drafting or critical revision of manuscripts, and approval of final versions. All individuals meeting these criteria should be offered authorship, while those making lesser contributions should be acknowledged appropriately (Rossouw et al., 2014).

The involvement of athletes, coaches, and other practitioners in research creates questions about appropriate recognition for their contributions. While their participation as subjects typically does not warrant authorship, substantial intellectual contributions to study design, data interpretation, or manuscript preparation may justify inclusion as authors. Collaborative relationships in sports science research often involve complex partnerships between academic researchers, applied practitioners, commercial entities, and sports organizations. These relationships must be managed transparently with clear agreements about roles, responsibilities, data ownership, and publication rights established before research begins (Madak, 1994).

Gift authorship, where individuals are included as authors despite not meeting authorship criteria, violates ethical publication standards and misrepresents the actual contributions to research. Similarly, ghost authorship, where individuals who meet authorship criteria are excluded from author lists, denies appropriate credit and may conceal conflicts of interest.

ETHICS OF TECHNOLOGY AND INNOVATION IN SPORTS RESEARCH

The rapid advancement of technology in sports science research has created unprecedented opportunities for understanding human performance while simultaneously introducing novel ethical challenges. Emerging technologies including artificial intelligence, genetic testing, virtual reality, and advanced sensing systems raise fundamental questions about privacy, consent, data ownership, and the appropriate boundaries of performance enhancement research (Ellapen & Paul, 2016).

Artificial intelligence and machine learning applications in sports science present complex ethical considerations related to algorithmic bias, data privacy, and the appropriate use of predictive models. AI systems trained on historical performance data may perpetuate existing biases related to gender, race, or other characteristics, potentially disadvantaging certain athletes or groups. The "black box" nature of many AI algorithms makes it difficult for participants to understand how their data is being used and what factors influence algorithmic decisions about their performance or development (Tonis et al., 2025).

Genetic testing and genomic research in sports science raise profound ethical questions about genetic privacy, discrimination, and the appropriate use of genetic information in athletic contexts. Genetic variants associated with performance, injury risk, or training response could have significant implications for athletes' careers and personal lives. The potential for genetic discrimination in team selection, scholarship decisions, or career counseling requires careful consideration of how genetic information is collected, stored, and used (Ellapen & Paul, 2016).

Wearable technology and continuous monitoring systems create new possibilities for understanding athletic performance and physiological responses while raising concerns about surveillance, autonomy, and the boundary between research and monitoring. Athletes may feel pressured to maintain optimal behaviour and performance when they know they are being continuously monitored, potentially affecting their wellbeing and natural behaviour patterns. Virtual and augmented reality technologies enable new forms of training research and performance analysis while creating questions about the psychological effects of immersive experiences, the potential for addiction or dependency, and the appropriate use of virtual environments in research contexts. The realism of virtual experiences may blur the boundaries between research and real-world activities, creating confusion about consent and the scope of research participation (Tonis et al., 2025).

Biometric identification and authentication technologies raise concerns about identity verification, data security, and the potential for unauthorized access to personal information. The use of facial recognition, fingerprint scanning, or other biometric methods in research contexts requires careful consideration of privacy implications and data protection measures (Loland, 2009).

The integration of multiple data sources through "big data" approaches creates comprehensive profiles of athletes that may be more sensitive and identifiable than any individual data source. Data fusion techniques that combine performance data, physiological measurements, behavioural information, and environmental factors require sophisticated approaches to privacy protection and consent management.

INSTITUTIONAL REVIEW BOARDS (IRBS) AND GOVERNANCE MECHANISMS

Institutional Review Boards serve as the primary mechanism for ethical oversight of human subjects research, but their application to sports science research presents unique challenges related to risk assessment, expertise requirements, and the evaluation of sport-specific ethical issues. The effectiveness of IRB review depends on board members' understanding of athletic populations, sports science methodologies, and the distinctive ethical considerations that arise in performance research contexts.

- **Composition and Expertise Requirements**

Traditional IRB composition requirements may be inadequate for effectively reviewing sports science research protocols. Standard IRBs typically include members with expertise in relevant scientific areas, ethicists, legal experts, and community representatives. However, sports science research often requires specialized knowledge of exercise physiology, biomechanics, sports psychology, and athletic training practices that may not be represented on general IRBs. The inclusion of individuals with sports science expertise is crucial for accurate risk assessment and evaluation of research protocols. Board members must understand the normal physical demands of athletic training and competition to appropriately evaluate whether research procedures present acceptable risks. Without this specialized knowledge, IRBs may either inappropriately approve risky protocols or unnecessarily restrict valuable research due to misunderstanding of sports contexts (Vitak et al., 2017).

Community representation on IRBs reviewing sports science research should ideally include individuals who understand athletic populations and sporting cultures. This might include former athletes, coaches, parents of young athletes, or others who can represent the perspectives and interests of research participants. Such representation helps ensure that ethical review considers the unique values, concerns, and decision-making processes characteristic of sporting communities. The international nature of much sports science research creates additional challenges for IRB composition and expertise. Research involving international athletes may require board members who understand different cultural contexts, legal systems, and sporting traditions. Language barriers, different consent practices, and varying ethical frameworks across cultures must be considered in protocol review and approval processes (Nurunnabi, 2014).

- **Risk Assessment in Sports Science Research**

Risk assessment represents one of the most challenging aspects of IRB review for sports science research. Traditional risk categories and assessment frameworks may be inadequate for evaluating the complex physical, psychological, and social risks associated with athletic performance research. IRBs must develop sophisticated approaches to risk assessment that account for the unique characteristics of athletic populations and sporting contexts.

The concept of minimal risk requires careful interpretation in sports science contexts. While minimal risk is typically defined as risks no greater than those encountered in daily life, the daily lives of athletes often involve significant physical stress and injury risk. This creates ambiguity about baseline risk levels and whether research procedures that might be considered high-risk for general populations are acceptable for athletic populations (Vitak et al., 2017).

Physical risk assessment must consider both acute risks associated with research procedures and longer-term risks that might affect athletes' health, development, or career trajectories. High-intensity exercise protocols may present immediate risks of injury, cardiovascular events, or metabolic disturbances, while also potentially affecting training adaptations, recovery patterns, or competitive performance over time. Psychological and social risk assessment requires understanding of the competitive pressures, identity issues, and social dynamics that characterize athletic populations. Research procedures that affect performance, provide comparative feedback, or alter training routines may have psychological implications that extend well beyond the immediate research context (Nurunnabi, 2014).

The cumulative nature of risks in sports science research presents additional assessment challenges. Athletes may participate in multiple research studies, train intensively while participating in research, or experience research-related stress in addition to competitive pressures. IRBs must consider how research participation fits within the broader context of athletes' lives and commitments.

- **Multi-Site and International Research Oversight**

The increasingly collaborative and international nature of sports science research creates complex oversight challenges that may exceed the capacity of traditional single-institution IRB review processes. Multi-site studies involving athletes from different institutions, countries, or sporting systems require coordinated oversight approaches that ensure consistent ethical standards while accommodating local variations in regulations and practices (Grady, 2015).

Central IRB models, where a single board provides primary ethical review for multi-site studies, can improve efficiency and consistency but may lack local expertise about specific populations or contexts involved in research. Local IRBs may better understand their particular athletic populations but may lack expertise in sports science research or may apply inconsistent standards across sites. International research collaboration requires navigation of different regulatory frameworks, cultural contexts, and institutional requirements across countries. What constitutes appropriate ethical oversight, consent procedures, or risk-benefit assessment may vary significantly across jurisdictions, creating challenges for ensuring both compliance and appropriate protection of participants (Vitak et al., 2017).

Data sharing and privacy protection requirements vary significantly across countries and may affect the design and conduct of international sports science research. European Union data protection regulations, for example, may impose requirements that affect how research data can be collected, stored, transferred, and used, with implications for multi-national research collaborations.

- **Ongoing Monitoring and Adverse Event Reporting**

The dynamic nature of sports science research, with its potential for changing participant circumstances, evolving competitive schedules, and cumulative effects of interventions, requires sophisticated approaches to ongoing monitoring and safety oversight that extend beyond traditional IRB review processes (Nurunnabi, 2014).

Adverse event reporting systems must be tailored to capture the range of potential negative outcomes associated with sports science research, including injuries, performance decrements, psychological distress, and social consequences. Traditional adverse event categories developed for biomedical research may be inadequate for capturing sport-specific harms or may misclassify normal training-related outcomes as adverse events. Data and safety monitoring boards (DSMBs) may be appropriate for larger or higher-risk sports science studies, providing independent oversight of study conduct and participant safety. However, DSMBs for sports science research require members with appropriate expertise in exercise science, sports medicine, and athletic populations to effectively evaluate safety data and make recommendations about study continuation or modification (van Stekelenburg, 2024).

Interim analyses and safety reviews must account for the seasonal nature of many sports, the varying demands of training and competition phases, and the potential for external factors to affect participant safety or study outcomes. Monitoring plans must be flexible enough to accommodate the realities of athletic participation while maintaining appropriate safety oversight.

CONCLUSION

The ethical landscape of sports and exercise science research is continuously evolving alongside technological innovations, shifting populations, and an expanding understanding of ethical principles. As this field advances, researchers face complex challenges that demand both scientific rigor and ethical sensitivity. Foundational principles such as respect for persons, beneficence, justice, and integrity remain central, yet their application in sporting contexts requires careful interpretation given the unique characteristics of athletic populations and the high stakes of competitive environments. Key ethical considerations include research design, participant selection, risk assessment, informed consent, confidentiality, and data privacy. The involvement of minors, international athletes, and those within hierarchical sporting structures makes voluntary participation particularly delicate. At the same time, the rise of wearable technologies, artificial intelligence, and genetic testing introduces new dimensions of privacy, autonomy, and fairness that call for innovative protections and responsible use.

Intervention studies raise some of the most pressing dilemmas, as they directly influence performance and welfare. Balancing scientific progress with participant safety, while navigating regulatory frameworks and competitive implications, demands vigilant oversight. Research integrity also remains a critical concern, with issues of selective reporting, conflicts of interest, and commercial influence requiring transparent governance and accountability. Moving forward, sports science research must embrace dialogue, collaboration, and education. Researchers, ethicists, athletes, coaches, institutions, and governing bodies share the responsibility of shaping ethical standards that safeguard participants while fostering innovation. Tailored training programs, discipline-specific guidelines, and adaptive review mechanisms are essential to build ethical capacity in this domain. Ultimately, the success of sports science research depends not only on methodological advancement but also on a steadfast commitment to human dignity, justice, and welfare. Ethical challenges should be viewed not as barriers but as opportunities to reinforce trust, integrity, and the responsible pursuit of knowledge.

REFERENCES

Bird, S. J. (1996). The Role of Science Professionals in Teaching Responsible Research Conduct. *Bioscience*, *46*(10), 783–786. DOI: 10.2307/1312856

Connelly, L. M. (2014). Ethical considerations in research studies. *Medsurg Nursing*, *23*(1), 54–56. https://go.gale.com/ps/i.do?id=GALE%7CA360608990&sid=googleScholar&v=2.1&it=r&linkaccess=abs&iss PMID: 24707669

Edwards, J. R. (2020). The peaceful coexistence of ethics and quantitative research. *Journal of Business Ethics*, *167*(1), 31–40. https://www.jstor.org/stable/45386639. DOI: 10.1007/s10551-019-04197-6

Ellapen, T. J., & Paul, Y. (2016). Innovative sport technology through cross-disciplinary research: Future of sport science. *S.A. Journal for Research in Sport Physical Education and Recreation*, *38*(3), 51–59. https://www.ajol.info/index.php/sajrs/article/view/149803

Grady, C. (2015). Institutional Review Boards: Purpose and Challenges. *Chest*, *148*(5), 1148–1155. DOI: 10.1378/chest.15-0706 PMID: 26042632

Harriss, D. J., & Atkinson, G. (2009). International Journal of Sports Medicine – Ethical Standards in Sport and Exercise Science Research. *International Journal of Sports Medicine*, *30*(10), 701–702. DOI: 10.1055/s-0029-1237378 PMID: 19809942

Harriss, D. J., Jones, C., & MacSween, A. (2022). Ethical Standards in Sport and Exercise Science Research: 2022 Update. *International Journal of Sports Medicine*, *43*(13), 1065–1070. DOI: 10.1055/a-1957-2356 PMID: 36495253

Horn, L. (2013). Promoting responsible research conduct in a developing world academic context. *South African Journal of Bioethics and Law*, *6*(1), 1. https://www.ajol.info/index.php/sajbl/article/view/90037. DOI: 10.7196/sajbl.256

Kang, E., & Hwang, H.-J. (2020). The Consequences of Data Fabrication and Falsification among Researchers. *Journal of Research and Publication Ethics*, *1*(2), 7–10. DOI: 10.15722/jrpe.1.2.202009.7

Ketefian, S. (2015). Ethical considerations in research. Focus on vulnerable groups. *Investigacion y Educacion en Enfermeria*, *33*(1), 164–172. http://www.scielo.org.co/scielo.php?pid=S0120-53072015000100019&script=sci_arttext. DOI: 10.17533/udea.iee.v33n1a19 PMID: 26148168

Laryeafio, M. N., & Ogbewe, O. C. (2023). Ethical consideration dilemma: Systematic review of ethics in qualitative data collection through interviews. *Journal of Ethics in Entrepreneurship and Technology*, *3*(2), 94–110. https://www.emerald.com/insight/content/doi/10.1108/JEET-09-2022-0014/. DOI: 10.1108/JEET-09-2022-0014

Lategan, L. O. K. (2012). The Building of a Responsible Research Community: The Role of Ethics. *The Journal of Research Administration*, *43*(1), 85–97. https://eric.ed.gov/?id=EJ976743

Loland, S. (2009). The Ethics of Performance-Enhancing Technology in Sport. *Journal of the Philosophy of Sport*. https://www.tandfonline.com/doi/abs/10.1080/00948705.2009.9714754

Macrina, F. L. (2007). Scientific Societies and Promotion of the Responsible Conduct of Research: Codes, Policies, and Education. *Academic Medicine*, *82*(9), 865–869. DOI: 10.1097/ACM.0b013e31812f7e58 PMID: 17726393

Madak, P. R. (1994). Ethical Considerations When Using Qualitative Methods in Evaluation. *The Canadian Journal of Program Evaluation*, *9*(2), 1–13. DOI: 10.3138/cjpe.9.001

McNamee, M., Anderson, L. C., Borry, P., Camporesi, S., Derman, W., Holm, S., Knox, T. R., Leuridan, B., Loland, S., Frias, F. J. L., Lorusso, L., Malcolm, D., McArdle, D., Partridge, B., Schramme, T., & Weed, M. (2025). Sport-related concussion research agenda beyond medical science: Culture, ethics, science, policy. *Journal of Medical Ethics*, *51*(1), 68–76. DOI: 10.1136/jme-2022-108812 PMID: 36868564

McNamee, M. J., Olivier, S., & Wainwright, P. (2006). *Research Ethics in Exercise, Health and Sports Sciences*. Routledge., DOI: 10.4324/9780203966853

Milanović, L. Ran\djelović, N., Živković, D., & Savić, Z. (2021). Ethics and sports, ethics in sports, sports ethics: Aspects of consideration of different authors. *Physical Education and Sport through the Centuries, 8*(2), 16–32. https://scindeks.ceon.rs/Article.aspx?artid=2335-05982102016M

Nurunnabi, A. S. M. (2014). Institutional Review Board (IRB): Its role and responsibility in making research ethical. *Bangladesh Journal of Bioethics*, *5*(1), 5–10. https://banglajol.info/index.php/BIOETHICS/article/view/18442. DOI: 10.3329/bioethics.v5i1.18442

Rao, K. N., Mair, M., Arora, R. D., Dange, P., & Nagarkar, N. M. (2024). Misconducts in research and methods to uphold research integrity. *Indian Journal of Cancer*, *61*(2), 354–359. DOI: 10.4103/ijc.ijc_4_23 PMID: 39016268

Resnik, D. B. (2014). Data Fabrication and Falsification and Empiricist Philosophy of Science. *Science and Engineering Ethics*, *20*(2), 423–431. DOI: 10.1007/s11948-013-9466-z PMID: 23982326

Rhodes, R. (2005). Rethinking Research Ethics. *The American Journal of Bioethics*, *5*(1), 7–28. DOI: 10.1080/15265160590900678 PMID: 16036651

Rossouw, T. M., van Zyl, C., & Pope, A. (2014). Responsible conduct of research: Global trends, local opportunities. *South African Journal of Science*, *110*(1–2), 1–6. http://www.scielo.org.za/scielo.php?script=sci_abstract&pid=S0038-23532014000100011&lng=en&nrm=iso&tlng=en. DOI: 10.1590/sajs.2014/20130103

Sachs, G. A., & Siegler, M. (1993). Teaching scientific integrity and the responsible conduct of research. *Academic Medicine*, *68*(12), 871–875. https://journals.lww.com/academicmedicine/abstract/1993/12000/Teaching_scientific_integrity_and_the_responsible.1.aspx. DOI: 10.1097/00001888-199312000-00001 PMID: 8259954

Steneck, N. H., & Bulger, R. E. (2007). The History, Purpose, and Future of Instruction in the Responsible Conduct of Research. *Academic Medicine*, *82*(9), 829–834. DOI: 10.1097/ACM.0b013e31812f7d4d PMID: 17726385

Stewart, R. J., & Reider, B. (2016). The Ethics of Sports Medicine Research. *Clinics in Sports Medicine*, *35*(2), 303–314. DOI: 10.1016/j.csm.2015.10.009 PMID: 26832979

Tonis, R. B.-M., Păun, D. G., Coelho, A. S., Urdes, L., Mihoreanu, L., & Martins, O. M. D. (2025). Ethical challenges using technology in sport competitions. *Intangible Capital*, *21*(3), 406–426. DOI: 10.3926/ic.3034

van Stekelenburg, M. H. M. (2024). *Navigating on the Ethical Compass.: What it Means and Implies to Equip Students of Universities of Applied Sciences with an Ethical Compass* [PhD-Thesis - Research and graduation internal]. DOI: 10.5463/thesis.782

Vitak, J., Proferes, N., Shilton, K., & Ashktorab, Z. (2017). Ethics Regulation in Social Computing Research: Examining the Role of Institutional Review Boards. *Journal of Empirical Research on Human Research Ethics; JERHRE*, *12*(5), 372–382. DOI: 10.1177/1556264617725200 PMID: 28831844

Waddington, I. (2013). Researching the World of Professional Football. In *Doing Real World Research in Sports Studies*. Routledge.

Watson, J. C., II, & Etzel, E. F. (2019). *Ethical and legal issues in sport and performance psychology*. https://psycnet.apa.org/record/2019-05400-041

Zhang, Y., Fu, J., Lai, J., Deng, S., Guo, Z., Zhong, C., Tang, J., Cao, W., & Wu, Y. (2024). Reporting of ethical considerations in qualitative research utilizing social media data on public health care: Scoping review. *Journal of Medical Internet Research, 26*, e51496. https://www.jmir.org/2024/1/e51496/. DOI: 10.2196/51496 PMID: 38758590

Chapter 7
Animal Research Ethics and Welfare Standards

Krutika Patre
http://orcid.org/0009-0001-7969-4409
Ajeenkya D.Y. Patil University, India

Sunil Sankathala
http://orcid.org/0009-0004-7809-7647
Ajeenkya D.Y. Patil University, India

Aryan Vasa
http://orcid.org/0009-0000-1837-8587
Ajeenkya D.Y. Patil University, India

Kritikka Kunte
Ajeenkya D.Y. Patil University, India

Devanshi Soni
Ajeenkya D.Y. Patil University, India

Susanta Das
http://orcid.org/0000-0002-9314-3988
Ajeenkya D.Y. Patil University, India

EXECUTIVE SUMMARY

Important scientific discoveries have been followed using animal research particularly in the field of medicine and biology. Nevertheless, animals have to be treated in an ethical manner. Ethics in animal research follow the 3S model whereby: Replacement (where possible, using alternatives), Reduction (involving the use of fewer animals) and Refinement (whereby procedures are altered to cause minimal suffering). The ethics committees must approve their studies on animals in order to make sure that research is necessary and that they take good care of the animals. Good welfare would entail adequate food, hygienic surroundings, medicine and the reduction of pain or stress. Almost all the countries have legislative acts that guard the research animals, but the enactments differ. Uniform care of animals There should be international standards and collaboration that bring about uniformity in taking care of animals. Education of researchers, regulation of laws, and education of alternatives present the important steps towards improved animal welfare in research all over the world.

DOI: 10.4018/979-8-3693-8302-5.ch007

1. INTRODUCTION

The use of animals in scientific studies dates back centuries, and it has contributed to the progress in medicine, pharmacology, and physiology. Animal research has been one of the most crucial factors in designing vaccines, antibiotics, and curative interventions that have led to better human health outcomes across the globe. Meanwhile, the application of animals in research has brought about a continuous ethical discussion mainly on the ethical duty that humans bear to the sentient. With the scientific advancement in animal cognition, emotional capacity, and other social behaviors, animal research has been subjected to increased ethical; demands have risen to make animal research more humane and transparent and more ethical in its methods (Iyalomhe, 2014; Kiani et al., 2022).

The most important aspect of this discussion is the idea of animal sentience and the status of morality. It is becoming scientifically more apparent that even the majority of animals can feel pain, suffering, and have other complex emotional states, whereas previous perceptions viewed animals as an experiment. The current trends have contributed to increased social interest in animal rights and welfare affecting the expectations, policies and ethical standards of science. Animal research ethics has therefore become a multidisciplinary area that can incorporate scientific necessity and moral responsibility which include the necessity of providing balance between the possible human advantage and the harm that animals may face (Kiani et al., 2022).

One of the most prominent ethical models of modern animal studies is the concept of Replacement, Reduction, and Refinement (3Rs) which was initially developed by Russell and Burch in 1959. The 3Rs were created to offer a more structured method of reducing animal use and suffering and still be scientifically sound and continue to play a fundamental role in animal research ethics worldwide (Iyalomhe, 2014; Clark, 2017). Replacement promotes using non-animal models wherever possible, such as in vitro testing, organ-on-chip technologies and artificial intelligence-based models that have been shown to be increasingly effective, efficient, and relevant to human biology (Clark, 2017). Reduction aims at reducing the number of animals used by designing better experiments, increasing the level of statistical rigor, sharing of data and reusing existing data (Festing et al., 1998). Refinement is done to minimize pain, distress, and suffering by enhancing housing localities, managing procedures, monitoring of welfare and the application of humane endpoints (Brink & Lewis, 2023).

The adoption and establishment of the 3Rs is different in regions yet has become a part of regulatory and ethical control measures in most regions of the world. Ethical review boards and authorities in various authorities, including the European Union, United States, and even parts of Asia demand that researchers explain the reasons behind animal use and also by demonstrating the 3Rs being followed during the research process (Martinez-Padilla et al., 2023). Ethics committees are important despite differences in their structure and implementation since they mediate between the scientific goals and the moral responsibility. The transparency, procedural justice, and ethical deliberation of animal research governance are facilitated by including various stakeholders, such as researchers, veterinarians, and representatives of the population in these committees (Pietrzykowski, 2021).

Simultaneously, due to the growth of open science practices, ethical initiatives to decrease the use of animals and enhance the integrity of research have grown. Preregistration, open data and collaboration research platforms are among the practices that enhance transparency and eliminate redundancy in animal research, thus promoting the Reduction principle (Nawroth & Krause, 2022). The application of common databases of findings of past animal research studies makes researchers add to the existing knowledge and avoid repeating the studies. The ARRIVE Guidelines 2.0 reporting standards are another tool that helps to increase reproducibility, transparency, and ethical accountability through the quality and completeness of reporting on animal research (Du Sert et al., 2020).

The scientific development of animal welfare has also been playing a key role of transforming the ethical demands of animal research. The modern welfare paradigms focus not on the prevention of suffering alone, but also on the positive welfare conditions. An example of this is the Five Domains Model, which is a holistic method of welfare assessment because both physical and psychological well-being are integrated through the evaluation of nutrition, environment, health, behavior, and mental state. These models represent a more general determination to move beyond the paradigm of purely utilitarian justifications into methods that appreciate animals as sentient beings, the welfare of which directly affect ethical acceptability and the scientific rigor (Littlewood & Mellor, 2016; Mellor, 2017).

Outside of laboratory studies, there are also some ethical issues in applied research, like conservation, aquaculture, biological management, and animal-assisted interventions (AAIs). These circumstances reflect why ethical conceptions should be sensitive to the situational complexity and emotional aspects of human-animal interactions (Bendorf et al., 2022; Nannoni & Mancini, 2024). Although AAIs may be useful to humans, especially those with special needs, the therapeutic value of these studies does not exclude consideration of significant ethical issues of animal welfare and the responsibilities of the caregiver and the clinician to avoid harm. These issues can only be solved by paying close attention to ethics to make sure that animal welfare is not sacrificed in the name of the human good (Bendorf et al., 2022; Nannoni & Mancini, 2024).

There has also been the rise of animal welfare concern in society which has also had an impact on education, policy, and the research culture. Some efforts like education are made through educational programs like the implementation of

Animal Welfare, Science, Ethics, and Law (AWSEL) in veterinary schools to ensure ethical responsibility in the future generation of scientists (De Briyne et al., 2020). Nonetheless, there are still huge international gaps in how animals are researched, and there have been areas which do not have detailed regulation systems. The international cooperation and academic integrity networks like the European Network of Academic Integrity (ENAI) are aimed at the solution of such inequalities to support ethical norms and the integrity of research across international boundaries (Sivasubramaniam et al., 2021).

In general, the current research ethics of animals are becoming more concerned with the necessity of the balance between scientific innovation and moral responsibility. Ethical research is not just regulatory compliance, but it should demonstrate a larger sense of compassion, transparency, and accountability. The culture of care is one of the concepts that emphasize the role of reflexivity, contextual judgment, and ethical sensitivity in research practice, urging scientists to consider animals not as experimental resources but as living beings that are ethically sensitive (Sankar, 2022; King, 2021).

This chapter is a discussion on these issues with reference to systematic study of animal research ethics. It begins with explaining the ethical foundation of animal research, animal sentience, either animal moral status, and ethical theories and principles that govern the research practices, including the 3Rs. The chapter moves to the regulatory and ethical control measures such as international regulations, regional regulations and ethics committees and harm benefit analysis. It subsequently addresses the issues of animal welfare and scientific rigor and puts emphasis on welfare assessment, humane treatment and experimental design and the culture of caring development.

This chapter further uses a systematic bibliometric review to examine trends in research in the field of integrating animal research ethics and animal welfare standards. PubMed and Lens.org dataset was selected as the sources of scholarly information to provide a broad and interdisciplinary spectrum of coverage. VOSviewer was used to visualize keyword co-occurrence networks and thematic clusters via which the presence of dominant and emerging research topics could be identified. The growth of publications, the country contribution and the most popular publishers were also discussed in the form of graphs and tables. The combination of these approaches gives a brief but sufficient summary of the history and the development of the research in the field of animal ethics and welfare standards. Finally, the chapter concludes by contemplating alternative and new solutions, including non-animal solutions to research, benefits and limitations of such research, and the implication of open science to do away with the use of animals. In this chapter we are going to discuss topics as follows:

2. ETHICAL PRINCIPLES OF ANIMAL RESEARCH

2.1. Sentience and moral status of animals

Ethical basis of animal research is strong and embedded in the acknowledgment of sentience in animals and moral status after the previous. The modern science shows that animals have evolved complex cognitive, emotional, and social abilities, which have challenged past beliefs that viewed them as natural mechanisms or simply reactive. Identification, social bonding, emotional reaction, critical thinking skills, and even future-oriented thinking have been studied in ethology, neuroscience, and behavioral science and have been demonstrated in animals. Such discoveries reinforce the ethical debate that animals can suffer not only physically but also psychologically hence being a matter of serious moral concern (Kiani et al., 2022; Muthu, 2016). In the ethical view of rights-based perspective, animals are considered as things that have basic value, that is, they should not be treated to an end irrespective of the gains that may be acquired due to a researched end (Iyalomhe, 2014). Proponents of this opinion claim that the deliberate harm of sentient beings is a kind of speciesism, which is the same as racism or sexism, the moral value is unfairly distributed on the basis of belonging to a species and not the ability to suffer. Practical approaches on the other hand tolerate the use of animals but on strict terms in that the harm inflicted on animals should be justified by the benefits that are expected to accrue to humans or other members of the society in a manner that the benefits accrue significantly (Kiani et al., 2022). The increased acceptance of animal sentience has impacted on the laws of welfare, ethics, and social expectations. Along with the growth of scientific knowledge about animal cognition, the ethical approaches shift steadily away from the bare minimum of welfare to the more holistic account of dignity, agency, and emotional well-being. This change highlights the changing ethical environment where animals are taken as important moral beings that should be treated with protection in the research setting (Muthu, 2016).

2.2. Ethics theories and principles of animal research

The combination of philosophical theories and applied ethical principles in animal research is used to give ethical direction in decision-making in animal research as a way of balancing scientific progress with moral accountability. Historically, utilitarianism has been dominating, and the justification of animal experimentation has been made when the potential benefits of the research, which are medical discoveries or benefits to society, are found to outweigh the potential harm caused to animals. This is a method based on minimization and proportionality, the foundations of numerous regulation schemes and ethical scrutinizes (Iyalomhe, 2014). In contrast with this idea, there are rights-based ethical theories, which state that animals have basic moral rights as they are sentient beings that can suffer. It is in this light that animals should not be treated as a tool to human desires irrespective of any positive gains. Such a conflict in utilitarian and rights-based ethics remains a current concern in terms of whether animal research can be morally justified (Sankar, 2022).

In everyday contexts, the concepts of generosity, justice, transparency, and procedural fairness are implemented in the form of ethics committees and regulative control strategies. Moral accountability and trust among the population require ethical review procedures to follow impartiality, fair hearing, and transparency. These guidelines enable the researchers to defend their research but at the same time make sure that proper attention is being paid to the welfare of the animals (Pietrzykowski, 2021). In more recent times, care based and relational ethics have been given more emphasis and focus, through empathy, reflexivity and contextual judgment as opposed to strict adherence to rules. The use of a culture of care that acknowledges both emotional and moral labor of animal research. Through this approach ethical sensitivity, responsibility, and responsiveness to lived experiences of animals and researchers are promoted and this is a move towards more reflective and humane ethical governance (King, 2021).

2.3. Replacement, Reduction, Refinement 3Rs principle

Ethical animal research rests on the 3Rs principal Replacement, Reduction and Refinement, which is a realistic combination of ethical theory and scientific responsibility. The 3Rs, which were originally proposed by Russell and Burch in 1959, are meant to reduce the use and suffering of animals at the expense of scientific validity (Iyalomhe, 2014; Clark, 2017). This framework has over the years gained wide use in ethical codes, legislation, and institutional review procedures globally (Martínez-Padilla et al., 2023). Replacement can be defined as the usage of the non-animal means wherever possible, which can be in the form of in vitro systems, computer-based simulations, organ-on-chip devices and artificial intelligence (AI) models. These options are also prone to offer more human-related information, less ethical considerations, and more effective experiments (Clark, 2017). Reduction aims at constructing the use of animals by enhancing experimental design, using rigorous statistical techniques, data sharing and reuse of available data with an aim of eliminating unwarranted duplication. Experiments that are poorly designed misuse scientific validity as well as lead to unethical wastage of the animals' lives (Festing et al., 1998). Refinement aims at reducing pain, suffering and distress caused to animals that should be utilized in research. This involves bettering housing, environmental enrichment, humane treatment, non-invasive treatment, and well-established humane endpoints. Refinement is an emerging concept of animal welfare which considers not only the physical but also the psychological health (Brink & Lewis, 2023). Many laws have the 3Rs enshrined in their rules, regulations, and laws, such as the 2010/63/EU directive in Europe, federal policies in the United States, and some developing Asian systems. The two of them form an internationally acknowledged ethical basis of responsible animal research (Martínez-Padilla et al., 2023).

3. ETHICAL AND REGULATORY SUPERVISION

3.1. Global rules and regulations

Global codes and practices are very important in streamlining ethical conduct in animal research studies in different jurisdictions. The Organization for Economic Co-operation and Development (OECD) and the World Health Organization (WHO) are bodies that support ethical research frameworks that lead to the use of alternatives to animal testing, especially in the fields of toxicology and pharmacology. These international organizations help in the sharing of best practices and promoting international cooperation in improving the standards of animal welfare (Martínez-Padilla et al., 2023). The ARRIVE Guidelines 2.0 are one of the most powerful international tools that promote transparency, methodological rigor, and ethical accountability on how the animal studies are reported. ARRIVE is expected to enhance reproducibility and decrease unnecessary animal utilization by mandating the description of the experimental design, characteristics of animals, welfare issues, and statistical procedures used (Du Sert et al., 2020). Although these have been made, large gaps in the implementation and enforcement of the ethical standards are still evident between the nations, and this is where the international ethical governance initiatives can play a role in addressing the gaps through capacity-building, education, and institutional development. The networks like the European Network of Academic Integrity (ENAI) encourage consistency of ethics across borders. International standards can, at the same time, be viewed as regulatory instruments as well as moral frameworks that capture the changing values of society in terms of animal welfare. These guidelines are important in promoting a more ethically consistent global research culture, even in areas where legal safeguards to this effect are still undeveloped (Sivasubramaniam et al.,2021).

3.2. Domestic and local regulation

The national and regional regulatory frameworks of animal research differ extensively on the levels of scope, enforcement and ethical rigor. Directive 2010/63/EU is one of the broadest in Europe, as a legal tool in scientific research, animal protection. It implies using the 3Rs, harm-benefit analysis, and obliges the researchers to explain why animals were used at each point of the experiment (Pietrzykowski, 2021; Martínez-Padilla et al., 2023). In the US, animal research is controlled mainly under the Animal Welfare Act and the National Institutes of Health policies. Although these frameworks encourage the 3Rs, the enforcement is decentralized and thus there is inconsistency in the ethical management within institutions (Martínez-Padilla et al., 2023). Some federal protections do not apply to certain species, including rats and mice, which can be ethical inconsistencies (Kiani et al., 2022). Asian regulating environments are more diverse. Some countries have already achieved some success in the formalization of legislation on animal welfare, such as Japan and South Korea, but many of the areas do not have a complete legal framework or effective enforcement practices. These differences emphasize the inequity in animal protection around the world and the necessity of worldwide coordination and collaboration. On the whole, though high-income regions are more likely to have more effective regulatory schemes, there are gaps in the world. To ensure that the ethics applied to animal research are applied in a consistent and fair way, it is necessary to strengthen national structures with the help of legal reforms, education, and international cooperation (Martínez-Padilla et al., 2023).

3.3. Ethics committee role and harm-benefit analysis

The ethical management of animal research is based on ethics committees also known as Animal Ethics Committee (AECs) or Institutional Animal Care and Use Committee. They mainly serve to critique proposed studies to determine that animal use is scientifically reasonable, ethically sound and legal. The principles of the procedural review encompassed in the key principles of ethical review refer to the principle of impartiality, transparency, fair hearing, and benevolence that lead to procedural justice and trust in the population (Pietrzykowski, 2021). Harm–benefit analysis is a fundamental role performed by ethics committees. This is done to determine whether the expected scientific or societal benefits of a study will supersede the harms put upon the animals, such as pain, distress, and long-term effects on welfare. Research that does not prove to be of adequate benefit compared to the harm is not ethical and cannot be passed (Nielsen & Rodenburg, 2025). In addition to formal assessment, the ethics committees develop awareness of ethics by facilitating the improvement of the experimental procedures, implementing alternatives, and following open science practices. As pointed out, transparency, data sharing, and preregistration minimize the unnecessary experimentation and justify the Reduction principle (Nawroth & Krause.,2022). There are however critics who proposed that ethical review is too bureaucratic and engages itself with the aspect of procedural compliance as opposed to substantive ethical reflection. By combining harm-benefit analysis with a spirit of care, ethics committees will be able to leave behind box-checking and make humane and context-aware decisions (King, 2021).

4. SCIENTIFIC RIGOR AND ANIMAL WELFARE

4.1. Humanity and welfare evaluation

The idea of animal welfare evaluation has changed quite significantly to be more holistic in terms of avoiding suffering and instead enhancing positive welfare conditions. The Five Domains Model is a popular model widely accepted to assess welfare through nutrition, environment, health, behavior, and mental condition This model recognizes the fact that the psychological well-being is equally significant with respect to the physical health in a determination of the overall welfare (Littlewood & Mellor, 2016; Mellor, 2017). Human treatment in research involves giving them the proper environment and housing, enriching the environment, socializing with them and letting species-specific behaviors to be exercised. Refinement strategies help to minimize pain and distress by enhancing the methods of handling, by using non-invasive monitoring and human endpoints, which are clearly defined Those practices are found to improve animal welfare as well as scientific validity because they decrease confounding variables of stress (Brink & Lewis, 2023). The literature reflects on the fact that welfare is contextual. Animal studies such as wildlife, conservation, aquaculture, or animal-assisted interventions have specific ethical considerations, such as ecological, behavioral, and relational factors (Bendorf et al., 2022; Nannoni & Mancini, 2024). Ethical welfare evaluation should be more malleable and receptive as opposed to strictly standardized. All in all, human treatment is an indicator of the change towards the acknowledgment of animals as living beings with emotional and psychological needs. This development moves the ethical responsibility to the level of scientific rigor, and this strengthens the argument that it is the good welfare practices that are core and not periphery to the high-quality research (Bendorf et al., 2022).

4.2. Reduction of harm, experimental design, and reproducibility

The scientific rigor and animal welfare are two closely related concepts because improper experimental design does not only damage the research validity but also causes animal wastage. Lack of blinding and poor randomization, reporting bias, p-hacking, and Harking are all identified by the author as major threats to internal validity in animal studies. The practices undermine reproducibility and add to the lives of wasted animals (Mason, 2023). The process of harm reduction starts at the design phase with sound statistical planning. The power analysis, resource equation methods, and sequential designs assist in the determination of the proper sample size so that neither underpowered nor overpowered studies take place. The cooperation with biostatisticians and compliance with the transparent reporting standards also contribute to the improvement of ethical and scientific results (Festing et al., 1998). The reproducibility is supported with standardized reporting systems like ARRIVE Guidelines 2.0 which implies clear reporting of the methodology, description of animal characteristics, welfare, and statistical results. Transparency in reporting also makes other researchers replicate studies or re-use data thereby limiting repetitive experiments. Finally, intense experimental design justifies the ethical demand of Reduction by increasing the amount of knowledge obtained per animal utilized. Ethical research cannot therefore be dissociated with greatness in methodology because the failures in science become failures in morality (Du Sert et al., 2020).

4.3. Research practices culture of care

Implementing a culture of care is a change towards animal research ethics based on compliance rather than being reflective and relational. The ethical review systems in most cases are too bureaucratic and emphasize procedural necessities instead of the experiences of researchers whose ethical life and the moral realities of research with animals. A caring culture is empathetic, responsible, reflexive, and contextual judging (King, 2021). This is because this method is aware of the emotional labor that accompanies animal research and demands institutional structures of assistance such as mentoring, ethical dialogue, and ethics training. A culture of care is an approach to embracing moral responsibility in the daily research activities instead of perceiving ethics as an administrative burden (King, 2021). Compassionate research culture also coincides with other wider welfare programs such as humane endpoints, harm benefit analysis and refinement strategies. It asks the researchers to constantly reconsider their ethical choices and to act responsively to the issues of welfare whenever they appear (Brink & Lewis, 2023). A culture of care improves animal welfare and scientific integrity by creating an ethical literate and accountable culture. It is an advanced system of ethics where compassion and rigor exist together that ensure sustainable and ethically justifiable research practices (Sankar, 2022).

5. OTHERS AND NEW STRATEGIES

5.1. Technological advances and non-animal research methods

Non-animal research methods development is one of the greatest ethical and scientific progressions in the modern research. The use of in vitro systems, such as cell cultures, tissue engineering, and organ-on-chip technologies, is increasingly being used as replacement strategies by researchers studying biological processes in controlled conditions without the need to use living animals (Clark, 2017). In some respects, organ-on-chip models in particular can recreate the structure and functioning of human organs and thus more accurately represent human physiology, disease pathophysiology, and drug effects than animal models (Iyalomhe, 2014; Martínez-Padilla et al., 2023). Replacement possibilities have been further increased with the help of computational biology and artificial intelligence. They involve computer modelling, machine learning and AI-based simulations to forecast toxicological consequences, pharmacokinetics and disease development and thus eliminate animal testing (Sankar, 2022). The international organizations, including the OECD and the WHO, support and promote such technologies, especially in regulatory toxicology and pharmacological research. Combined, these developments show a paradigm shift of research methodologies that are human relevant, efficient and morally responsible (Martínez-Padilla et al., 2023).

5.2. Advantages and shortcomings of alternatives

Alternatives to animals have a tremendous ethical benefit due to providing the elimination or a big reduction in animal suffering, which is in line with the Replacement principle of the 3Rs (Clark, 2017). These techniques can be more reproducible and human-specific in a scientific sense, which has long been an issue with animal models due to their lack of translational validity (Sankar, 2022). Nevertheless, literature does realize significant limitations as well. Other complicated biological processes like immune responses of an entire organism, lasting behavioral impacts, and interactions between different systems are hard to reproduce completely with a non-animal system (Iyalomhe, 2014). Also, there is disparity in the use of alternatives. Such obstacles as a lack of financing, technical skills, and poor regulatory implementation are common to low- and middle-income countries and make the transition to non-animal use slow (Sankar, 2022). Regardless of these restrictions, the literature reveals that the restrictions are not used as an excuse to conduct irrelevant animal experiments. Instead, they emphasize the necessity of further investment, certification, and dissemination of other technologies across the world to increase their relevance and access (Martínez-Padilla et al., 2023).

5.3. Open science role in the reduction of animal use

The concept of open science is being claimed as a potent ethical practice that can alleviate the use of animals by increasing the level of transparency, collaboration, and efficiency in research. There are measures to minimize unnecessary duplication of animal studies preregistration of studies or open-access publication, sharing of datasets and experimental protocols. The Reduction principle is directly supported where researchers can use existing data and are therefore less likely to repeat the already conducted research (Nawroth & Krause, 2022). Open science enhances ethical scrutiny and harm to benefit analysis as well. Common repositories of the results of animal research enable ethics boards and investigators to determine whether proposed research will involve truly novel discoveries or a repetition of an existing study. Moreover, an increased transparency of reporting guidelines like ARRIVE Guidelines 2.0 contributes to the increase of reproducibility and reliability by which the data gained after animal studies are used to the fullest and is ethically justified (Du Sert et al., 2020). Besides reduction, open science promotes faith among the population, because research is more transparent and responsible. Not only will openness help in accomplishing great jobs to animals and researchers, but it will also play a key role in society by fostering ethical accountability and scientific integrity. By so doing, open science becomes a methodological and moral tool to the betterment of a more humane research culture, more efficient, and more socially responsible (Nawroth & Krause, 2022).

6. RELATION BETWEEN HUMAN RESEARCH AND ANIMAL RESEARCH

The interdependence between human research and animal research is offered in the chapter as scientifically dependent and ethically controlled in the past, with animal study is the basis of the biggest improvements on human health and needs to be provided with moral justification at the same time. The chapter elaborate that animal research has contributed to the discovery of vaccines, antibiotics, and curative interventions that have resulted in better human health outcomes in the world, which make animal research one of the pillars of biomedical and physiological research that benefits humanity (Iyalomhe, 2014; Kiani et al., 2022). Yet, this contribution does not seem to be unconditional; instead, animal research ethics is defined as a multipurpose system that combines the principles of scientific necessity with the ethical responsibility of being able to balance the potential benefits, as well as the physical and psychological suffering of animals (Kiani et al., 2022). It explains a utilitarian rationale of animal research in which experimentation is ethically acceptable only when the anticipated gains on humans or society outweigh the harms to animals, which is the foundation of harm-benefit analysis and ethical review processes (Iyalomhe, 2014; Sankar, 2022). Also, it mentions that the scientific rigor and ethical acceptability of research results are directly dependent on animal welfare because a higher quality of welfare and ethical standards like the 3Rs increases the reliability of data and the validity of results that are to be applied by humans. Generally, the chapter describes human and animal research as two related fields where human scientific advancement should be attained with ethical responsibility towards animals maintained at any given moment (Littlewood & Mellor, 2016; Mellor, 2017).

7. BIBLIOMETRIC ANALYSIS ON ANIMAL RESEARCH ETHICS AND WELFARE STANDARD

This section examines the relationship between animal research ethics and welfare standards by the help of bibliometric analysis in the form of research records improved in the PubMed and Lens.org database. VOSviewer is used to visualize and analyze bibliometric networks as the appearance of essential terms, current topics of study and their evolution over time. More so, the systematic analysis of Lens.org data is also conducted to identify the trends in the publications, geographical distribution of the development of the research articles, as well as the pattern of the type of document, larger publishing company and the coverage of the field in terms of discipline. The research is outlined in graphs and tables to obtain a general idea of the research on the ethics and welfare standard in the Animal research.

Figure 1 shows the VOSviewer network of keyword co-occurrences is shown. It was developed on PubMed literature on animal research ethics and standards of animal welfare. In this visualization, the nodes denote each key word, and the size of the node corresponds to the frequency in which the key word is observed in the dataset. The connecting lines (links) demonstrate the frequency of the occurrence of the keywords in the same publications, whereas the various colors indicate clusters, signifying significant thematic areas in research.

The map identifies the keywords of the highest rank animals and animal welfare as the most prevailing and central ones, which proves that the sphere is highly organized around welfare-based research and the ethical control of animal-based experiments. Related terminologies like animal care committees, guideline adherence, government regulation, ethical review, and reference standards are a sign that a sizeable fraction of publications deals with formal mechanisms of oversight, such as institutional review boards and regulatory compliance systems of using laboratory animals. This cluster is an indication of the relevance of the ethical governance frameworks that guarantee research accountability, humane treatment, and adherence to national and international standards.

Figure 1. VOSviewer network visualization of keyword co-occurrence on Animal research ethics and welfare standard, generated using the PubMed database (visualized using VOSviewer) (Data may vary slightly).

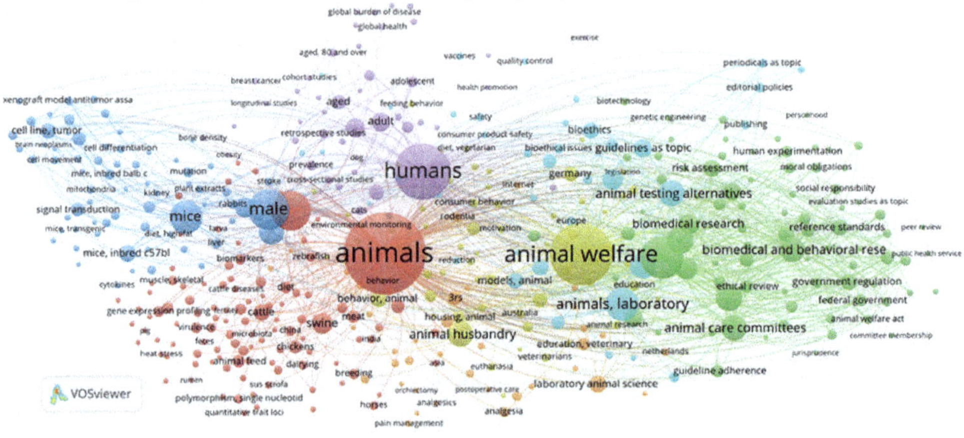

The other notable domain in the network is the connectivity of animal welfare with terms of biomedical and behavioral research including biomedical research, behavioral research, models, animal and animals, laboratory. This demonstrates that the topic of welfare is highly entrenched in the framework of experimental design and laboratory operations and is not to be viewed as a separate issue. The occurrence of references such as risk assessment and bioethical issues also indicates that literature is becoming more judgmental on welfare based on organized ethical or scientific risk-benefit models.

Another group is around the term's humans, male and mice which is indicative of the methodological organization of biomedical research in which animal models are directly related to human health outcomes. The abundant usage of species-specific names, including "mice," rats, swine, and chicken, shows that the problem of ethical welfare is expanded among a variety of animal species in the laboratory and on the farm.

Overall, Figure 1 shows that animal ethics are not confined to moral philosophy, but are intricately connected with regulatory systems, laboratory welfare guidelines, biomedical experimentation, and institutional governance hence the multidisciplinary character of the field.

The VOSviewer overlay visualization of keyword co-occurrence on animal research ethics and welfare standards is represented in Figure 2 based on PubMed records.

An overlay visualization adds a time dimension (as opposed to a normal network map where clusters are separated by color groups only). In this case the node size is again the frequency of occurrence of a keyword, and the links are again the strength of co-occurrence of keywords. Nevertheless, the color of the node shows the mean of the year of publication of the papers containing that keyword. The terms in blue/purple category are the older ones whereas the terms in green/yellow category are the new research themes (color bar 2000-2020).

The most central and most common keywords are of course wild animals, animal welfare, and humans, which demonstrates the fact that literature always relates animal-based experimentation to welfare and the ethical implications of this issue concerning biomedical studies in humans. Nevertheless, the overlay coloring shows the change in the research focus with time.

Closely linked to institutional and regulatory frameworks are old research themes (blue to purple). The keywords like government regulation, ethical review, animal care committees, reference standards, and adherence to guidelines are placed in cooler colors, indicating that the initial work in this area focused on the development of formal ethical oversight systems, institutional responsibility, and standardized welfare requirements.

Themes more recent (green to yellow) display a trend of refinement and modernization of welfare science. Such keywords as animal testing alternatives, bioethics, risk assessment, genetic engineering, and biotechnological applications are presented in warmer colors, which shows the growing number of attentions to animal use reduction by alternatives and the enhancement of animal's welfare based on the progressive scientific approaches of the last few years. This is in line with the increasing trend of the world in supporting the 3Rs concept (Replacement, Reduction, Refinement) in which ethical research is more focused on replacing animal experiments where feasible and refining procedures to ensure that suffering is minimized.

Species-specific research evolution is also pointed out in the overlay map. There are still common terms like mice, rats, swine, and other laboratory animals, which prove that the ethics of animal welfare is still strongly associated with experimental biomedical models. Simultaneously, more recent keywords in association with the complexity of experiments (e.g., genetic engineering and advanced biomedical research) indicate that the issues regarding welfare are growing along with the current methods of research.

In general, it can be seen in Figure 2 that animal ethics and welfare studies have now advanced to the stage of ensuring regulatory compliance and regulation, but is now more of a technology-focused, risk-focused, and alternatives-focused structure, despite keeping at the center of the game the animal and welfare.

Figure 2. VOSviewer overlay visualization of keyword co-occurrence on Animal research ethics and welfare standard, generated from the PubMed database using VOSviewer (Data may vary slightly).

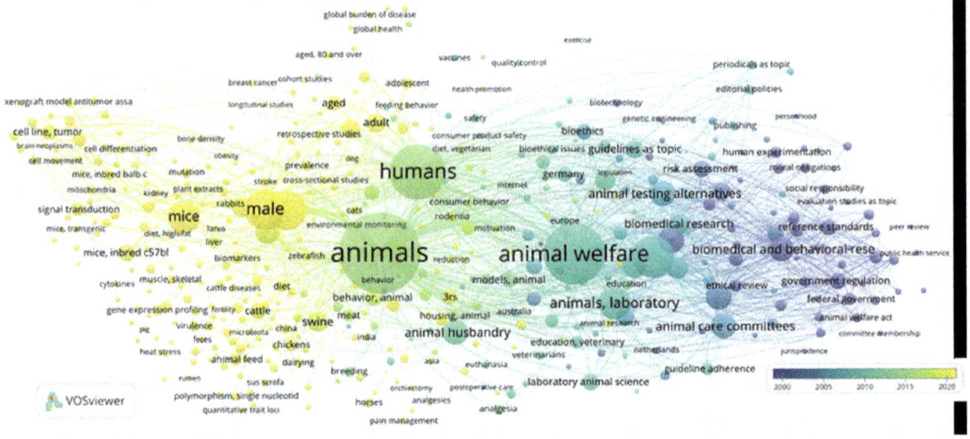

Figure 3 is a VOSviewer density diagram of keyword co-occurrence of research on animal research ethics and welfare standards created with the PubMed data. In a density map, there is no emphasis on individual clusters of data such as in a network map; rather, emphasis is placed on the concentration of keywords. The colors depict the concentration (frequency + strength of co-occurrence) of keywords: yellow color suggests the highest concentration, green color suggests the moderate density, and blue color suggests the areas of low density. As such, the most prevalent and recurrent themes of the data set are the ones that are the brightest.

The main area with the highest intensity (yellow) demonstrates the keywords of animals, animal welfare, and human beings. This makes it clear that the gist of the literature is highly based on research based on animals and ethical and welfare concerns linked to experiments that eventually translate to human health outcome. The popularity of these keywords indicates that the issue of animal welfare is not an independent subject, but rather it is closely combined with the aims of biomedical research.

A second large clump contains the keywords laboratory and experimental context like the keywords: animals, laboratory, animal husbandry, models, animal, and biomedical research. This shows that much of the literature that is on welfare is specifically on the laboratory settings to do with housing, handling, and the experimental settings. Use of such terms as animal care committees, ethical review, guideline adherence, and government regulation further indicates that welfare discourses are often placed in the context of institutional compliance and regulatory management.

Species-based experimental dominance is also shown by density visualization. The use of words like mice, male, rats, swine, and similar biomedical keywords are in the central hot spot, which indicates that rodents and other conventional laboratory models are still the most popular in terms of the welfare related research. The abundance of the word male implies that reporting and biological variation in animal research concerning sex is an issue of recurring concern in welfare and experiment design debates.

A moderately dense area is located on the right side of the map, and it contains keywords like animal testing alternatives, bioethics, risk assessment, and reference standards. Herein is an emerging powerful accent on ethical modernization, especially the formulation of alternatives and the consideration of the research risks and welfare burdens.

Overall, Figure 3 shows that the research area is centered around a very narrow central theme of animals, welfare, and humans, and backed by powerful secondary themes of laboratory practice, regulatory compliance and ethics review mechanisms and animal testing alternatives. Such an arrangement of densities shows that animal research welfare is a well-established area of research that has a solid core but continues to be growing in the new directions of ethical and methodological innovations.

Figure 3. VOSviewer density visualization of keyword co-occurrence on Animal research ethics and welfare standard, generated from the PubMed database using VOSviewer (Data may vary slightly).

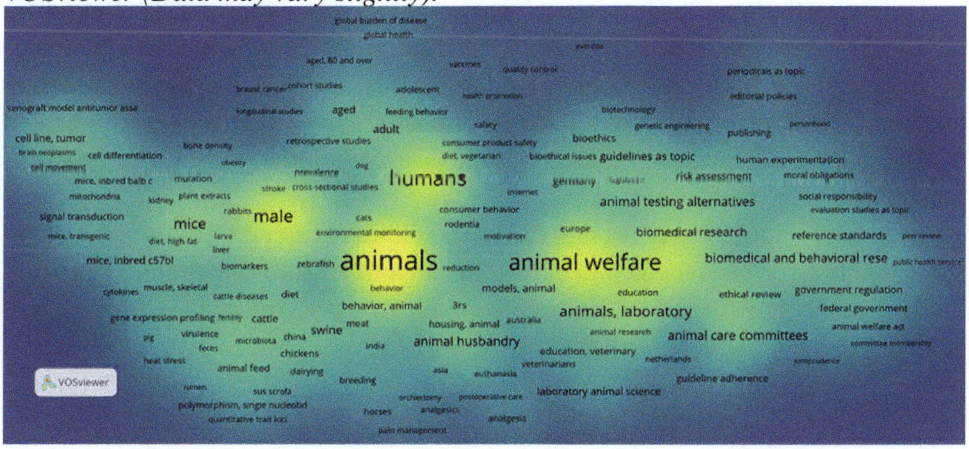

The trend of the yearly publication of research on animal research ethics and standards of animal research welfare is presented in Figure 4 to illustrate the yearly record of the research that was retrieved in the Lens.org database. The graph shows clearly that the publications on this topic have been rising consistently over the years, which is an indication that the world is becoming more concerned with ethical regulation, animal welfare practices, and responsible utilization of animals in research.

Between the year 2000 and about 2005, there are very few publications where there are limited studies that have been published annually. This implies that animal welfare and ethics were a research topic in the early 2000s, but the issue was not yet a core and well-developed research issue.

The number of publications is starting to increase gradually between 2006 and 2012. It is associated with the start of more academic and institutional interest perhaps because of a growing awareness of welfare issues, a more refined system of ethical review, and the availability of standardized recommendations regarding laboratory animal care and use.

Furthermore, a noticeable rise begins in 2013 where consistent yearly output of publications is observed. The period of 2016-2019 presents a great upward trend, which means that ethics and welfare of animal research became a more active research field. The increase can be associated with the intensification of welfare regulations, development of animal ethics committees, and the global spread of the concept of the 3Rs (Replacement, Reduction & Refinement).

The peak is noticed in the year 2021 where the number of publications is maximum. This peak is an indication that the topic received exceptional focus in this period. This can be because it is during the time of COVID-19 that more animal models are being used and more ethical debates are taking place concerning animal experimentation, alternative tests and humane research.

Following the year 2021, the graph indicates that there are decreasing numbers of publications in the period 2022-2025 but still significantly higher than the initial years. It means that the field might have left the peak but still, it remains the active and significant sphere of study. The exceptionally low value that it has indicated in 2026 can be attributed to the fact that the year at the time of data mining is still incomplete in the database.

In general, Figure 4 verifies that the topic of animal research ethics and welfare standards is a very advanced and expanding area of research, with the significant increases in its growth since 2015 and the maximum activity since 2021.

Figure 4. Year-wise publication trend on Animal research ethics and welfare standard (Source: Lens.org database) (Data may vary slightly).

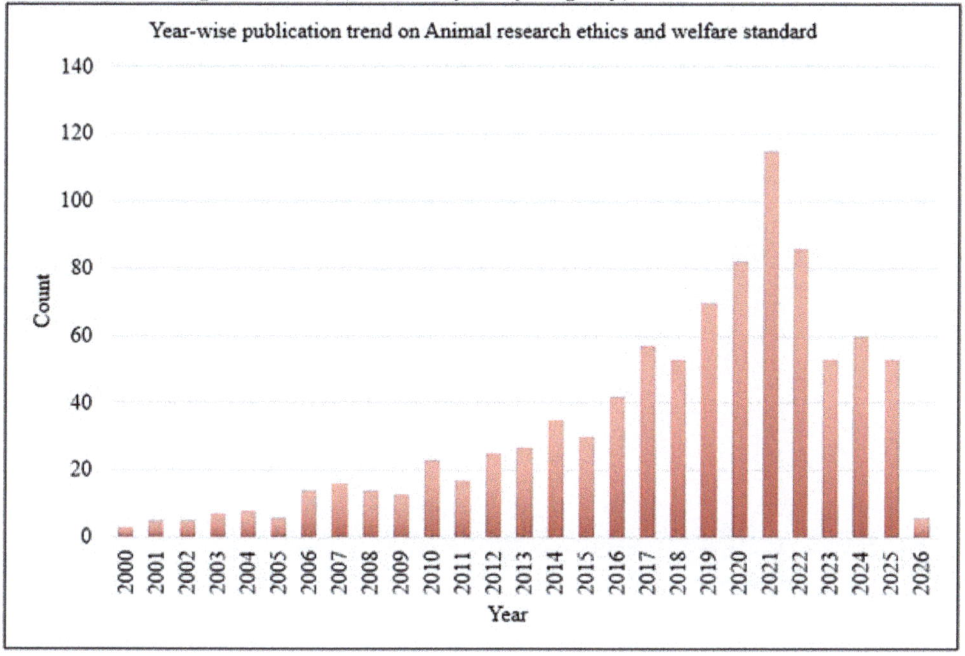

Figure 5 shows a country-based publication distribution map of the research on animal research ethics and welfare standards, on the Lens.org database. It is a choropleth world map of visualization, which is colored by the number of publications of countries. The darker the shade, the greater the number of publications and the light shade denotes that there is less research activity.

The United States is the most dominant contributor as indicated in the dark shade by the map. This points to the fact that USA has the most research on animal ethics, welfare control and laboratory animal standards in the world. This prominence is attributable to the high quality of biomedical research infrastructure in the country coupled with rigid institutional ethical review systems (including IACUC-based governance) and preclinical studies that use animals on large scale.

A publication activity is also notable in several European countries, particularly the United Kingdom, Germany, France and other Western European countries, which are characterized by the long-established ethical frameworks and welfare-based research policies. The production of European research is usually inspired by internationally accepted welfare laws, animal testing substitute, and advocation of the 3Rs principle.

Figure 5. Country-wise publication distribution on Animal research ethics and welfare standard (Source: Lens.org Database; visualized using geographic mapping tools) (Data may vary slightly).

Moreover, other countries like Canada, Australia as well as Japan demonstrate moderate research production. Such countries usually possess the established regulatory framework and well-established academic input into the field of laboratory animal science, veterinary ethics, and governance of biomedical research.

There is some activity with publication in India and some of Asia, pointing to the fact that the topic is spreading internationally. The smaller coloration on much of Africa and some areas of South America, however, implies that the publication output of these areas is still somewhat less, perhaps because of variations in research funding, reduced number of indexed publications, or less assimilation of animal welfare policy research into academic products.

In general, Figure 5 shows that the studies on the ethics of animal research and welfare standards are distributed on a global level but they are highly concentrated in the countries where the biomedical research ecosystem is highly developed and where animal ethics governance structures are established.

Figure 6 shows the heatmap of publication-type distribution of the research on animal research ethics and welfare standards, which were extracted in the Lens.org database. This visualization has every block as a type of document and the intensity of the color (lights to dark red) is the number of publications that are published with darker colors most commonly representing the most commonly published categories of documents.

The findings are also indicative in that the leading mode of scholarly communication in animal ethics and welfare study is dominated by journal articles of which 64,398 journal articles were found. This emphasizes the fact that the subject has been primarily written using peer-reviewed scientific publications, such as the research of animal model refinement, welfare assessment, and ethical governance, and policy-based discussions on bioethics.

Books (3,020) and, right behind, preprints (3,150) and dissertations (2,801), represent the second and third largest categories respectively. The fact that there are numerous books is evidence that animal welfare and animal ethics are also highly covered in the academic textbooks, ethical theories, veterinary science textbooks, and literature on regulatory guidelines. Equally, the existence of dissertations points to the fact that the field is a still developing area of graduate and doctoral dissertation research, usually with experimental ethics, animal behavior or institutional regulation as their themes.

In other types of publications like book chapters (1,113) and conference proceeding articles (222), the topic is also covered in the collaborative academic environment, but conferences do not play such a key role as journals. The dataset (77) indicates a new interest in systematic data sharing, such as welfare scoring systems, datasets of behavior and documents of ethical review.

The lesser document types indicate that policy commentaries, reviews, and report writing are also part of the field, including editorials (49), letters (31), reviews (68), and reports (487); although these are more infrequent.

Figure 6 demonstrates that the field of animal ethics and welfare standards is a greatly journal-oriented field that has been long-established in the academic tradition and shows further evidence of an increased research rate.

Figure 6. Heatmap showing publication-type distribution for Animal research ethics and welfare standard (Source: Lens.org Database) (Data may vary slightly).

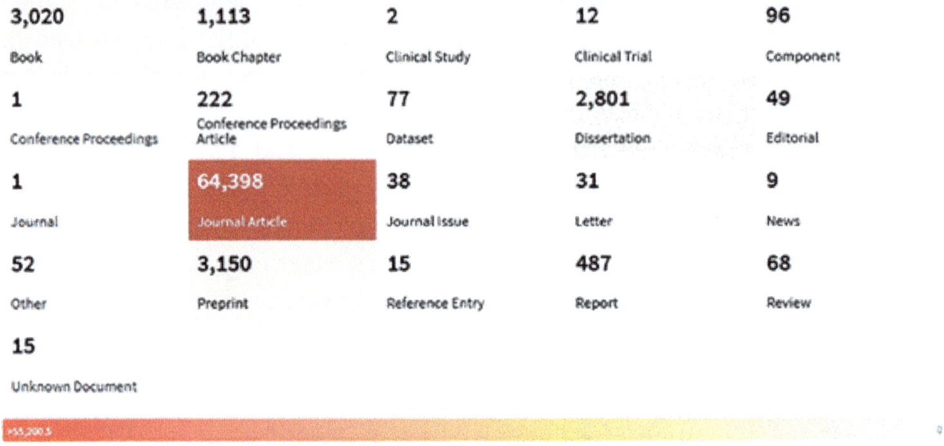

Table 1 displays the leading publishers in publications on the subject "Animal research ethics and welfare standards" according to the data found in the Lens.org database (data might change slightly). The following table has the list of publishers and the number of documents indexed by each publisher.

Wiley-Blackwell (135 publications) and Wiley (105 publications) are the strongest publishers in the field of this research, according to the table. This means that a considerable amount of high-impact academic research on animal welfare ethics including ethical frameworks, laboratory animal care policies, veterinary ethics, and biomedical research ethics is published via these international large-scale academic publishers.

Furthermore, the European academic and scientific publishers are also very supportive through their contributions made by the publishers. As an illustration, Walter de Gruyter GmbH (62 publications) and WellBeing International Publications (67 publications) note the existence of both the mainstream academic publishing and welfare-oriented specialized journals to facilitate the research in the field of animal ethics, welfare policies, and humane care guidelines. One of the observations is that there are several university libraries and academic institutions like the University of Wollongong Library (24), university of Queensland Library (12), Victoria University of Wellington Library (37), and University of Sao Paulo (13). It is indicative of the fact that much of the output of research is being circulated by and through institutional repositories, theses and university-based publications, and in these areas, open-access academic archiving plays a significant role.

In addition, it is estimated by the presence of discipline-specific journals that animal welfare ethics is highly connected with veterinary science, animal health policy, and laboratory animal management, including Veterinary World (30) and the Veterinary Journal of Ankara University (27).

Overall, Table 1 confirms that animal research ethics and welfare standards research is backed by a mix of mega-international publishers, specialized welfare publishers and university repositories which suggests both international and institution-wide research interest.

Table 1. Top publishers contributing publications on Animal research ethics and welfare standard (Source: Lens.org Database) (Data may vary slightly).

Publishers	Count
Wiley-Blackwell	135
Wiley	105
WellBeing International Publications	67
Walter de Gruyter GmbH	62
Wageningen University and Research	48
Wageningen Academic Publishers	43
Victoria University of Wellington Library	37
Veterinary World	30
Veterinary Journal of Ankara University	27
VEB Verlag Volk und Gesundheit	26
University of Wollongong Library	24
University of Toronto Press Inc. (UTPress)	21
University of Southern California	14
University of Sao Paolo	13
University of Queensland Library	12

8. CONCLUSION

Conclusively, animal use in scientific studies has played a significant role in the advancement of medicine, biology, and the general health of people, but it also has some serious ethical implications. As explained in this chapter, increasing awareness of the sentience of animals, their cognitive skills, and emotional welfare has altered the approach towards the use of animals in experiments. This consciousness has resulted in the evolution of ethical concepts like the 3Rs concept, harm and benefit analysis, check and balances of regulation, welfare evaluations, and others, which seeks to strike a balance between scientific advancement and caring, conscientiousness, and respect to animal life.

The chapter shows that effective ethical animal research does not solely rely on institutional review committees and formal regulations but also on scientific rigor, transparency and professional integrity. Sound experimental design, effective statistical planning and compliance with reporting guidelines like the ARRIVE 2.0 will help limit the number of animals used in research and increase the reliability of the research. Furthermore, by fostering the spirit of a robust culture of care, the researchers will be motivated to be ethically thoughtful and attuned to the issue of animal welfare in their everyday activities. The one significant positive change in this area is technological advancements and new research techniques. Such effective opportunities to substitute or minimize animal use and enhance the research to human health include in vitro systems, organ-on-chip technologies, artificial intelligence models, and open science practices. These inventions testify to the fact that scientific excellence and ethical responsibility can co-exist and support each other.

Although such improvements are made, there are still critical issues. Ethics and regulations are heterogeneous with respect to countries and institutions with gaps in animal protection. The poor condition of funding, lack of training, and legal frameworks remain the major barriers to the application of humane practice in most other parts. To solve these gaps, international cooperation should become stronger, the ethical education should be extended, and alternative technologies need to sustain investment. In general, the chapter shows that no humane treatment of animals is an obstacle to scientific discovery, but a source of credible, trustworthy, and socially responsible research. The research community can make sure that scientific progress in the future will be conducted with respect to the dignity, welfare, and intrinsic value of all sentient beings because the scientific research will be effective and sound through the incorporation of ethical principles, good governance, scientific rigor, and technological innovation.

LIST OF ABBREVIATIONS

- 3Rs: Replacement, Reduction, and Refinement
- AAIs: animal-assisted interventions
- AECs: Animal Ethics Committee
- AI: Artificial Intelligence
- AWSEL: Animal Welfare, Science, Ethics, and Law
- ENAI: European Network of Academic Integrity
- OECD: Organization for Economic Co-operation and Development
- WHO: World Health Organization

REFERENCES

Bendorf, C. M., Yun, S. C., Kurath, G., & Hedrick, R. P. (2022). Comparative susceptibilities of selected California Chinook salmon and steelhead populations to isolates of L genogroup Infectious Hematopoietic Necrosis virus (IHNV). *Animals (Basel)*, *12*(13), 1733. DOI: 10.3390/ani12131733 PMID: 35804631

Brink, C. B., & Lewis, D. I. (2023). The 12 RS framework as a comprehensive, unifying construct for principles guiding animal research ethics. *Animals (Basel)*, *13*(7), 1128. DOI: 10.3390/ani13071128 PMID: 37048384

Clark, J. M. (2017). The 3Rs in research: A contemporary approach to replacement, reduction and refinement. *British Journal of Nutrition*, *120*(s1), S1–S7. DOI: 10.1017/S0007114517002227 PMID: 29081302

De Briyne, N., Vidović, J., Morton, D. B., & Magalhães-Sant'Ana, M. (2020). Evolution of the teaching of animal welfare science, ethics and law in European veterinary schools (2012–2019). *Animals (Basel)*, *10*(7), 1238. DOI: 10.3390/ani10071238 PMID: 32708281

Du Sert, N. P., Ahluwalia, A., Alam, S., Avey, M. T., Baker, M., Browne, W. J., Clark, A., Cuthill, I. C., Dirnagl, U., Emerson, M., Garner, P., Holgate, S. T., Howells, D. W., Hurst, V., Karp, N. A., Lazic, S. E., Lidster, K., MacCallum, C. J., Macleod, M., & Würbel, H. (2020). Reporting animal research: Explanation and elaboration for the ARRIVE guidelines 2.0. *PLoS Biology*, *18*(7), e3000411. DOI: 10.1371/journal.pbio.3000411 PMID: 32663221

Kiani, A. K., Pheby, D., Henehan, G., Brown, R., Sieving, P., Sykora, P., Marks, R., Falsini, B., Capodicasa, N., Miertus, S., Lorusso, L., Dondossola, D., Tartaglia, G. M., Ergoren, M. C., Dundar, M., Michelini, S., Malacarne, D., Bonetti, G., Dautaj, A., & Bertelli, M. (2022). Ethical considerations regarding animal experimentation. *Journal of Preventive Medicine and Hygiene*, *63*(Suppl. 3), E255–E266. DOI: 10.15167/2421-4248/jpmh2022.63.2S3.2768 PMID: 36479489

King, G. (2021). Towards a culture of care for ethical review: Connections and frictions in institutional and individual practices of social research ethics. *Social & Cultural Geography*, *24*(1), 104–120. DOI: 10.1080/14649365.2021.1939122 PMID: 36712287

Lens, (2025). *Scholar search: Animal research ethics and welfare standard* https://www.lens.org/lens/search/scholar/list?q=Animal%20research%20ethics,%20welfare%20standards

Littlewood, K., & Mellor, D. (2016). Changes in the welfare of an injured working farm dog assessed using the five domains model. *Animals (Basel)*, *6*(9), 58. DOI: 10.3390/ani6090058 PMID: 27657140

Martínez-Padilla, L., Escribano, F., Tizon-Couto, D., & Gil-Borrelli, C. (2023). The 3Rs principle in animal experimentation: A legal review in Europe, the Americas, and Asia. *Animals (Basel)*, *13*(1), 1–16. DOI: 10.3390/ani13010077 PMID: 38200732

Mason, G. J. (2023). Animal welfare research is fascinating, ethical, and useful—But how can it be more rigorous? *BMC Biology*, *21*(1), 302. Advance online publication. DOI: 10.1186/s12915-023-01793-x PMID: 38155349

Mellor, D. (2017). Operational details of the five domains model and its key applications to the assessment and management of animal welfare. *Animals (Basel)*, *7*(8), 60. DOI: 10.3390/ani7080060 PMID: 28792485

Muthu, S. S. Miguel Angel Gardetti, & Subramanian Senthilkannan Muthu (Eds.). (2016). *Environmental footprints and eco-design of products and processes*. Springer. https://doi.org/DOI: 10.1007/978-981-10-0245-8

Nannoni, E., & Mancini, C. (2024). Toward an integrated ethical review process: An animal-centered research framework for the refinement of research procedures. *Frontiers in Veterinary Science*, *11*, 1343735. Advance online publication. DOI: 10.3389/fvets.2024.1343735 PMID: 38694478

Nawroth, C., & Krause, E. T. (2022). The academic, societal and animal welfare benefits of open science for animal science. *Frontiers in Veterinary Science*, *9*, 810989. Advance online publication. DOI: 10.3389/fvets.2022.810989 PMID: 35498731

Nielsen, B. L., & Rodenburg, T. B. (2025). Ethical approval of studies involving humans and other animals. *Animal Welfare (South Mimms, England)*, *34*, e12. Advance online publication. DOI: 10.1017/awf.2025.5 PMID: 40028438

Pietrzykowski, T. (2021). Ethical Review of Animal Research and the Standards of Procedural Justice: A European perspective. *Journal of Bioethical Inquiry*, *18*(3), 525–534. DOI: 10.1007/s11673-021-10111-5 PMID: 34283345

Sankar, P. (2022). Ethical and welfare issues in the use of animals in biomedical research. In Muthu, S. S. (Ed.), *Animal Welfare* (pp. 15–24). Springer., DOI: 10.1007/978-981-16-8508-8_2

Sivasubramaniam, S., Dlabolová, D. H., Kralikova, V., & Khan, Z. R. (2021). Assisting you to advance with ethics in research: An introduction to ethical governance and application procedures. *International Journal for Educational Integrity*, *17*(1), 14. Advance online publication. DOI: 10.1007/s40979-021-00078-6

VOSviewer. (https://www.vosviewer.com/) PubMed. *Scholar search: Animal research ethics and welfare standard* https://pubmed.ncbi.nlm.nih.gov/?term=Animal+research+ethics%2C+welfare+standards+

ADDITIONAL READING

Animal-Assisted Interventions in the Classroom—A Systematic Review. (2017). In the International Journal of Environmental Research and Public Health [Journal-article]. DOI: 10.3390/ijerph14070669

Cohen, C. (1986). The case for the use of animals in biomedical research. *The New England Journal of Medicine*, *315*(14), 865–870. DOI: 10.1056/NEJM198610023151405 PMID: 3748104

Hecht, E. E., Mahovetz, L. M., Preuss, T. M., & Hopkins, W. D. (2017). A neuroanatomical predictor of mirror self-recognition in chimpanzees. *Social Cognitive and Affective Neuroscience*, *12*(1), 37–48. DOI: 10.1093/scan/nsw159 PMID: 27803287

Macleod, M., & Mohan, S. (2019). Reproducibility and Rigor in Animal-Based research. *ILAR Journal*, *60*(1), 17–23. DOI: 10.1093/ilar/ilz015 PMID: 31687758

Plannthin, D. (2016). Animal ethics and welfare in the fashion and lifestyle industries. In Environmental footprints and eco-design of products and processes (pp. 49–122). DOI: 10.1007/978-981-10-0245-8_3

KEY TERMS AND DEFINITIONS

Animal Care: Animal care refers to the giving of animal's good food, water, shelter, medical care, and environmental enrichment in order to facilitate their health and well-being, particularly, laboratories or research centers.

Animal welfare: Animal welfare is the health and physical and psychological nature of animals, their freedom to express natural behaviors, and freedom of pain, suffering, and distress, proper housing, nutrition and care.

Ethical Research: Ethical research provides that the studies are undertaken with utmost integrity, and ethical approaches to human (ethical) subjects, and within the sphere of moral and legal regulations, such as consent and a way of reducing harm.

Humane Treatment: The humane treatment of animals is treating the animal with respect, empathy and understanding, without causing un-necessary distress or pain and taking great care to provide the best possible care and methods of euthanasia, when it is needed.

International Standards: The international standards are international guidelines and rules (e.g. WHO, OIE, CIOMS) that control when animals are being used, looked after, handled and in what ethically sound manner and animal research needs to be conducted in a unified and accountable method.

Principle of 3Rs: The 3Rs refer to a group of ethical guidelines of animal research.

Replacement: where possible avoid using animals by using other alternatives.

Reduction: have the minimal number of animals.

Refinement: Reduce pain, suffering and enhance animal welfare through experiments.

Research Principles: The regulations of a scientific inquiry that centers on honesty, transparency, reproducibility, and ethical behavior, particularly about animal and human subject use, are considered as the research principles.

Chapter 8
Ethical Guidelines in Descriptive Accounting Research

Siriyama Kanthi Herath
http://orcid.org/0000-0002-6443-9739
Clark Atlanta University, USA

EXECUTIVE SUMMARY

The concept of ethical integrity permeates descriptive accounting research, because, although this type of research is observational in its design, it nonetheless holds considerable responsibility because it relies on live financial information. Inaccuracy or prejudice to any degree can lose stakeholder credibility, as well as taint reputations and reputation of the research and the profession. In this chapter, the philosophy-based moral values including professional and philosophical theories of ethics are discussed with reference to the principles of integrity, objectivity, accountability, and confidentiality. It examines the real-life data managing, methodological Openness, and responsible reporting and deals with contemporary technological issues. The moral questions that could appear to researchers during their work find an illustration in a case study, confirming theoretical research and practical moral choice.

1. INTRODUCTION

Ethics is the foundation of all plausible and rigorous research. Without ethical integrity, even technically sound studies will lose their legitimacy and will not be able to fulfill their purpose. Even though descriptive accounting research may initially appear to be ethically less complex than experimental, intervention-based

DOI: 10.4018/979-8-3693-8302-5.ch008

or behavioral research, it has its own ethical concerns (Bernardi and Bean, 2006a). This type of research is aimed at observing, summarizing, and analyzing patterns of real accounting data. Such trends may affect the choice of investors, corporate governance policies, influence regulators, and inform the general discussion of financial issues.

Since descriptive accounting studies tend to work with empirical evidence in terms of financial statements, regulatory reports, or proprietary databases, the stakes are high. The associated high stakes of financial reporting are replicated in the literature on earnings management, which states that manipulations of financial statements can undermine the stakeholder trust and jeopardize any long-term sustainability, even when the manipulations are used to meet short-term objectives. However, omission or even the hint of bias can be misleading to decision-makers, can be damaging to reputations and undermining to collective trust in the work of the researcher and the scientific institution in general. Fowler (2023) notes that the manipulations, even minor in the eyes of the outsiders in financial reports may weaken the confidence and viability, thus when the data remain based on real-world accounting figures, the ethical stakes associated with descriptive research are quite high. The consequences of such ethical slip-ups, intentional or not may thus impact the industry and the policymaking process at a systemic level (Carberry et al., 2018). This ethical ideal, however, is constantly challenged by the intense pressures within the academic environment, such as the competition for funding and the demanding "publish or perish" culture.

The responsibility of ethical awareness proceeds beyond the institution review boards and the academic supervisor as the plan to check and monitor this behind the scenes of the graduate and doctoral learners is the obligation of ethical awareness that lasts. Academic studies are building on one another; thus, a study produces a body of knowledge on which future studies, professional practice and policy recommendations are built. Once such a foundation is weakened through ethical breaches, it can be a hard-said task trying to fix it. The chapter will focus on how the tension between ethical duties and institutional pressures (e.g., career incentives) drives many of the problems discussed.

1.1 Ethical Foundations in Descriptive Accounting Research

According to the International Ethics Standards Board for Accountants (IESBA, 2023), observance of the fundamental principles and fulfilment of the requirements peculiar to the Code help professional accountants ensure that they honor their duty to uphold the best interest of the general population. This claim shows the most important purpose of ethical models in guiding the actions of professionals, including accounting researchers. Ethics incorporates such principles as integrity, objectivity, professional competence, confidentiality and professional conduct that are the pillars which guarantee integrity in and rigor in research.

Ethics does not merely prove to be an aspect of a mere procedural formality; it is the pillar that public trust is founded on. According to IESBA (2023), ethical behaviors are at the core of the trust of the people in the work of every professional accountant. Business and societal opinion-making and decision-making depends on interpretation of the data, which is why descriptive studies, in which the information is viewed by people, are credible and reliable and the credibility and reliability of it is its ethics directly. The inability to comply with ethical requirements may have a far-reaching adverse impact on firms, clients, and the community in general.

Application of these principles introduces a roadmap to the research process at every step, including finding the research question through the presentation of the findings (Poje and Groff, 2021). Being ethical implies reporting results without distortion and manipulation, whereas being objective implies that there is no bias (intentional or unintentional) influencing interpretation. Responsibility can be used to assure the citizens that the practice is valid. This trend in accounting ethics research reflected by Bernardi and Bean (2006) indicates an increasing appreciation of the fact that the primary research is characterized by ethical standards, which makes descriptive research provide unambiguous, faithful and reliable data on accounting phenomena to various stakeholders.

Moreover, accountability enables the establishment of responsibility between the researchers to their methods and the findings and provides the population with confidence in the validity of their practice. The confidentiality dilemma protects privacy of sensitive information and assures individuals or groups involved the rights, and this again strengthens ethical research.

All these aspects of ethics can be employed to contribute towards strengthening the overall goal of descriptive accounting research that is to provide clear, faithful and reliable information about the phenomena in accounting that would satisfy the expectations of various stakeholders including businesses, regulators, investors and the average citizens. The adoption of these principles can help the researcher preserve the image of the profession and promote the development of clear and trusted financial systems.

1.2 Importance of Ethics in Descriptive Accounting Research

Ethics Credibility and reliability of accounting research reside in its ethics particularly in descriptive research where interpretation of data can influence the decisions that are based on businesses and perception of the masses. As noted by the International Ethics Standards Board of Accountants (IESBA, 2023), not living up to ethical standards may have far-reaching adverse consequences on the firms, employing organizations, their clients and other interests such as the general community, even though they put accountants at a risk of having an adverse impact on accountants professionally or in regulatory terms. That is the dramatic significance of introducing ethics in preserving the trust at any of the levels and reflects the need to follow ethical standards not just to the individual researcher but also to the entire accounting ecosystem. Maintaining the descriptive research, making good decisions, and protecting the interests of all the involved stakeholders will be preserved by the defense of these principles.

Ethics is not a mere process of accountancy research, but the core on which the trust of the people is founded. The International Ethics Standards Board of Accountants (IESBA, 2023) argues that Ethics is central to the trust people have in the performance of every accountant, and it should always be central to their judgments, decisions, and actions when engaging in professional activities or the provision of professional services. This assertion points out the severe importance of ethical behavior in ensuring that descriptive research studies in the field of accounting are believable, reliable, and helpful to stakeholders such investors to policymakers.

Ethics does not represent a voluntary accessory to accounting research; it constitutes an essential ingredient in the generation of knowledge that can be trusted (Bernardi & Bean, 2006). Ethics is core in the descriptive study of accounting research often conducted with the processes of gathering, interpretation, and reporting of information of how organizations are practiced, financial disclosures and stakeholder behaving. Descriptive research also differs with prescriptive research since its purpose is to describe reality because it must give a commitment to do so with accuracy, neutrality, and transparency.

In this sense, as Bernardi and Bean (2006) observe, the expansion of accounting research in ethics-oriented journals during the last decades is related to the rising recognition that ethical principles cannot be an accessory interest but the hallmark of fundamental research. Ethical standards, in terms of honesty in the presentation of data, compassion to confidentiality, and the prevention of conflicts of interests, keep researchers and help sustain morals of the academic document as well as the entire accounting profession.

Critical Synthesis and Evolving Trends in Accounting Ethics Research

The development of accounting ethics literature identifies an important turn towards the attention of systemic and methodological ethical issues, rather than being to individual professional behavior (that the prior codes, such as that of the IESBA, were primarily concerned with).Many early studies were conducted on the topic of personal integrity and fraudulent reporting in the firm, but the modern discourse is laid on the ethical aspect of the research process itself. This also encompasses the fair usage of Big Data, the disclosure of proprietary archival information, and the possible biases of algorithmic analysis (IESBA, 2023). One of the main gaps that have become observable in the literature is that it has not established stiff ethical codes of conduct specific to the new challenges of technologies such as AI and machine learning when it comes to the development of descriptive financial models and interpretation. The changing emphasis here highlights why the current discussion requires reconciliation of traditional approaches on ethics with the current studies in research.

Lastly, ethics is applied to uphold the validity of descriptive findings that can be employed to inform informed decisions and can ensure the principles of accountability and transparency which the accounting field is founded on. To understand how these responsibilities can be preserved in a practical manner we must first be made to understand how the prescribed theories and principles of ethics are adhered to in bringing about sound judgment of research.

2. ETHICAL FOUNDATIONS: THEORIES AND GUIDING PRINCIPLES

Philosophical traditions form the basis of ethical researching practice that developed the moral choice-making in the centuries. They are the intellectual structure of these traditions in making decisions of what to do, as well as why to do it as researchers. The philosophical traditions that have governed the process of making moral decisions over centuries form the foundation of research practice. Such traditions give intellectual structure in ascertainment of what actions the researchers ought to perform and the reasons why they ought to perform those actions. The explanations of these theories are critical to the accounting researcher in that each provides a distinct perspective through which to assess the complex situation that arises where there are competing demands of the sponsors, regulator or the people. These frameworks are used in descriptive accounting studies whose purpose is often faithfully to describe the findings of financial or organizational data to ensure that findings are accurate and ethically defensible. The recent studies indicate that adopted ethical practices within the accounting profession can not only boost the corporate governance levels but also elevate the compliance patterns, which strengthen the parties of interest and corporate ethics (Jejeniwa et al., 2024).

2.1 Major Ethical Theories

(a) Utilitarianism

Utilitarianism, which is the most vocalized by Jeremy Bentham and John Stuart Mill, is based on the weighing of an action according to the outcomes of an action in the effort to advance the greatest good to the greatest number of people. In the context of accounting research, this can be evaluated by putting into consideration the role the findings will have on the various stakeholders, including the shareholders, regulators, employees, and society and ensuring that the findings will have better than evil when they are published. Study of the view is now more concentrated on the aspect of accounting as the mode of enhancing the well-being, and sustainability in society and best aligned with the principles of utilitarianism (Nørreklit et al., 2024).

To illustrate, when a descriptive study reveals the inefficiencies of a government spending program, the utilitarian approach parallels the need to report those results, since the advantages to taxpayers and policymakers would outnumber any embarrassment to agencies. But the strict application of utilitarian logic needs to be brought with a lot of care; at times, over attention to the consequences of actions may result in neglect of rights or responsibilities that are not supposed to be broken, in the best interests of a greater creature.

(b) Deontology

Deontology is a problem concept which is founded on the philosophical beliefs of Immanuel Kant which focuses on existence of certain duties and rules of moral actions that need to be adhered to irrespective of the consequences. This is a fundamental principle of Kantian ethics (Misselbrook, 2013), according to which the researchers must follow the accepted ethical procedures, including those defined by the American Accounting Association (AAA) Ethics Guidelines, even in cases where the violation of rules would yield more impressive or popular results (American Accounting Association, n.d.).

Deontology is an ethical model established by Kant based on the ideology that human beings through their ability to reason can interpret and then act responsibly in all circumstances in response to universal moral laws (Gelman, 2020; Misselbroek, 2013). This implies that, in accounting research, the researcher is required to adhere to all ethical considerations although neglecting them could result in more impressive or popular findings.

For example, in descriptive accounting work, a deontological perspective would require a researcher to disclose data limitations, even if doing so might weaken the perceived strength of their conclusions. Knowingly, omitting a relevant variable from an analysis to simplify or enhance findings would be considered unethical, regardless of the potential appeal of the results.

(c) Virtue Ethics

Virtue ethics, which gained its roots with Aristotle, lays more stress on character development than simply following the rules or concentrating on the consequences (Mixon, 2024). It turns the question not into: what rule should be used? or, what would be the best outcome? to: what would a virtuous man (or woman) do? Honesty, prudence, diligence, humility, and fairness are virtues that are critical in ethically practicing in accounting research (Bretag & Green, 2014; Mintz, 2021). A researcher working with the guidance of virtue ethics would not only see the manipulation of data as a violation of the rules but also as the acts that have no connection with the personality of a dependable researcher (Bretag & Green, 2014). As an illustration, facing ambiguous or unfinished information, a virtuous researcher would act out of transparency and integrity as the scope of professional reputation can be established over time and can be destroyed with ease (Mintz, 2021; Mixon, 2024).

These three theories are not mutually exclusive, but they are different. Practically, ethical deliberation will frequently fund its judgment based on each of these three: the attention of consequences (utilitarian theory), duties (deontology), and virtues (virtue theory). In total, they offer an equilibrium between toolkits on approach to the moral complexity of descriptive type of accounting research. To understand better the premise on which ethical reasoning in descriptive accounting research is based, Table 1 highlights the differences between the three major ethical theories, i.e., utilitarianism, deontology, and virtue ethics. Both theories have clear principles and considerations that will guide the researchers in their handling of the ethical dilemmas and decision-making.

Table 1. Comparison of Major Ethical Theories

Ethical Theory	Core Principle	Application in Descriptive Accounting Research	Key Considerations
Utilitarianism	Maximize overall goods and minimize harm	Ensure findings benefit the public and professionals; avoid misinformation	Balancing stakeholder interests; long-term consequences
Deontology	Follow rules, duties, and codes of conduct	Adhere strictly to AAA Ethics Guidelines; avoid rule-breaking even if beneficial	Importance of honesty, fairness, and accuracy
Virtue Ethics	Develop moral character and virtues	Cultivate honesty, diligence, humility in data handling and reporting	Focus on researcher's integrity beyond formal rules

2.2 Foundational Principles in Practice

Ethical theories provide a general philosophical cloud; however, researchers also require practical principles that empower them to make day-to-day decisions. With respect to the various theoretical approaches, four guiding concepts recurrently arise in the accounting studies:

1. **Integrity** –Willingness to be accurate, honest, and truthful in all research facets. This encompasses reporting honestly and ethically, admitting mistakes, and avoiding pressures, whether by the funding agencies, employers, or colleagues to bend the results.
2. **Transparency-** Clear declaration on the sources of data, methods used in compilation, analytical tools, and study limitations. To use an example, when a specific set of data is restricted to publicly traded companies per region, then such a constraint should be explicitly stated to prevent overgeneralization.
3. **Confidentiality**- The privacy of the individuals and organizations, the data of which is used, is treated confidentially. In the case of aggregated data, researchers might be forced to prevent reverse engineering of sensitive information regarding reporting.
4. **Accountability** – Being responsible to the repercussions of his/her work and how it can be understood or used. This implies that it will be about foreseeing the misuse of results and putting them in context so they can be used properly.

These principles are an applied gap linking a theoretical philosophical approach and a social research reality. They are the points of the conscience that directs descriptive accounting researchers to make their daily choices however small to big, and these may include data collecting to producing the write-up in a descriptive accounting inquiry maintaining that quest to achieve knowledge without compromising ethical responsibility.

Table 2 provides an overview of the essential ethical principles that are often followed in accounting studies, along with the main methods of practical implementation and the dangers of violating them. The framework assists the researcher to convert the abstract values into practice through the research undertakings.

Table 2. Ethical Principles and Practical Examples

Ethical Principle	Description	Practical Implementation	Potential Risks if Violated
Integrity	Accuracy and truthfulness in research	Accurate data recording; honest reporting of results	Fabrication, falsification, bias
Transparency	Openness about methods and limitations	Clear documentation; disclosure of conflicts	Misleading findings; loss of trust
Confidentiality	Protecting confidential information	Anonymizing data; secure storage	Privacy breaches; legal consequences
Accountability	Responsibility for research impact	Peer review participation; correction of errors	Damage to reputation; harm to stakeholders

Accounting and financial results are hugely affected by legal, regulatory, and political changes, which determine the degree of transparency and accountability that firms and researchers need to comply with (Taylor et al., 2025). Awareness of these institutional frameworks is invaluable to accounting researchers to make their research relevant to the changing compliance needs as well as to reportable financial matters that are dependable and ethically acceptable in the world of accounting.

2.3 The Academic Environment and Ethical Pressures

Although even though the philosophical foundations and general ethical principles (Integrity, Transparency) are applied to inform best research practice, the descriptive accounting research in the real world is situated within a problematic academic ecosystem that often creates powerful incentives to breach ethical principles. Most ethical violations have their roots in these institutional forces, which transform abstract moral problems into life-and-death career decisions.

Key Systemic Drivers of Ethical Risk

Academic competition and research environment create some pressures that question the purpose of the researcher:

The "Publish or Perish" Mandate: One of the secrets of career development, tenure, and even promotion is the quality of journal publications. This puts a tremendous strain on researchers to deliver meaningful or good results in a short duration of time. This institutional effect can promote questionable research practices, such as p-hacking (modifying analysis until a significant finding is obtained) or Hypothesis After Results Are Known (HARKing), thereby influencing methodological transparency.

Competition for Funding and Resources: The business of descriptive accounting research may be expensive because of proprietary data (e.g., Compustat, Audit Analytics) or databases that are expensive to maintain. It is this lack of these resources, coupled with the external grants, which causes researchers to prioritize the projects which may be of interest to the corporate sponsors or funding agencies and therefore the implicit and even explicit conflict of interest which can lead to bias in the research design or interpretation.

Career Incentives and Self-Censorship: This can be said to be the case with junior researchers and more so doctoral students. The motive of self-censorship could be the need to have a good dissertation and good recommendation by their supervisors or corporate sponsors which fail to deliver good results when the lines of inquiry are pursued. The anxiety of risky future employment opportunities usually dominates over the necessity to report a bad or uncertain outcome.

The Impact on Integrity

These forces are central to the principles of integrity and accountability. The breach of ethics does not have to be founded on malice, but on a utilitarian calculation where the immediate good of the researcher (publication, funding) is higher than the long-term, diffused evil of loss of trust and scientific integrity. These systemic forces are important to identify since they will not only work towards identifying individual failures but also the institutional protection and open governance systems will be needed to enable ethical decision-making.

Table 3. Common Researcher Conflicts of Interest and Mitigation Strategies

Type of Conflict	Specific Example in Accounting Research	Potential Ethical Risk	Practical Mitigation Strategy
Financial Interest	Researcher holds significant stock in a publicly studied firm or industry.	Biased selection of variables or methodology to ensure favorable results that protect personal investment.	**Mandatory Disclosure:** Full and explicit disclosure of all relevant financial holdings in the publication's COI statement.
Sponsorship/Funding	Research is funded by a large corporation that grants data access contingent upon a review of the findings prior to publication.	Self-censorship or omission of critical results to avoid jeopardizing the funding stream or data access.	**Firewalling:** Institutional separation between the researcher and sponsor regarding data analysis and publication decisions. **Blind Data Analysis** by a third party.
Career/Academic Incentive	Doctoral student finds results contradictory to their advisor's prior published work, or a junior faculty member needs "significant" results for tenure.	P-hacking, HARKing, or selective reporting to generate publishable (positive) results, sacrificing scientific rigor for career gain.	**A Priori Registration:** Pre-registering research design, hypotheses, and analytical plan to prevent post-hoc manipulation.
Consulting Ties	Researchers actively consult for a firm being included in the sample of a descriptive study (e.g., on corporate governance).	Interpretation of ambiguous data points in a manner favorable to the consulting client or omission of the client from the "troubled" subsample.	**Recusal/Separation:** The researcher must recuse themselves from analyzing data related to consulting clients or implement a **cooling-off period** after the consultancy ends.

Table 3 translates the theoretical concept of the Conflict of Interest (COI) into actionable strategies for descriptive accounting researchers, ensuring a commitment to the principles of integrity and transparency (IESBA, 2023).

2.4 The Researcher's Conflict of Interest

While ethical principles generally focus on rules and data integrity, a subtle yet profound source of bias in descriptive accounting research stems from the researcher's own interests, known as a Conflict of Interest (COI). A COI exists when a researcher's professional judgment regarding a study's primary interest (e.g., the validity of the research) may be unduly influenced by a secondary interest (e.g., financial gain, career advancement, or personal relationships). These conflicts may be conscious or unconscious in descriptive accounting, and may bias the choice of samples, methods of analysis, or interpretation of findings.

Common Conflicts in Accounting Research

The access to and use of corporate data is a central conflict among descriptive accounting researchers:

Funding Sources and Sponsorship: When a study is sponsored by a certain corporation or industry group, there is an unspoken pressure to come up with results that are favorable to the sponsor. To illustrate, an earnings management researcher who is paid by an industry lobby may be reluctant to report systemic, wide-ranging non-compliance.

Consulting and Financial Ties: When researchers are consultants to companies or have a strong financial interest in the company they are researching (e.g. owning stock in the company), they may have an incentive to interpret ambiguous information in a way that benefits their consulting client or personal investment.

Data Access and Non-Disclosure Agreements (NDAs): NDAs are especially common in the case of corporate sponsors who provide access to proprietary data by strict NDAs or a privilege to view the findings prior to their publication. This control may cause researchers to self-censor or evade sensitive lines of inquiry so that they can have access to valuable data.

Career Ambition and Publication Bias: Academic pressure to make a significant and positive contribution can cause methodological flexibility, including p-hacking (flexibility in the way research is carried out), or the reporting of research biased to confirm a hypothesis, even when the overall study is weak.

Case Illustration: The Conflict of the Hidden Dip

The abstract idea of a conflict of interest tends to turn into a harsh reality of young scholars. Take the case of the so-called "Hidden Dip in Profitability": A doctoral student who is doing a descriptive study of an organizational sponsor finds out that the profitability of the company is decreasing dramatically- a fact that is in direct opposition to the official story of the sponsor. The student is aware that revealing this discovery correctly would put the sponsor relationship at risk, which may result in loss of funding, access to future data and vital career opportunities. Nevertheless, the minimization or exclusion of the data would be against the obligation to the duty of integrity to the population and the academic community. This scenario is a clear indication of the conflict between the individual interest of the researcher to advance his/her career and duty to tell the truth.

Practical Mitigation Strategies

The management of COIs is essential because of maintaining integrity and transparency (as the former is defined in Section 2.2). Mitigation needs to be institutionally monitored and personally committed. These ideas are converted into practical steps leading to action as directly presented in Table 3 (in Section 2.3), which further supports the key ethical obligations existing in Deontology and the development of the Virtue Ethics of honesty and prudence in the researcher.

COIs are also managed through integrity and transparency (as provided in Section 2.2). Controlling requires institutional discipline and individual commitment.

2.5 Laws and Regulations

The major role in the formation of accounting research is played by legislation and regulations, such as the Sarbanes-Oxley Act (SOX), the Securities Exchange Act, and various tax codes that lay down compulsory requirements of financial reporting transparency, accuracy, and accountability (Eilifsen et al., 2014). Adherence to these bodies of law means that research activities will be in synchrony with the demands of the industry and the interests of the stakeholders. Ethical research in accounting should abide by numerous regulations and laws that are aimed at enhancing transparency, accuracy, and safeguarding of interests of the stakeholders. Legal guidelines that are important include financial reporting, data privacy, and corporate governance to inform the researcher how limited their operations can be. Knowledge and use of these laws ensure neither legal consequences nor impairment of research occurring in accounting studies.

Table 4 summarizes important laws and regulations accounting under consideration and their definite implications.

Table 4. Key Accounting Laws and Regulations: Overview and Definitive Implications

Law/Regulation	Relevance to Accounting Research
Sarbanes-Oxley Act (SOX)	Enhance corporate financial disclosures and protect against fraudulent reporting. Crucial for transparency in research data.
General Data Protection Regulation (GDPR)	Governs data privacy and protection in the EU, critical for handling personal or sensitive data in research (**GDPR [European Union, 2016]**).
Securities Exchange Act of 1934	Regulates securities transactions and financial disclosures, affecting research relying on publicly traded company data.
Foreign Corrupt Practices Act (FCPA)	Addresses anti-bribery and accounting transparency standards in multinational contexts; relevant to cross-border research.

The accounting researchers are required to remain within the domain of these laws to make their work legal and respectful towards the law and ethical practices that are paramount in sustaining the interest of the populace. The legal, regulatory, and political changes play a key role in influencing accounting and financial results as the transparency standards and accountability that firms and researchers are required to follow can be altered through a legal, regulatory, and political intervention (Taylor et al., 2025). Awareness of these institutional frameworks is invaluable to accounting researchers to make their research relevant to the changing compliance needs as well as to reportable financial matters that are dependable and ethically acceptable in the world of accounting.

3. DATA MANAGEMENT AND ETHICAL RESPONSIBILITIES

Ethical descriptive accounting studies start by acknowledging that data management is not just a technical exercise, but it is a fundamental ethical obligation that determines validity, reliability, and trustworthiness of research results. Since it must be studied through the lens of ethics, researchers must follow ethical principles throughout the course of data acquisition, storage, and analysis.

3.1 Data Acquisition and Bias

The ethical process begins with data collection. Acquirers of data should be able to stop in the best way without violating the law, or breaching any privacy, or violation of restriction laws or privacy breaches. In this regard, an illustrative example is borrowed by the proprietary financial data sets where owners must acquire the rightful permission to use the proprietary financial data thus respecting intellectual property and confidentiality agreements. Also, selection of samples is an important consideration that should be conducted by researchers without being biased enough to mislead stakeholders. Representativeness matters and using only convenient samples or performing a fragmented one can result in systematic errors and thus compromise the quality of the study. Lastly, there should be complete disclosure of any shortcomings or constraints to the data like absence of variables or constrictive time bound boundaries to be able to interpret results.

3.2 Data Cleaning, Analysis, and Manipulation

After the data is obtained, it has been noted that the information tends to need cleaning so that it can be made accurately in case of mistakes or inconsistencies. This, however, requires a delicate matter of ethics: in sterilizing the data, no harm is to be done to the truth of the matter. Any manipulation that is done, be it the correction of outliers or adjustment of missing data- must be well and properly documented to facilitate transparency and repeatability. There are also unethical research strategies, like p-hacking wherein the analysis or reporting of results is skewed to give positive results so it can accommodate the hypothesis initially formed. Ethical researchers can take into consideration ethical practices. Analysis must be done according to a pre-determined plan when it is possible and all discoveries made by exploration methods must be clearly marked as such. Having an unambiguous audit trail where raw data and code are included in the analysis enhances accountability as it will allow others to analyze and check the outcomes.

3.3 Data Storage and Security

After Ethical considerations go as far as the data storage and security after the analysis. Secured secrets or sensitive data: could this include data that belongs within the specific company, or financial? Data that is not supposed to be made public. Among the best practices are encryption, role-based access, and back-ups according to the protocol of the institutional review board (IRB) or equivalent supervisory organizations. These precautions prevent leakage of data, which can lead to risks to the participants or organizations and maintain confidence regarding the research process. Furthermore, the archival work would involve that the data can be checked in the future or be utilized in the secondary analysis, but the unnecessary disclosure would be a balance to find. An ethical requirement to have a sustainable stewardship of data in the future would therefore be beneficial to the credibility and sustainability of academic work.

Based on these basic ethics and theoretical theories, it is important to consider how these theories manifest in the context of managing data which is one of the most important studies of descriptive accounting research. Data ethics is a multi-dimensional task, and it includes acquisition, cleaning, analysis, and storage. An example of such allegations can be demonstrated by Sayed (2023), in which the researcher points to cases of lacking data security in the context of big data analytics because of the breach of confidential data in different industries, which sparked the urgent necessity of implementing sufficient ethical protection of data storage and privacy.

Table 5 outlines the ethical requirements of each step of data work, the best practice examples, and after-thoughts of the unethical action to emphasize vigilance and transparency.

Table 5. Data Management Ethical Responsibilities

Stage	Ethical Responsibility	Example Actions	Risks of Unethical Behavior
Data Acquisition	Obtain data lawfully; avoid bias	Secure permissions; representative samples	Unauthorized use; sampling bias
Data Cleaning & Analysis	Document changes; avoid selective reporting	Maintain audit trails; avoid "p-hacking"	Data manipulation; irreproducibility
Data Storage & Security	Protect data privacy; ensure secure storage	Encryption; restricted access; IRB compliance	Data leaks; identity exposure

4. ETHICAL CONSIDERATIONS IN SPECIFIC DESCRIPTIVE METHODOLOGIES

Various descriptive accounting studies have different methodologies, each of which holds its share of ethical challenges and responsibility. Although ethical principles are universal, adherence differs in relation to the data collection and analysis method that an individual may apply (Guragai et al., 2017). In this section, a focus is made on the ethical insights of certain very prevalent descriptive methodologies, such as surveys and interviews, analysis of archival data, and case study.

4.1 Surveys and Interviews

The surveys and interviews continue to be popular methods of collecting both qualitative and quantitative data directly with the help of the participant; yet the methods demand serious ethical considerations to ensure the rights and well-being of the participant are not violated (Engel, 2016). One of the most important factors is the matter of informed consent: the subjects must know perfectly what the research is all about; the risks involved, disclosure to be made to their data, and their inerrant right to pull out at any time without reprisal. Such openness creates trust and upholds the autonomy of individuals. Accounting ethical dilemma can be met especially in gray areas when professional judgment is conducted to reconcile between the rule of regulation, and the business practicality by appreciating robust ethical frameworks and continuous watchfulness (Park University, 2024).

Also, participation must be voluntary, there should be no coercion, and participants must never feel excessively pressured to participate. Researchers need to avoid the issue of power obliviousness particularly where the participants could be worried about the consequences. Kang and Hwang (2023) stated that it would be essential to make a clear distinction between anonymity and confidentiality of survey research. Anonymity refers to the fact that the identity of participants is unknown to the researcher, whereas confidentiality entails the fact that the identity is known, although it is held in confidence and will not be released. The involvement of the participants in informed decision-making and protection of their privacy can be achieved through the presence of clear policies on how the data is overseen.

4.2 Archival Data Analysis

In accounting, archival research refers to the use of existing documents including records of previous financial statements, regulatory reports, or business databases. Although the latter sources may be perceived as publicly available, there are ethical issues to think about when it comes to intellectual property and the context of the information data. As an example, Parry and Mauthner (2004) consider the practical, legal, and ethical issues of archiving qualitative research data, and as a result, the ownership of the data and the impact of its openness to others raise a concern. According to U.S. copyright office (2005), the researchers need to be cautious when using the proprietary databases in the research because not all the available data can be used freely as the copyright regulations and licensing terms of use are reproduced. To avoid breaching the law and ethics, it is necessary to distinguish between the publicly accessible (data is available but with restrictions) and public domain (data that can be used at will).

Moreover, archival documents were produced to be used in original intents that may not necessarily be the same as those of the investigator. Ethical researchers should be allowed to identify the original context and limitations and not misinterpret and overgeneralize the findings. This consideration of the data provenance helps to maintain academic integrity and gives the reader a better picture of the extent to which the research has been done.

4.3 Case Studies and Observations

Case studies are elaborate and descriptive accounts of organizations or incidents, which in most cases are founded on personal observation and examination of records. Case studies are extremely enlightening and descriptive, but such ethical issues as organizational privacy and prevention of harm must be addressed carefully (Mökander and Floridi, 2023). The researchers ought to take caution in exposing sensitive strategic information that could harm the organization either on its competitive edge or image. Even though description may help understand it better, it is a thin line between transparency and security of confidential information. Another problem is fair representation. As Otley and Berry (1994) note, case studies of management accounting must be realistic, and not biased or exaggerated, and should not generalize beyond the context being studied. As Maddi and Teixeira da Silva

(2024) point out, ethical publication practices presuppose good and clear identifications of intellectual contributions, which means that the concept of authorship is a real representation of substantial contribution to the research process. The present method-related ethical issues discussion demonstrates the importance and challenge of ethical practices adjustment based on the nature of information and approaches used (Hassan et al., 2021). With these considerations in mind, we now move to the ethics of communicating research findings and publishing results where new issues of transparency and responsibility emerge.

Expanding on the main principles of ethical data management and researcher accountability, it is important to note that the application of such sophisticated tools as big data analytics can not only contribute to transparency but also present new ethical issues in descriptive accounting research (Pham et al., 2025). Various descriptive methodologies are associated with different ethical issues. Table 6 includes the most common research methods like surveys, analysis of archival data, and case studies, as well as the ethical considerations, the main practices to maintain integrity, and the challenges that researchers might encounter.

Table 6. Ethical Considerations by Research Methodology

Methodology	Ethical Considerations	Key Practices	Common Ethical Challenges
Surveys & Interviews	Informed consent; voluntary participation	Clear consent forms; anonymity assurances	Coercion; privacy breaches
Archival Data	Compliance with copyright/ licensing; data context	Check licenses; contextualize data use	Misuse of proprietary data; misinterpretation
Case Studies & Observations	Protect sensitive information; fair representation	Confidentiality agreements; balanced analysis	Overgeneralization; confidentiality risks

4.4 Cross-Cultural Ethical Challenges in International Accounting Research

Cultural differences exist in the expectations of ethics, what is expected of a particular country in terms of confidentiality or consent may be very different in another. Multinational researchers must be aware of such differences, and they must strive to attain an ethical level that does not violate the local practices but adheres to the universal ones. It is important that the multinational accounting firms, as well as the researchers that might anticipate the influence of cultural values on ethical sensitivity and decision-making, predict the international differences in ethical behavior (Cohen et al., 1996).This emphasizes the need to appreciate cultural aspects and apply it in the ethical systems to make the research relevant and considerate in a wide spectrum.

As an example, the European Union (2016) has laws involving greater requirements surrounding data privacy through regulations such as the General Data Protection Regulation (GDPR), whereas other authorities may lack more formal data protection laws. These variations have consequences on how researchers collect, store, and share information. In a comparable manner, culture related regulations governing hierarchy and authority might affect consent process and willingness of people to express themselves freely during interviews or surveys.

Gupta (2025) notes that cross-cultural differences have a dominant role in shaping the practices of the ethical and accounting codes based on principles followed in other countries regarding international research, and hence, careful consideration should be given to the cross-cultural differences in a manner that does not affect the local traditions and customs and that uphold the universal code of ethics. Similarly, Vinson et al. (2020) specify that cultural preferences along with risk-taking inclinations are susceptible to the national environment of countries and condition the understanding and finding solutions to the moral ambiguities of the accounting profession.

To address these concerns, the accounting researchers ought to introduce a culturally sensitive research strategy that has mediative powers between adhering to local values and fundamental ethics. This would imply that you will conduct sufficient background research into the legal and ethical climate of the host country, seek the assistance of local partners or ethics committees and apply ethical rules which are flexible and at the same time uniform. It is possible to conclude that tolerating cultural diversity in ethical decision-making process is the best option available to improve the quality and relevance of global accounting research studies, safeguard the rights of the subjects involved, and maintain the confidence of the population.

5. ETHICAL ISSUES IN PUBLICATION AND DISSEMINATION

5.1 Authorship and Attribution

Proper and equal intellectual credit is the key to the first step of ethical publication. The authors ought to have representation of individuals who have made significant contributions to the work in conceptualization, data procurement, analysis of the data, or even authorship. Misra et al., (2018) refers to the issues and the opportunities of improving the soundness of the modeling and peer-reviewing practices. Malpractice of suppressed authorship or falsification of attribution to members leads to the harm of scholarly communication and this can only harm confidence in the research activity (Maddi & Teixeira da Silva, 2024). This tradition of such practices undermines the credibility of any academic studies and demands more delineated authoring monitoring rules. Furthermore, crediting any source of help, even financial supporters, owners of databases, and technical support is a crucial step to the reputation and integrity of the study conditions. In more than 81,000 publications published in PLOS ONE, Maddi & Teixeira da Silva (2024) examine the truth and justice rate of authorship assignments. The authors found out that among the articles, 9.14 percent were connected to at least one instance of inappropriate authorship, hence the more prevalent cases of misattributions. The paper reveals the significance of candid and fair acknowledgment of everyone to sustain integrity of the academic communication process (Maddi & Teixeira da Silva, 2024).

Analysis of Pressure: The high stakes publish or perish culture in academia is a direct cause of authorship disputes and unethical inclusion/exclusion of names. The tenure, promotion, and scarce research funds can be a motivating factor that encourages researchers to misuse their authority in pressuring subordinates to give them authorship or, on the other hand, withhold junior researchers to amass credit, which is a complete breach of the principle of fair attribution.

5.2 Authorship Peer Review Process

Peer review is based on confidence, privacy, and objectivity. The researchers as the reviewers are bound to access shared work without prejudice to enable ethical guidelines and practice of reviewing manuscript without any bias. Similarly, manuscript presenters should indicate a related study and duplicated study to avoid duplication of works and attain fair judgment. Ethical violations can cripple scientific progress and reputations in this process.

In a special journal issue on fertility, pregnancy, and mental health, McIntosh and Vitale (2023) explore the problem of conflicts of interest between authors, editors, and peer reviewers. The authors found that there were undisclosed conflicts of interest between authors, an editor and several peer reviewers who were members of anti-abortion advocacy groups, which undermined objectivity and transparency. Their research highlights the need to ensure confidentiality and impartiality in peer review process to promote scientific integrity and avoid propagation of biased or misguided research.

5.3 Responsible Dissemination of Findings

It is the duty of the researchers to report and describe their findings in an appropriate manner. This means that overgeneralization and overstating of the results should not be done beyond the evidence at hand. The clear definition of limitations and the spheres where errors can contribute to the prevention of the inaccurate interpretation by other researchers, the Cabinet, or the general population. In the case of dealing with the media or publishing public summaries, researchers should ensure that the nature and depth of their research are not lost, which would otherwise lead to false reports or unrealistic ideas. Dempster et al., (2022) underline that the depth and subtlety of a research project should not be lost in the production and release of public statements or involvement in the media, which can subsequently potentially lead to misinformation and expectations that are not representative of reality.

With the changing nature of the research dissemination landscape, to introduce the tools and technologies that define how data is gathered, analyzed, and distributed, introduce new ethical issues and opportunities into the limelight (Schreyer et al., 2022) The integrity of research goes beyond data collection and analysis to publication and dissemination. Table 7 lists the key ethical challenges that were faced when writing, reviewing, and communicating findings to the public, and the best practices to follow to become a responsible and credible scholar.

Table 7. Ethical Issues in Publication

Issue	Description	Best Practices	Consequences of Misconduct
Authorship & Attribution	Recognize genuine contributions	Follow journal guidelines; disclose funding	Plagiarism; disputes; loss of credibility
Peer Review	Maintain confidentiality and impartiality	Avoid conflicts of interest; honest feedback	Breach of confidentiality; bias
Responsible Dissemination	Present accurate findings with limitations	Include caveats; avoid hype in press releases	Public misinformation; reputational damage

5.4 Ethical Training and Education

Bérubé and Gendron (2023) stress that accounting researchers need to be provided with ethical training and education to be able to resolve complex moral dilemmas that they face in their practice. Special workshops on ethics, case studies, and discussion groups, which place students in a real-life scenario where they can practice and apply their direct ethical principles in real life, should be introduced as a part of graduate programs. This background exposure helps in fostering a high ethical attitude at an early stage in the careers of the researchers.

In addition to initial training, life-long professional training contributes to ensuring that researchers are informed of any changes in ethical norms, regulatory code, and emerging issues in the area. The need to keep up with the changes in ethical regulations and governmental regulations make continuing professional education important to accounting professionals (Ghosh, 2025). Lifelong learning in terms of refresher courses, seminars and certification programs allows professionals to maintain integrity, transparency, and accountability all through their career. Ethics training in each level empowers the accounting research community to sustain its integrity, and promotes the faithful, accountable scholarship (Mardawi et al., 2021).

There is research in support of the importance of practical and ongoing ethical education. As Winterberg et al. (2021) note, inclusion of ethical dilemmas experienced in the real world during the training period makes the researchers more effective moral navigators. In the same way, Parks-Leduc et al. (2021) observed that constant and customized ethics training elevate the educational quality of ethical decision-making with respect to the consideration of individual differences. According to Andersson et al. (2022), teaching the learners reflective practice and training with real-world cases is the crucial step to become ethically competent and face complicated problems.

Undergraduate training and education in ethics can be valuable because it increases the ethical awareness and decision-making skills of students and forms a basis on which future researchers should be built (Okougbo, 2021). Poje et al., (2025) further state that achieving the above goals requires comprehending the ethical predispositions of students to develop effective programs that cultivate integrity in students at the beginning stages of their academic careers. The fact that ethics can be learned through case-based learning and active discourse is also emphasized by Eaton et al. (2023) to equip students with the keys of how they are going to manage ethical issues in their careers.

With the changing dynamic of ethics with the advancement in technology, it is also pertinent to understand the role of technology in ethical trends as well as issues of research and this is what the subsequent section attempts. The rising use of technology in accounting research gives rise to novel ethical issues and opportunities which ought to be resolved to support the integrity of the profession.

6. THE ROLE OF TECHNOLOGY IN ETHICAL RESEARCH

6.1 Artificial Intelligence and Algorithmic Bias

The use of artificial intelligence (AI) in the accounting has made it possible to detect more advanced patterns and make predictive models. Adelakun et al. (2024) examine the way AI methods (machine learning, natural language processing, and data mining) are used to find trends and abnormalities in financial data and that, in numerous ways, improve fraud-detecting abilities in accounting. Nevertheless, in the case of AI systems, they tend to be found on historical data, which are also subject to existing biases and may support inequalities or unjust interpretations. These tools should be audited carefully by scientists, they should announce their restraints, and they should not blindly trust the results of algorithms. This implies that ethical use of AI should be practiced ensuring that automated processes are estimated to advance rather than harm the integrity of the research. According to Forbes (2024), even unstructured documents such as PDFs and emails can be successfully processed by AI systems that specialize in certain fields since they can convert them into a structured format to enable reliable analysis and decision-making.

6.2 Big Data and Privacy

The advent of big data provides new opportunities to handle large and complex volumes of accounting information. Even though big data analytics offers a baseline benefit to data-driven decision-making and strategic expansion, organizations need to emphasize the strength of data governance and ethical guidelines to avoid data abuse and ensure the trust of stakeholders (Galvez, 2024). However, there is also the risk of invasion of privacy of individuals or organizations with this power. Big data technologies have changed the volume and size of data that can be used in accounting research and allow conducting more detailed analysis, but they also pose serious questions about data management, privacy, and ethical use (Abdalla, 2022). Although datasets are anonymized, a combination of sources may accidentally disclose sensitive data. Ethical research practice requires ongoing risk evaluation, strict anonymization procedures, and data aggregation strategies to ensure privacy and permit meaningful analysis. In the age of big data, transparency is not a choice, as Ikezuruora (2023) claims it is the foundation of keeping the population trustful and making sure that people know how their data is gathered and processed. The author underlines that the absence of disclosure and open communication may lead organizations to lose credibility and become the subject of increased ethical scrutiny in their data-driven activities (Ikezuruora, 2023). The disclosure that Amazon has thousands of workers across the globe to listen to voice recordings on Echo devices underscores the ethical issues in data collection and privacy, the importance of transparency and user consent in research practices (Day, Turner, and Drozdiak, 2019). Recent developments emphasize that professional accountants have a changing environment of ethical issues, which are influenced by technological changes and more stringent regulation, and they need to be more vigilant and flexible to maintain integrity in their practice (Lane, 2024). The emergence of big data analytics has brought about intricate ethical issues, especially related to privacy, data accuracy, accessibility, and algorithmic bias, which require the development of strong ethical guidelines that can be used to regulate responsible data use (Dorn et al., 2024).

6.3 Blockchain and Data Trust

The blockchain has potential solutions in data integrity and transparency (not possible to modify) and the audit trails can be verified, and it has a promise of achieving this through its technology. It may further confidence in data provenance and minimize potential manipulation when it is applied to accounting research. However, according to Maddi and Teixeira da Silva (2024), there are issues of scalability, regulation compliance, and underexperienced with traditional systems that need to be brought to fruition to realize the potential in the accountancy sector of blockchain. There are three points which researchers must consider when incorporating the blockchain tools, and the technological advantages must not be supported by unacceptable social and ecological costs.

Technological advances have brought new forms of ethics to research. In Table 8, we discuss in detail the ethical issues of artificial intelligence, big data, and blockchain technology, as well as the measures that should be taken to avoid risks and adhere to moral norms of this rapidly changing environment (Pabel & Akther, 2025).

Table 8. Ethical Considerations with Emerging Technologies

Technology	Ethical Challenge	Mitigation Strategies	Impact on Research Ethics
Artificial Intelligence	Algorithmic bias; lack of transparency	Audit AI tools; disclose limitations	Risk of reinforcing bias; reduced trust
Big Data	Privacy breaches via data linkage	Anonymization; aggregation; risk assessment	Unintended data exposure
Blockchain	Environmental impact; scalability	Evaluate sustainability; weigh costs	Balancing transparency with ecological responsibility

6.4 Ethical Challenges in Data Visualization and Reporting

In accounting research, ethical considerations go beyond the data collection and analysis to incorporate the means of presenting the findings in a visual format. Maddi and Teixeira da Silva (2024) note that data visualization techniques are the most important in the process of presenting results properly and correctly, yet they also comprise specific ethical dilemmas. Misinterpretation of data can be misleading like graphs which are plotted on truncated axes, selective scaling or distorted graphics plotted on exceedingly small or large scale, respectively. This undermines transparency and can mislead the interpretations of research findings by the readers either intentionally or unintentionally.

The portrayal of data between researchers is an ethical or moral obligation whose accuracy should be veracious and interpretable by others. This also involves the application of uniform scales, no biased exclusion of data points and proper nomenclature and subtext of every figure and graph. According to Hussien, Alshaketheep, and Al-Ahmed (2025), transparent crashes of information increase the level of honesty, as well as effectiveness of sustainability reporting in the research of accounting.

The ethical issues of data visualization need to be considered closer due to the increased usage of digital tools and dashboards. Scholars need to be educated to identify and prevent visualization pitfalls as well as follow good practices that would facilitate honest and accurate reporting (Niel et al., 2024). Ensuring they are upheld helps to preserve the confidence of the public and facilitate good decision-making on the available evidence that is consistent and clearly communicated.

7. RESPONDING TO MODERN ETHICAL DILEMMAS: A DECISION FRAMEWORK

Ethical dilemmas are complex, and, in the environment of the modern rapidly changing research, it can happen in unpredictable and diverse ways. Researchers need to possess a systematic decision-making process that will be consistent, accountable, and reflective to find their way in these dilemmas (Maddi & Teixeira da Silva, 2024). The five-step approach provided will give certain guidelines on how to contextualize descriptive accounting research by addressing ethical issues.

7.1. Identify the Ethical Issue

The initial move in solving any ethical dilemma is to state the problem or the possible risk. In accounting research, the most important thing is to pay close attention to the facts and be ready to acknowledge that a situation is associated with competing values or conflicting responsibilities, which is what Kraus, Mikes, and Véliz (2024) refer to as the conceptualization of accounting as a moral interlocutor that helps to establish reflective equilibrium in organizational decision-making situations. As an example, a researcher may be faced with a dilemma on whether to reveal some sensitive information that may be detrimental to the reputation of a company but is essential in terms of transparency. Awarenessining the ethical aspects involved, it is possible to prevent hasty or uninformed choices and precondition the consideration of ethical issues in accounting, since the awareness of ethical dilemmas in accounting is essential to preserve the trust of the population and guarantee the reliability of financial data (Zhang, 2024).

7.2. Consult Guidelines and Principles

When an ethical problem has been identified, consultation of ethical framework relevant to the problem should follow. This comprises both professional codes of conduct, i.e., AICPA Code of Professional Conduct, and academic and institutional advice (American Institute of Certified Public Accountants, n.d.; Park University, 2024). These codes describe the major principles that should guide most decisions and such principles include integrity, objectivity, confidentiality, and professional conduct. By adhering to them, one can make sure that these actions will be related to the legal demand and accepted ethical norms, which will lead to proper moral and professional decision-making.

7.3. Consider Stakeholders and Consequences

Ethical decision-making involves a broad perspective of the parties that are going to be impacted. Research participants, organizations, fellow scholars, regulators, and the public may be some of the charter stakeholders. A recent study by Owusu and Korankye (2023) uses bibliometric tools to analyze the research on ethical decision-making (EDM) of accountants between 1987 and 2022, pointing out that the study in accordance with ethical decision-making shows development in studies and their focus on both individual and firm-based problems. Both short-term and long effects should be considered to clarify advantages and drawbacks. Incorporating information communication technology (ICT) and artificial intelligence (AI) into auditing could be a revolutionary step in the advancement of auditors in terms of being able to process large amounts of data, detecting possible risks and overall quality of an audit in the era of digitization (Thottoli, 2024). This step motivates researchers to have an integrated approach where competing interests must be affordable.

7.4. Explore Alternative Actions

As Mökander and Floridi (2023) observe, there are hardly any effortless ways out of ethical dilemmas, and discussing various lines of action also assists organizations in cautiously considering the advantages and the hazards that each action can entail. Alternatives can be total disclosure, partial anonymization, post-publication, GR, seeking mediation or further consultation. Such a creative process of solving perspective promotes adaptability and creativity, which are highlighted in modern research concerning creative accounting and external audition where a due criticality to the ethical standards is stressed as the aspect of adjustment to the complex financial operations (AlShanti et al., 2024).

7.5. Make and Justify the Decision

Once analyzed, the researcher should make a clear and well-justified decision which reflects the ethical principles and not convenience or the pressure (Gong, 2017). This determination must clearly be made and recorded in the form of a written explanation and reasonable consultations. The rationale of the choice breeds accountability and may be used whenever a problem arises in future. It is also eager to practice ethical rigor to support the trust among peers, participants, and the public. Through this systematic approach, scholars are more prepared to deal with subtle and dynamic ethical aspects of descriptive research in accounting which eventually leaves their research in the realms of credibility and social usefulness.

8. CASE STUDY: THE HIDDEN DIP IN PROFITABILITY

8.1 Scenario

A doctoral student engaged in the study of descriptive accounting research on a dissertation has uncovered a possible and important decline in profitability in the company under study in the materials of an organizational sponsor. The finding contradicts the official story of the organization whereby the picture is of steady or improving financial performance. The student may be in a severe situation whereby revealing such a discovery will be a risk in the continuity of the relationship with the sponsor which will impact future funds or career prospects. Conversely, it can be a disadvantage on the study if findings are not reported or flattened, as it would not wish to strain the relationship.

Pressure Analysis: This is due to the extreme pressure the student is under due to careers ambition and dependency on funding. The connection with the sponsor would most likely determine the access to future data, good professional references, and even employment, which is why the risk of losing the sponsor becomes a serious risk factor to the career path and dissertation writing of the student.

8.2 Ethical Dilemma

The student poses an ethical dilemma that is grave; the question is whether the student ought to give truthful data and make the sponsor happy about the decline in profitability despite the blowback or neglect data respectively. The professional allegiance conflict, the individual interest in professional growth and the obligation to speak out the truth in the interest of investors, policy and the academic world is brought out in the dilemma. The result of the decision reached will reflect what the researcher intends to his/her core moral values and superiority as far as the descriptive accounting research is concerned.

Underlying Conflict: It is a classic conflict between the Deontological obligation to tell the truth (integrity), and the Utilitarian need to maximize personal good (career stability) and sustain a relationship with the researcher to work in the future (funding/data access).

8.3 Discussion Questions

A deliberate analysis is needed to help resolve this dilemma the clash between the integrity requirement and the career inducement on the doctoral student. The ethical theories, principles, and decision models explored in this chapter are applied to the following questions:

- **Ethical Foundations:** Which ethical principles (Integrity, Transparency, Accountability, Public Interest) are most relevant, and how do they conflict with the pressures of the scenario?
- **Consequences of Misconduct:** What specific violations of research ethics and academic codes could be performed by omitting, changing, or downplaying the finding?
- **Systemic Pressures:** How does the culture of "publish or perish" and face-off competition over funding aggravate the dilemma faced by the student and also can tend to mar objectivity?
- **Conflicts of Interest:** What are the potential conflicts of interest (financial, career-based and academic) that may affect the judgement of the student or cause bias?
- **Candid Communication:** How then should the researcher communicate this issue to the advisor, to have candid conversation, but at the same time avoid conflicting obligations to report the truth and obligations to sponsor agreements?
- **Scholarly Autonomy:** What are some of the practical activities or steps (e.g., detailed audit trails, transparent reporting) that might be evidence of scholarly autonomy and methodological independence?

The ethical theories, principles and decision models that are discussed in this chapter have a practical application in this case study. It makes the researcher go beyond the rules of abstractions and think out to the full extent of the practical consequences of his/her decisions, particularly in the case of high professional pressure.

9. EMERGING TRENDS AND FUTURE CHALLENGES

There is no strict ethics of descriptive accounting research, it thrives over time and with technological advancements, regulatory setting and in social demands. These are new trends that not only offer an opportunity to researchers but also ethical concerns.

9.1 The Rise of Artificial Intelligence and Big Data

On the same note, as AI-driven analytics and machine learning in accounting studies continue to rise, these fields become more complex particularly when financial forecasting, credit risk analysis, and fraud detection are involved (Gao et al., 2024). Even though the use of these tools can handle exceptionally big amounts of data and expose patterns that are impossible to see by human eyeglasses, they equally raise questions that are related to the following:

- **Algorithmic Opacity and the Veneer of Objectivity:** Most AI models are black boxes, and it is hard to understand how decisions or classifications arrive at. This generates a very pernicious issue in accounting research: the statistical, quantitative production of AI gives an illusion of absolute, objective objectivity of the results. The output of the black box could be perceived as undeniable reality by the researchers and obscure a set of implicit biases and promote required human doubt.
- **Bias Amplification:** Predictive algorithms have the potential to preserve any historical bias within financial or organizational data or even increase it. In case historical data is biased (e.g. in loan decisions), the AI will internalize the bias and enhance it, so that the wrong habit would look scientifically reasonable.
- **Data Ownership and Consent:** Since big, integrated data tends to blur the boundaries between data ownership and the inability to secure an informed consent, the privacy and confidentiality principles are challenged.

Future Priority: The researchers are supposed to develop the transparent audit frameworks of AI and adhere to ethical principles, which would support fairness, explainability and accountability in the automated analysis.

9.2 Increasing Globalization and Cross-Border Research

Globalization means data often originates from multiple authorities with differing ethical and legal standards. Challenges include:

- **Varying Privacy Laws:** Regulations like the EU's GDPR, Brazilian LGPD (Brazil, 2018), and emerging data protection acts in other regions create compliance complexity.
- **Cultural Norms and Ethics:** What is considered ethically acceptable in one region may be viewed as invasive or inappropriate in another.

Future Priority: Ethical frameworks must evolve to incorporate context-sensitive guidelines that respect local laws and cultural norms while maintaining universal principles of transparency and integrity.

9.3 Real-Time and Continuous Data Reporting

The shift toward real-time financial reporting and continuous auditing introduces pressure for speed, which can compromise ethical safeguards such as:

- Thorough consent procedures.
- Proper data anonymization.
- Comprehensive disclosure of limitations.

Future Priority: Researchers need to establish real-time ethics procedures, ensuring that enhanced reporting does not sacrifice ethical rigor.

9.4 Sustainability and ESG (Environmental, Social, Governance) Metrics

As organizations integrate ESG factors into their reporting, descriptive research now involves non-financial data such as environmental impact and social responsibility metrics. Ethical questions arise around:

- Data Authenticity and Verification: How to verify the integrity of sustainability disclosures, which are often less regulated and standardized than financial data.
- Greenwashing Risks and Legitimacy: Ensuring research does not unintentionally legitimize misleading corporate narratives (greenwashing). A descriptive study relying uncritically on self-reported, unaudited ESG data risks

providing a stamp of academic approval to potentially misleading corporate claims, thereby misinforming stakeholders and policymakers.

Future Priority: Develop robust ethical standards for ESG research, emphasizing verification, impartiality, and social accountability.

9.5 Cybersecurity and Data Protection

With the development of cloud-based accounting systems and distant research environments, data breaches pose serious ethical risks. Researchers have a duty to:

- Implement strong encryption and access controls.
- Comply with evolving cybersecurity regulations.
- Plan for ethical responses to breaches.

Future Priority: Establish ethical contingency protocols for managing breaches, ensuring transparency and minimal harm to stakeholders.

9.6 The Human Dimension in a Digital Era

While technology directs the conversation, ethical research must remain human-centered. Overreliance on automatic tools risk:

- Reducing professional skepticism.
- Overlooking qualitative context in favor of quantitative outputs.

Future Priority: Encourage critical human oversight, ensuring that automation supplements—not replaces—ethical judgment.

The research in the future of ethical descriptive accounting is found in the adaptive frameworks that would sustain a balance between innovation and responsibility. With AI, big data, and globalization redefining the field of research, scholars should be highly anticipated to foresee ethical dilemmas, establish clear governance systems and principles of fairness, accountability and stakeholder respect.

10. CONCLUSION

Ethical integrity of descriptive research in accounting is not merely an incidental overlay to the research process but rather an integral part as it cannot be separated into the research process, data selection and acquisition, analysis, interpretation, and eventual reporting of the results. These central moral principles such as integrity, transparency, confidentiality, and accountability ensure that the investigators not only preserve the integrity of their research findings but also underpin the credibility that they have argued about in the same by the scientific community, the professionals practicing the discipline, and society.

It would assist in good decision making in case of conflicts or strains, through utilitarianism, deontology, and virtue theory using the major theories of ethics. Better still, the emerging technological landscape alongside all its innovations including AI, big data, and blockchain demonstrate new realities and issues that demand ethical mindfulness at any given moment. The principles are a moral compass and guard to ensure that research does not only add to knowledge but also benefit society.

Limitations

Despite the clear explanation of the ethical considerations, there exist some limitations. The first is that ethical dilemmas occur in subtle cases where strict rules are not easily applicable, it must be judgmental and discretional on the part of the researchers. Second is that institutional and cultural variations can influence interpretations and practical use of ethical standards and dissimilarities may appear during implementation and anticipations between different settings. Third, the frequency of the appearance of the new technologies surpasses the pace of the establishment of the ethical frameworks of their dealing which are currently in place, and the gaps that require continuous innovation and adaptations are still present.

Future Directions

The areas that should be given some more attention in future to strengthen ethics in descriptive accounting research include the following:

- **Development of Adaptive Ethical Frameworks:** Adaptive ethical frameworks should be developed to address emerging risks such as algorithmic bias, data privacy breaches, and so on due to modern technologies and approaches.
- **Enhanced Training and Awareness:** Doctoral and graduate level courses must address training focused on incorporating ethics with the discourse of descriptive accounting studies and offers scholars skills of applying it in practice and evidence-based learning.
- **Cross-Disciplinary Collaboration:** Ethicists, data scientists, legal experts and accounting professionals should be integrated to work together in creating systemic solutions to complex ethical problems.

- **Institutional Support and Oversight:** Universities, professional associations and publishers must intervene to establish ethical norms and provide procedures to define rules and policies, provide explicit system of review and procedures that can deal with ethical breaches.
- **Global Harmonization of Standards:** Harmonization of research ethical expectations, both nationally and internationally, can enhance uniformity and equitability in cross-border research.

To summarize, ethical excellence as a journey is a matter of constant pursuit or interest that involves commitment, reflection, and flexibility in descriptive accounting research. Willingness to accept these principles and preplanning to meet future challenges would help researchers to exist credible, transparent, and socially responsible knowledge base to promote and serve society.

REFERENCES

Abdalla, H. B. (2022). A brief survey on big data: Technologies, terminologies and data-intensive applications. *Journal of Big Data*, *9*(1), 107. DOI: 10.1186/s40537-022-00659-3

Adelakun, B. O., Onwubuariri, E. R., Adeniran, G. A., & Ntiakoh, A. (2024). Enhancing fraud detection in accounting through AI: Techniques and case studies. *Finance & Accounting Research Journal*, *6*(6), 978–999. DOI: 10.51594/farj.v6i6.1232

AlShanti, A. M., Al-Azab, H. A. H., Humeedat, M. M., & AlQudah, M. Z. (2024). Exploring the evolution of creative accounting and external auditors: Bibliometric analysis. *Cogent Business & Management*, *11*(1), 2300500. Advance online publication. DOI: 10.1080/23311975.2023.2300500

American Accounting Association. (n.d.). *Publications ethics for academic research* [Guide for authors of Behavioral Research in Accounting]. Retrieved August 2025, from https://aaahq.org/Research/Journals/Behavioral-Research-in-Accounting/Guide-for-Authors

American Institute of Certified Public Accountants. (n.d.). *AICPA Code of Professional Conduct*. Wikipedia. https://en.wikipedia.org/wiki/AICPA_Code_of_Professional_Conduct

Andersson, H., Svensson, A., Frank, C., Rantala, A., Holmberg, M., & Bremer, A. (2022). Ethics education to support ethical competence learning in healthcare: An integrative systematic review. *BMC Medical Ethics*, *23*(1), 29. DOI: 10.1186/s12910-022-00766-z PMID: 35305627

Bernardi, R. A., & Bean, D. F. (2006). The debate on ethics in accounting education: The views of our forgotten stakeholders. *The CPA Journal*, *76*(7), 56–57. https://ssrn.com/abstract=2250592

Bernardi, R. A., & Bean, D. F. (2006). The growth of accounting research in ethics journals. In Jeffrey, C. (Ed.), *Research on professional responsibility and ethics in accounting* (Vol. 11, pp. 1–28). Emerald Group Publishing., DOI: 10.1016/S1574-0765(06)11003-1

Bérubé, J., & Gendron, Y. (2023). Developing ethical sensitivity in future accounting practitioners: The case of a dialogic learning for final-year undergraduates. *Journal of Business Ethics*, *183*(3), 763–781. DOI: 10.1007/s10551-021-04947-5

Brazil, (2018). *Lei n° 13.709, de 14 de agosto de 2018. Lei Geral de Proteção de Dados Pessoais (LGPD)* [Law No. 13.709, of August 14, 2018. General Law for the Protection of Personal Data (LGPD)]. Diário Oficial da União. Retrieved from https://www.planalto.gov.br/ccivil_03/_ato2015-2018/2018/lei/l13709.htm

Bretag, T., & Green, M. (2014). The role of virtue ethics principles in academic integrity breach decision-making. *Journal of Academic Ethics, 12*(3), 165–177. DOI: 10.1007/s10805-014-9209-z

Carberry, E. J., Engelen, P.-J., & Van Essen, M. (2018). Which firms get punished for unethical behavior? Explaining variation in stock market reactions to corporate misconduct. *Business Ethics Quarterly, 28*(2), 119–151. DOI: 10.1017/beq.2017.46

Cohen, J. R., Pant, L. W., & Sharp, D. J. (1996). A methodological note on cross-cultural accounting ethics research. *The International Journal of Accounting, 31*(1), 55–66. DOI: 10.1016/S0020-7063(96)90013-8

Copyright Office, U. S.U.S. Copyright Office. (2005). U.S. Copyright Office report on legal protection for databases. *Bulletin of the American Society for Information Science, 24*(2), 25–29. DOI: 10.1002/bult.82

Day, M., Turner, G., & Drozdiak, N. (2019, April 11). Thousands of Amazon workers listen to Alexa users' conversations. *TIME.* https://time.com/5568815/amazon-workers-listen-to-alexa/

Dempster, G., Sutherland, G., & Keogh, L. (2022). Scientific research in news media: A case study of misrepresentation, sensationalism, and harmful recommendations. *JCOM, Journal of Science Communication, 21*(1), A06. DOI: 10.22323/2.21010206

Dorn, M. F., Haemapun, H., Jin, Z., Sgandar, T. O., Pillai, D., Sandeep, A., & Yilmaz, A. (2024, September 16). *Ethical considerations in big data analytics*. OxJournal. https://www.oxjournal.org/ethical-considerations-in-big-data-analytics

Eaton, T. V., Hazel, M., Pyzoha, J. S., & Zhang, J. (2023). How do the accounting profession and higher education currently approach ethics training? *Issues in Accounting Education*, •••, 1–18. DOI: 10.2308/ISSUES-2023-104

Eilifsen, A., Messier, W. F., Glover, S. M., & Prawitt, D. F. (2014). *Auditing and assurance services: A systematic approach* (10th ed.). McGraw-Hill Education.

Engel, C. J. (2016). Ethical standards of scientific research for beginning accounting scholars. *Business and Economic Research, 6*(1), 129–136. RePEc:mth:ber888:v:6:y:2016:i:1:p:129-136

European Union. (2016). Regulation (EU) 2016/679 of the European Parliament and of the Council of 27 April 2016 on the protection of natural persons with regard to the processing of personal data and on the free movement of such data (General Data Protection Regulation). *Official Journal of the European Union, L119*, 1–88. https://eur-lex.europa.eu/eli/reg/2016/679/oj

Forbes Technology Council. (2024, May 24). How AI can unlock the power of unstructured data. *Forbes*. https://www.forbes.com/councils/forbestechcouncil/2024/05/24/how-ai-can-unlock-the-power-of-unstructured-data/

Fowler, D. S. (2023). Ethical implications of deceptive earnings management practices. *Business Ethics and Leadership*, *7*(3), 84–96. DOI: 10.61093/bel.7(3).84-96.2023

Galvez, R. (2024, January 16). Data governance trends for 2024. *Precisely*. https://www.precisely.com/datagovernance/data-governance-trends-for-2024/

Gao, H., Kou, G., Liang, H., Zhang, H., Chao, X., Li, C.-C., & Dong, Y. (2024). Machine learning in business and finance: A literature review and research opportunities. *Financial Innovation, 10*(1), 86. DOI: 10.1186/s40854-024-00629-z

Gelman, M. (2020). Kantian ethics—duty, reason, and the grounding of moral law. *Essays in the Philosophy of Humanism, 32*, Article 7. https://americanhumanist.org/what-we-do/publications/eph/journals/volume32/gelman/

Ghosh, A. (2025). Navigating ethical dilemmas at the workplace: A study of Indian professional management accountants in business. *Journal of Human Values*, *0*(0), 09716858251352091. Advance online publication. DOI: 10.1177/09716858251352091

Gong, J. (2017). Ethics in accounting: A decision-making approach. *Journal of Business Ethics*, *142*(3), 621–623. DOI: 10.1007/s10551-016-3386-1

Gupta, M. (2025). Cross-cultural influences in international accounting: A comparative study. *International Journal of Academic Excellence and Research*, *01*(01), 14–20. DOI: 10.62823/MGM/IJAER/01.01.37

Guragai, B., Hunt, N. C., Neri, M. P., & Taylor, E. Z. (2017). Accounting information systems and ethics research: Review, synthesis, and the future. *Journal of Information Systems*, *31*(2), 65–81. DOI: 10.2308/isys-51265

Hassan, S., Pandey, S., & Pandey, S. K. (2021). Should managers provide general or specific ethical guidelines to employees: Insights from a mixed methods study. *Journal of Business Ethics*, *172*(3), 563–580. DOI: 10.1007/s10551-020-04442-3

Hussien, L. F., Alshaketheep, K., Al-Ahmed, H., Shajrawi, A., Alghizzawi, M., Zraqat, O., & Deeb, A. (2025). The impact of big data analytics on the sustainability reports quality. *Discover Sustainability*, *6*(1), 776. DOI: 10.1007/s43621-025-01678-9

Ikezuruora, C. (2023). *Unveiling the truth: Why transparency is essential in the age of big data*. PrivacyEnd. https://www.privacyend.com/transparency-essential-age-big-data/

International Ethics Standards Board for Accountants. (2023, July 12). *IESBA emphasizes the critical importance of ethical behavior for all professional accountants*. International Federation of Accountants. https://www.ethicsboard.org/news-events/2023-07/iesba-emphasizes-critical-importance-ethical-behavior-all-professional-accountants

Jejeniwa, T. O., Mhlongo, N. Z., & Jejeniwa, T. O.Temitayo Oluwaseun Jejeniwa-Noluthando Zamanjomane MhlongoTitilola Olaide Jejeniwa. (2024). The role of ethical practices in accounting: A review of corporate governance and compliance trends. *Finance & Accounting Research Journal*, *6*(4), 707–720. DOI: 10.51594/farj.v6i4.1070

Kang, E., & Hwang, H.-J. (2023). The importance of anonymity and confidentiality for conducting survey research. *Journal of Research and Publication Ethics*, *4*(1), 1–7. DOI: 10.15722/JRPE.4.1.202303.1

Kraus, K., Mikes, A., & Véliz, C. (2024). Bringing morality back in: Accounting as moral interlocutor in reflective equilibrium processes. *Accounting, Organizations and Society*, *113*, 101570. Advance online publication. DOI: 10.1016/j.aos.2024.101570

Lane, S. (2024, October 14). *The new era of ethical challenges for professional accountants*. ACCA Professional Insights. https://www.accaglobal.com/us/en/professional-insights/global-profession/ethical-challenges.html

Maddi, A., & Teixeira da Silva, J. A. (2024). Beyond authorship: Analyzing contributions in PLOS ONE and the challenges of appropriate attribution. *Journal of Data and Information Science*, *9*(3), 88–115. DOI: 10.2478/jdis-2024-0015

Mardawi, Z., Seguí-Mas, E., & Tormo-Carbó, G. (2021). Rethinking the accounting ethics education research in the post-COVID-19 context. *Cogent Business & Management*, *8*(1), 1984627. Advance online publication. DOI: 10.1080/23311975.2021.1984627

McIntosh, L. D., & Vitale, C. H. (2023). *Safeguarding scientific integrity: Examining conflicts of interest in the peer review process*. ArXiv preprint arXiv:2308.04297. https://doi.org//arXiv.2308.04297DOI: 10.48550

Mintz, S. M. (2021). *Accounting Ethics Education.* The CPA Journal. https://www.cpajournal.com/2021/10/19/accounting-ethics-education/

Misra, D. P., Ravindran, V., & Agarwal, V. (2018). Integrity of authorship and peer review practices: Challenges and opportunities for improvement. *Journal of Korean Medical Science, 33*(46), e287. DOI: 10.3346/jkms.2018.33.e287 PMID: 30416407

Misselbrook, D. (2013). Duty, Kant, and deontology. *The British Journal of General Practice, 63*(609), 211. DOI: 10.3399/bjgp13X665422 PMID: 23540473

Mixon, K. (2024). The role of virtue ethics in modern moral dilemmas. *International Journal of Philosophy, 3*(4), 14–28. https://www.researchgate.net/publication/382246077_The_Role_of_Virtue_Ethics_in_Modern_Moral_Dilemmas#:~:text=Ultimately%2C%20virtue%20ethics%20was%20found,more%20just%20and%20sustainable%20world. DOI: 10.47941/ijp.2094

Mökander, J., & Floridi, L. (2023). Operationalising AI governance through ethics-based auditing: An industry case study. *AI and Ethics, 3*(3), 451–468. DOI: 10.1007/s43681-022-00171-7 PMID: 35669570

Niel, S., Florence, A., John, A., & Berky, B. (2024). Investigating the effectiveness of data visualization in communicating financial information. *Journal of Financial Reporting and Analysis, 9*(2), 45–62.

Nørreklit, L., Nørreklit, H., Cinquini, L., & Mitchell, F. (2024). Accounting for a better world: Towards a conceptual framework to enable corporate reporting to contribute to the sustainability of the good life. *Meditari Accountancy Research, 32*(5), 1608–1640. DOI: 10.1108/MEDAR-05-2023-2015

Okougbo, P. O., Okike, E. N., & Alao, A. (2021). Accounting ethics education and the ethical awareness of undergraduates: An experimental study. *Accounting Education, 30*(3), 258–276. DOI: 10.1080/09639284.2021.1888135

Otley, D. T., & Berry, A. J. (1994). Case study research in management accounting and control. *Management Accounting Research, 5*(1), 45–65. DOI: 10.1006/mare.1994.1004

Owusu, G. M. Y., & Korankye, G. (2023). The state of ethical decision-making research in accounting: A retrospective assessment from 1987 to 2022. *Business Ethics, the Environment & Responsibility, 32*(2), 419–434. DOI: 10.1111/beer.12519

Pabel, A. S., & Akther, W. (2025). Navigating the ethical landscape: A systematic review of challenges in AI adoption within the accounting profession. *Multidisciplinary Reviews, 8*(11), 2025357. DOI: 10.31893/multirev.2025357

Park University. (2024, September 23). *Ethical considerations in accounting: Navigating the gray areas*. https://www.park.edu/blog/ethical-considerations-in-accounting/

Parks-Leduc, L., Mulligan, L., & Rutherford, M. A. (2021). Can ethics be taught? Examining the impact of distributed ethical training and individual characteristics on ethical decision-making. *Academy of Management Learning & Education*, *20*(1), 30–49. DOI: 10.5465/amle.2018.0157

Parry, O., & Mauthner, N. S. (2004). Whose data are they anyway? Practical, legal and ethical issues in archiving qualitative research data. *Sociology*, *38*(1), 139–152. DOI: 10.1177/0038038504039366

Pham, V. A. T., Nguyen, L. A., Dellaportas, S., Phan, D. H. T., & Nguyen, Q. H. (2025). How does big data analytics impact accounting manipulation? *Accounting and Finance*, *65*(3), 2918–2934. Advance online publication. DOI: 10.1111/acfi.70024

Poje, T., & Groff, M. Z. (2021). Mapping ethics education in accounting research: A bibliometric analysis. *Journal of Business Ethics*, *179*(2), 451–472. DOI: 10.1007/s10551-021-04846-9 PMID: 34127872

Poje, T., Ličen, M., & Groff, M. Z. (2025). Who enrolls in accounting? Ethical predispositions of undergraduate accounting students. *Journal of Accounting Education*, *71*, 100966. Advance online publication. DOI: 10.1016/j.jaccedu.2025.100966

Sayed, R. (2023). Ethical considerations in the use of big data for business analytics: A cross-industry perspective. *Journal of Innovations and Sustainability*, *7*(4), 08. DOI: 10.51599/is.2023.07.04.08

Schreyer, M., Sattarov, T., & Borth, D. (2022). Federated and privacy-preserving learning of accounting data in financial statement audits. *Proceedings of the Third ACM International Conference on AI in Finance*, 105–113. DOI: 10.1145/3533271.3561674

Taylor, G., Cumming, D., Li, D., & Shi, J. (2025). The role of legal, regulatory and political reforms and institutions in influencing accounting and financial outcomes. *The British Accounting Review*, *57*(3), 101647. DOI: 10.1016/j.bar.2025.101647

Thottoli, M. M. (2024). Leveraging information communication technology (ICT) and artificial intelligence (AI) to enhance auditing practices. *Accounting Research Journal*, *37*(2), 134–150. DOI: 10.1108/ARJ-09-2023-0269

Vinson, J., Curtis, M., Conover, T., & Chui, L. (2020). Ethical relativism in accounting: A cross-cultural examination of the influence of culture and risk-taking propensity on ethical decision-making. *Journal of International Accounting, Auditing & Taxation*, *41*, 100350. DOI: 10.1016/j.intaccaudtax.2020.100350

Winterberg, C. A., Brummel, B. J., Ayres, T. B., & Kerr, A. J. (2021). Applying ethical dilemmas to professional ethics training and education: Practical implications and future directions. *Industrial and Organizational Psychology: Perspectives on Science and Practice*, *14*(3), 325–330. DOI: 10.1017/iop.2021.73

Zhang, W. (2024). Ethical dilemmas in accounting: A comprehensive analysis of professional ethics. *Academic Journal of Business & Management*, *6*(2). Advance online publication.

APPENDIX: ETHICAL SELF-ASSESSMENT CHECKLIST FOR DESCRIPTIVE ACCOUNTING RESEARCH

This checklist is a useful guide to the researchers to establish their work compliments with the utmost ethical standards of descriptive accounting research. It is possible to be done in research design, collection of data and reporting.

Table 9. Ethical Self-Assessment Checklist

Ethical Step	Yes	No	Notes
Transparency and Disclosure			
Have all data sources been fully disclosed?	☐	☐	
Have potential conflicts of interest been declared?	☐	☐	
Are funding sources and sponsorships clearly stated?	☐	☐	
Informed Consent and Privacy			
Has informed consent been obtained from participants?	☐	☐	
Are participants aware of how their data will be used?	☐	☐	
Has data been anonymized where needed?	☐	☐	
Are data storage and security measures adequate?	☐	☐	
Integrity and Accuracy			
Are limitations and assumptions clearly stated?	☐	☐	
Is the interpretation of results free from bias?	☐	☐	
Have errors or inconsistencies been disclosed?	☐	☐	
Compliance and Accountability			
Does the study comply with institutional review board (IRB) or ethics committee guidelines?	☐	☐	
Have applicable legal and regulatory requirements been followed (e.g., data protection laws)?	☐	☐	
Is there a clear process for addressing ethical breaches if identified?	☐	☐	

Chapter 9
Ethical Dilemmas in Social Science Research: Case Studies and Solutions

Andi Asrifan
http://orcid.org/0000-0002-9934-6129
Universitas Negeri Makassar, Indonesia

Mardyawati Mardyawati
Universitas Islam Makassar, Indonesia

Ranita Rope
Universitas Muhammadiyah Maluku Utara, Indonesia

Adi F. Mahmud
Universitas Muhammadiyah Maluku Utara, Indonesia

Suratman Dahlan
Universitas Khairun, Indonesia

EXECUTIVE SUMMARY

This chapter discusses social science researchers' ethical dilemmas as scientists and social agents. Social scientists must provide correct data and ethical issues arising from their involvement in social movements or politically sensitive situations. Insufficient studies have examined how researchers manage ethical issues in practice, especially when their findings contradict existing structures or affect disadvantaged people. This article examines how deontological, utilitarian, and virtue ethics apply to real-world ethical situations. Our findings show that research ethics must balance scientific rigor and ethical responsibility. The chapter suggests that researchers need stronger frameworks and training to overcome ethical issues and

DOI: 10.4018/979-8-3693-8302-5.ch009

ensure research integrity and social relevance. These findings join conversations on ethical research practices and their effects on study validity and researcher and participant well-being.

INTRODUCTION TO ETHICAL DILEMMAS IN SOCIAL SCIENCE RESEARCH

1.1 The Ethical Tensions in Social Science Research

The intricate interaction between researchers' dual responsibilities as scientific investigators and social agents raises ethical issues in social science research (Drolet et al., 2023; Brown and colleagues, 2020; Sartoretto & Custódio, 2020. Social scientists are dedicated to revealing and presenting factual information from meticulous empirical data. Nevertheless, their obligations extend beyond that; they are also expected to behave in manners that mirror the ideals and standards of society. These double responsibilities give rise to inherent conflicts, particularly when researchers are engaged in social movements or work in politically delicate settings. For instance, they may have ethical quandaries concerning their moral responsibilities to reveal certain facts that could harm particular groups or maintain the privacy of susceptible populations. Moreover, ethical quandaries frequently arise when researchers' discoveries question established political and economic systems, suggesting their involvement in processes of social regulation (Bhatt, 2022; Hota et al., 2023; Williams et al., 2020. These ethical dilemmas are theoretical and have significant consequences for the probity of social science research and the personal integrity of the researchers.

Social science research, which centers on human behavior, social systems, and cultural contexts, possesses intrinsic ethical challenges due to its distinctive nature (Bhatt, 2022; Pritchett et al., 2022; Henrickson et al., 2020). Furthermore, the field's epistemological focus on comprehending social realities influenced by power dynamics and methods of social control intensifies this complexity.

Ethical quandaries have become increasingly prominent in modern circumstances, especially in the years after the events of 9/11. Academic researchers may encounter ethical dilemmas when selecting research subjects that may result in adverse outcomes, employing potentially misleading or intrusive techniques, and managing the unexpected repercussions of their research discoveries. When the research overlaps with heightened security concerns or involves politically heated themes, these dilemmas become very difficult to resolve since the boundaries between ethical research practice and societal control become invisible. Hence, social scientists face a complex ethical terrain that requires a meticulous equilibrium between scientific rigor and ethical obligation (McGowan, 2020; Burgess et al., 2023; Shilton et al., 2021).

While ethical issues in social science research have been extensively studied, there is still a significant lack of knowledge on how these challenges directly affect the validity and integrity of study findings and the professional and personal lives of researchers. Current scholarly works often concentrate on either the ethical principles that support research methodologies or the pragmatic challenges encountered during research activity.

Nevertheless, there is a dearth of thorough investigation of how social scientists manage these challenges in practical environments, their ethical compromises, and the possible ramifications of these choices on the research process and the general acceptance of their work. Addressing this gap is essential to building a more sophisticated understanding of ethical decision-making in social scientific research.

Through an analysis of both the theoretical foundations and practical reality of ethical issues in social science research, this paper seeks to close this gap. Our discussion encompasses prominent ethical theories such as deontological ethics, which centers on obligations and principles; utilitarianism, which highlights the outcomes of activities; and virtue ethics, which takes into account the character and intentions of the researcher. Through the analysis of case studies and real-world situations, we illustrate the application of different ethical frameworks in resolving intricate challenges. Our analysis demonstrates that ethical decision-making in social science research is a fluid process that necessitates the careful consideration of conflicting ethical principles with practical limitations. The primary objective of this study is to enhance the ongoing discussion on ethical research procedures by offering valuable perspectives on how social scientists might better manage ethical dilemmas, therefore assuring the ethical integrity and social responsibility of their findings. Building on these foundational ethical responsibilities, we now explore specific ethical frameworks that guide decision-making in research

THEORETICAL FRAMEWORKS IN RESEARCH ETHICS

2.1 Key Ethical Frameworks for Decision-Making in Research

Understanding and solving ethical issues in social science research require theoretical frameworks (Taquette & Borges, 2022; Hosseini et al., 2022; Putra, 2023). These frameworks provide systematic methods for examining and resolving difficult ethical concerns, ensuring moral and professional research practices. Social science researchers employ deontological, utilitarian, and virtue ethics to make ethical decisions and conduct their research. With these dilemmas in mind, we now explore ethical frameworks that guide balancing responsibilities

Immanuel Kant's deontological ethics emphasizes morality (Fardid et al., 2023). This concept holds that activities are right or wrong regardless of their results. Deontological ethics demands that social science researchers get informed permission, maintain confidentiality, and avoid deception (Ess, 2020; Nii & Ogbewe, 2023; Knöpfle et al., 2024). A deontological approach would deem a study unethical if it deceived or harmed participants without consent, even if the findings could benefit society. This ethical perspective is crucial to research integrity and compliance with universal ethical standards.

Jeremy Bentham and John Stuart Mill's utilitarianism bases morality on outcomes, particularly happiness and pain (Savulescu et al., 2020). Utilitarian research ethics considers an action ethical if it benefits the most people. A study with minimal risk to a small group of volunteers may be ethical if it could benefit a larger community. Utilitarianism allows for ethical flexibility when weighing competing interests or assessing the pros and cons of alternative research methods. However, this approach may justify compromising the rights or well-being of a few for the benefit of many, making it contentious.

Aristotelian virtue ethics emphasizes the researcher's character and morality rather than laws or consequences (Darr, 2020; Karimova et al., 2020; Queloz & van Ackeren, 2024). According to this concept, being a “good” researcher requires honesty, empathy, integrity, and justice. Virtue ethics is crucial in social science research, as researchers face ethical dilemmas and complicated social situations (Baerøe et al., 2022; Poulton, 2023; Robinson et al., 2022). Virtue-ethical researchers prioritize participant trust, autonomy, and cultural sensitivity. This ethical framework helps researchers to develop a strong moral code that guides their research.

These ethical theories offer different viewpoints yet are compatible. Researchers use many frameworks to solve complicated ethical problems. For example, a researcher may use deontological principles to meet ethical requirements, utilitarian reasoning to evaluate their work's impact, and virtue ethics to retain personal integrity and participant trust. Social science researchers can address ethical issues using deontological, utilitarian, and virtue ethics. These theoretical frameworks help ensure ethical research and advance respectful, just, and socially responsible knowledge. By considering these frameworks, researchers can better manage ethical issues and improve research quality and impact.

HISTORICAL PERSPECTIVES ON RESEARCH ETHICS

Historical events and cultural transformations have affected social science research ethics by emphasizing participant protection and ethical integrity. Understanding these historical viewpoints helps explain present human subject research ethics.

Unethical research techniques like the Tuskegee Syphilis Study (1932-1972) established the necessity for specific ethical principles in the early to mid-20th century (Paunova, 2023; Flynn, 2021). This study left African American men with syphilis untreated to track the disease without their consent. Outrage over such ethical transgressions led to the creation of foundational ethical norms. In reaction to Nazi medical experiments during World War II, the Nuremberg Code (1947) established voluntary consent and the duty to avoid unnecessary suffering (Cosman, 2023; Aurey, 2020).

Reforms in the mid-20th century According to Petkov et al. (2022) and Resnik & Hecking (2020), the 1964 Declaration of Helsinki strengthened ethical norms by emphasizing informed consent and oversight in human subject research. In this time, Institutional Review Boards (IRBs) were established to assure ethical research ideas. These advancements were especially important for social sciences, which often study sensitive themes and vulnerable individuals and require careful ethical considerations to avoid harm and maintain justice.

The 1979 Belmont Report established three core research ethics principles—respect for humans, beneficence, and justice—that continue to drive ethical practices today (Nagai et al., 2022; Bitter et al., 2020). Research methods and technologies have changed, requiring ethical rules to handle digital privacy, data security, and cultural sensitivity. Modern ethical standards prioritize participant rights and research's social impact.

Research ethics evolved over time, becoming the basis for social science research ethics. Researchers can better handle ethical issues and ensure their work is ethical, respectful, and socially responsible by reflecting on these historical viewpoints.

REGULATORY FRAMEWORKS AND GUIDELINES IN SOCIAL SCIENCE RESEARCH

Regulations and guidelines are essential for ethical social science research. These frameworks protect study subjects, maintain research integrity, and ensure ethical compliance. Regulatory organizations, ethical guidelines, and institutional procedures have evolved to control social science research ethics, reflecting changing ethical norms and societal expectations.

The Institutional Review Board (IRB), also known as Research Ethics Committees (RECs) or Human Research Ethics Committees (HRECs), is a key regulatory mechanism in social scientific research (Resnik, 2021; Serpico, 2024; Lynch et al., 2022). Research institutes have IRBs to examine human subject research proposals for ethical compliance. IRBs assess the risks and benefits of planned study, get informed consent, and supervise the research process for ethical compliance to protect research participants. Federal regulations like the Common Rule (45 CFR 46 in the US) govern IRB evaluation of human subject research (Corneli et al., 2021). These standards mandate minimal participant risks, recorded informed permission, and privacy and confidentiality protection.

IRBs and other professional organizations have ethical standards of conduct for researchers (Ferretti et al., 2020; Corneli, 2021). These discipline-specific codes include research ethics issues like participant permission, confidentiality, and data management. The American Sociological Association and American Psychological Association have standards of conduct that promote confidentiality, informed consent, and participant safety. These ethical guidelines guide social science researchers through complicated ethical challenges and help them make ethical decisions that meet professional standards.

International and national rules dictate ethical principles as well as institutional and professional associations (Brenner et al., 2023; MacArthur, 2020). The Declaration of Helsinki and the Belmont Report shaped global ethics. The Declaration of Helsinki, developed by the World Medical Association, emphasizes informed consent and ethical evaluation of medical and social research procedures. The 1979 Belmont Report, which outlines three core ethical principles—respect for persons, beneficence, and justice—guides ethical guidelines across many fields, including social sciences (Nagai et al., 2022; Serpico, 2024).

Social science research ethics and regulations change with study methods and technologies. Digital data collecting and big data analytics have raised ethical questions about data privacy, security, and informed permission in online environments (Favaretto et al., 2020). The regulatory frameworks progressively address these developing challenges, ensuring that research procedures keep pace with technical and societal advances. Cross-cultural and vulnerable population research increasingly emphasizes ethical considerations, requiring researchers to use culturally sensitive and context-specific ethical norms.

Social science research must follow ethical and regulatory requirements (Head, 2020; Stilgoe et al., 2020). IRBs, ethical codes of conduct, and international guidelines protect study participants, maintain research integrity, and build public trust in social science research. As social science research evolves, these regulating mechanisms must adapt to new ethical concerns to keep it ethical and relevant. By following these criteria, researchers can progress knowledge while preserving the highest ethical standards.

INFORMED CONSENT AND PARTICIPANT AUTONOMY

Social science research ethics require informed consent, transparency, and participant autonomy (Tamminen et al., 2021; Akpa-Inyang & Chima, 2021). It entails giving potential participants detailed information about the research purpose, procedures, risks, and benefits to help them decide to participate. Legal formalities aside, informed consent is a moral imperative to respect autonomy and protect individuals from harm.

Participant information, comprehension, and voluntariness are key to informed consent (Pietrzykowski & Smilowska, 2021). To ensure permission without coercion, researchers must present clear, accessible information and evaluate comprehension. Working with vulnerable populations or in hierarchical environments has power imbalances, making this crucial. Respecting participant autonomy includes continuing communication and permission, especially in long-term investigations or when the research scope changes (Wallace & Miola, 2021). Cultural sensitivities and local conventions must be respected to ensure the consent process is relevant in varied circumstances.

Informed consent and methodological requirements of covert observations or deception studies are difficult to balance (Podschuweit, 2021). After debriefing participants, ethical guidelines recommend seeking retrospective agreement to utilize the data. Authentic, voluntary, and informed consent that respects participant autonomy is crucial to ethical research, building trust, and protecting participant rights and dignity. Ethical research hinges on informed permission, which requires careful consideration of ethical imperatives, cultural settings, and social scientific power dynamics.

CONFIDENTIALITY AND ANONYMITY IN SOCIAL SCIENCE RESEARCH

Confidentiality and anonymity are crucial ethical norms in social science research to preserve participants' privacy and maintain data integrity (Moriña, 2021). The researcher must keep participant data confidential. However, anonymity requires concealing participants' names, even from researchers, or removing identifying information from data. These principles are essential for fostering researcher-participant trust in sensitive study domains like personal, societal, and political topics. Researchers must use secure storage, encrypted communications, and restricted access to sensitive material to maintain secrecy (Hassan et al., 2022). This is essential to preventing social shame, legal issues, and psychological distress from illegal personal information publication.

Data is more private when participants are anonymous. In tiny or close-knit groups, demographic facts may indirectly reveal identities, making anonymity difficult (Lowe & Ridgway, 2023; Ho, 2024). Researchers must carefully examine anonymization and eliminate or generalize indirect identifiers.

Data sharing and storage make confidentiality and anonymity difficult in the digital age (Keshta & Odeh, 2021; White et al., 2022). Researchers must keep up with data protection laws and technologies to protect participant data. Legal or organizational demands can conflict with confidentiality, such as when law enforcement requires study data.

In conclusion, ethical social science research requires confidentiality and anonymity. To ensure participant privacy and study integrity, researchers must combine ethical commitments with practical factors and regularly review their processes. These criteria ensure participant safety, boosting study validity and trustworthiness.

ETHICAL ISSUES IN DATA COLLECTION AND ANALYSIS

Adhering to ethical principles in data gathering and analysis is crucial for carrying out responsible social science research (Hosseini et al., 2022; Nii & Ogbewe, 2023). Researchers must adhere to ethical norms in all aspects of data collection, processing, and interpretation to safeguard the well-being of participants and maintain the integrity of the research findings.

During the process of collecting data, ethical quandaries can emerge concerning the selection of subjects, obtaining informed consent, and assessment of potential harm. Researchers must meticulously select sampling techniques that are equitable and inclusive while taking care to avoid any biases that may skew the research results. Effective communication of the study's objectives, the nature of involvement, and any possible hazards is crucial for researchers to adhere to the principle of informed consent (Klykken, 2022; Pietrzykowski & Smilowska, 2021). Researchers must navigate the dilemma of reconciling inconspicuous techniques with the ethical obligation to safeguard participants' rights in situations where clandestine data gathering is required, such as in investigations of sensitive or unlawful activities.

Statistical analysis also poses ethical dilemmas, namely with the precise and truthful depiction of results (Taquette & Borges, 2022; Kyngäs et al., 2020). Researchers must refrain from manipulating data to obtain specific results, as this represents a violation of scientific integrity. Transparency in data management, which involves recognizing constraints and possible biases, is crucial for preserving confidence and credibility. Furthermore, ethical concerns encompass the use of software and algorithms in data analysis, which have the potential to introduce biases if not meticulously supervised and verified.

Furthermore, ethical dilemmas may arise from the power dynamics within the relationship between researchers and participants, especially in qualitative research, where the researcher's interpretation holds considerable importance. Researchers must actively acknowledge their biases and take measures to prevent any misrepresentation or exploitation of participants' experiences in their analysis (Ivey, 2023). Safeguarding the privacy of participants throughout data analysis is of utmost importance, especially when dealing with sensitive or identifiable data.

Ultimately, ethical standards in the gathering and analysis of data are crucial for the conscientious execution of social scientific research. Researchers must ensure constant vigilance and introspection, strictly following ethical principles and procedures that foster openness, respect, and impartiality throughout the study. By employing this method, the safety of participants is ensured, and the reliability and validity of the research results are improved.

7.1. Privacy and Data Security

In an age of data breaches and personal data exploitation, social science research ethics must incorporate privacy and data security (Labrecque et al., 2021; Rafiq, 2022). Participant privacy involves protecting their identity and sensitive data during the study. Data security protects data from unwanted access, loss, or tampering so researchers can control it.

Privacy begins with data collection, storage, and management (Wieringa et al., 2021). Researchers need informed consent to use, preserve, and distribute data and protect participants' privacy. This includes anonymizing data whenever possible, especially in studies with vulnerable populations or sensitive themes, to prevent participant identification. Accidental sharing of anonymized data requires vigilance.

Data security protects data integrity and confidentiality by technical and administrative measures (Thapa & Camtepe, 2021). This comprises safe servers and databases, data encryption, and frequent security audits to uncover and fix flaws. Researchers should provide authorized workers with access to sensitive data only and securely dispose of it when no longer needed. Online surveys and social media data are more vulnerable to breaches and unlawful access in the digital age (Jain et al., 2021).

Cloud storage and third-party data processing raise privacy and security concerns (Razaque et al., 2021). Researchers must carefully evaluate these platforms' security and ethical and legal compliance to secure participant data. Legal or institutional requirements for secret data may conflict with participant pledges, creating ethical issues. Privacy and data security are legal and ethical obligations to respect and safeguard participants. Social science researchers preserve trust and integrity by strictly adopting data protection measures and being transparent about their processes, ensuring participant safety.

ETHICAL RESPONSIBILITIES OF RESEARCHERS

Researchers in social sciences have significant ethical obligations beyond informed permission and data protection (Moriña, 2021; Favaretto et al., 2020). These duties include integrity, respect, and fairness in all research processes to protect participants' rights, dignity, and welfare. Avoiding participant damage is a key ethical obligation in research (Taquette & Borges, 2022). This covers bodily and psychological injury and protects study participants from social or economic consequences. Researchers must carefully design research to reduce hazards and ensure that benefits outweigh costs (Favaretto et al., 2020; Jamrozik et al., 2021). This requires sensitivity to cultural contexts and power dynamics that may affect participant consent and experience. Research honesty and transparency are also important. Reporting study findings honestly, acknowledging limits, and avoiding plagiarism and data manipulation are examples. Researchers must report data honestly without modifying them to meet hypotheses or external pressures (Elaad, 2022; Bishop, 2020). Integrity ensures authentic and dependable knowledge that benefits the field and society.

Ethics require researchers to respect participant autonomy and privacy. This includes obtaining informed consent, preserving participants' right to withdraw from the study without penalty, and delivering the promised secrecy and anonymity. In an age of digital vulnerabilities, researchers must protect participants' data and follow data protection rules and best practices (Shilton et al., 2021; Andrew & Baker, 2021).

Justice and inclusion in research are also ethical obligations. Researchers must avoid exploiting vulnerable people and properly spread research gains and liabilities (Bitter et al., 2020; Saleh, 2020). Be careful not to utilize research findings to reinforce preconceptions or injustices. In conclusion, researchers' ethical duties are complex and need the highest ethical standards. By following ethical guidelines, researchers maintain the trust of participants and the community, ensuring that social science research advances knowledge while protecting human dignity and rights.

CASE STUDIES IN ETHICAL DILEMMAS

Social science ethical concerns are best explored through case studies. These case studies present real-world ethical dilemmas that require careful navigation. They demonstrate the complexity of ethical decision-making and the difficulties researchers encounter in combining ethics and practicality.

Informed permission and participant autonomy are common ethical issues (Taquette & Borges, 2022; Laurijssen et al., 2022). In a study on sensitive themes like domestic violence or mental health, full disclosure of the study's purpose may affect participants' willingness to participate or cause distress. The ethical requirement to provide full information and the desire to avoid harm or distress force researchers to make tough judgments about how much to disclose and when to debrief.

As longitudinal research collects more precise and individualized data, confidentiality breaches grow more difficult. Researchers must balance longitudinal data richness with participant privacy (Sperling, 2022; Di Feliciantonio, 2024). Collecting fresh, potentially identifiable data demands ongoing vigilance and maybe renegotiating consent. Analysis and distribution of research findings often present ethical issues (Newman et al., 2021) (Head, 2020; Pietilä, 2020). A case may involve the unintended repercussions of releasing findings stigmatizing a community. Researchers must balance scientific progress with community harm. They may need to change the presentation of the findings or consult community stakeholders to assess the consequences and address objections. These case studies emphasize ethical reflexivity—reflecting on one's ethical commitments and research decisions' potential effects. Ethical decision-making frameworks like the IDEA model help researchers discover ethical challenges, assess their implications, explore alternatives, and act ethically. These case studies help researchers and students understand social science research ethical concerns and build the skills to handle them properly.

9.1. Case Study 1: Informed Consent Violation

Case Study 1 addresses a major ethical issue in social science research: informed consent violations. This study examined military academy student experiences and recruited worldwide conference volunteers for interviews. After promising confidentiality and anonymity, the research initiative violated military policy (Bray, 2020; Bloemraad & Menjívar, 2022). The initiative continued after agreeing to limit certain portions of the analysis, particularly those critical of academic culture.

This instance illustrates ethical issues regarding informed consent. First, changing the study scope without alerting participants broke the confidence between researchers and participants. Under institutional pressure, changing the conditions under which volunteers agreed to a study breaches voluntary and informed consent. The increased constraints and potential biases in the research did not allow participants to reconsider their participation. Second, the case shows the ethical challenge of institutional demands versus research integrity. Military rules forced the researchers to change their goals, raising problems regarding academic independence and how researchers might balance external requirements with ethics.

Researchers must be clear about research design adjustments and obtain fresh consent to overcome such issues (West & Schill, 2022; Rahman et al., 2021). Ethical review boards assess such cases to ensure researchers stay ethical despite external influences. This instance emphasizes the significance of informed consent throughout the research process and ethical integrity despite external challenges. Researchers learn the importance of protecting participant rights and transparency throughout the research process.

9.2. Case Study 2: Data Falsification

Case Study 2 involves data manipulation by Samantha, a first-year PhD student, who was under pressure to achieve project deadlines on a privately sponsored study on urban secondary school student learning. Samantha found data inaccuracies that could delay project completion due to stress and workload (Parker, 2023; Edmondson, 2023). She changed the data to achieve the required results in a desperate attempt to rescue the research and her academic position.

This instance raises the ethical issue of data manipulation, a major research integrity breach. Data fabrication impairs research credibility and public, financial, and researcher trust in the scientific community (Haven et al., 2022). Samantha may reach incorrect conclusions that influence policy and education, creating widespread harm.

Samantha's tight deadlines, personal challenges, and lack of support show how academic structural flaws can lead to unethical actions. This raises the question of whether institutions and supervisors should encourage and mentor early-career researchers to avert ethical lapses. This issue requires research organizations to promote ethical behavior and provide support for researchers to manage stress and ethical issues (Drolet et al., 2023). Strong research ethics and integrity training and support systems can reduce pressures that could lead to misconduct. A culture of openness and transparency, where mistakes may be shared and corrected without penalties, is also necessary for ethical behavior. This case emphasizes the necessity of preventing data manipulation and researchers' and institutions' ethical obligations. It reinforces scientific endeavor and society's trust by emphasizing the importance of ethical research in all areas.

9.3. Case Study 3: Ethical Implications in Emerging Areas of Artificial Intelligence and Social Media Research

Technological innovations create new ethical issues, especially in AI and social media research. These domains provide extraordinary study potential but contain considerable ethical dangers that social science researchers must overcome.

1. **Artificial Intelligence and Machine Learning**

AI-driven social science research presents significant ethical problems about data privacy, prejudice, and accountability, yet it also offers valuable findings. Machine learning models require massive volumes of personal data, frequently without agreement, raising privacy and consent concerns. AI algorithms may also accentuate biases in their training data, resulting in discriminating results. Thus, AI research ethics must incorporate data protection, algorithmic transparency, and bias detection and mitigation.

2. **Social Media Research**

Social media platforms provide useful data on social habits, trends, and public opinion. This research raises ethical issues about privacy, consent, and injury. Social media users may not realize their data can be utilized for study, causing a privacy risk. Researchers must balance user privacy with openness on these platforms, possibly anonymizing and getting consent where possible. Researchers must also evaluate the psychological impact of sensitive topics to avoid stigmatizing or harming specific populations.

3. **Transparency and Accountability in Data Usage**

As AI and social media research progress, ethical norms must prioritize data usage transparency and accountability in data collection, analysis, and sharing. AI may use "explainable AI" principles to explain algorithmic conclusions to non-specialists. Social media research openness might include openly disclosing objectives and methods when using public data.

4. **Dynamic Consent Models**

Traditional informed consent processes may need to change due to the fluidity of AI and social media research. Participants can alter their consent preferences over time in dynamic consent models, giving them more data control. This strategy protects participant autonomy and helps researchers to be ethical in quickly changing technology situations.

5. **Implications for Future Research**

Future research in these domains will need flexible ethical frameworks to handle emerging technology's problems and complexities. Interdisciplinary guidelines incorporating data science, ethics, and social sciences may improve ethical standards. By proactively addressing these challenges, researchers can ensure ethical technological developments in social science.

ETHICAL DECISION-MAKING MODELS

Social science research involves complex moral challenges that require ethical decision-making models. Researchers can identify, analyze, and resolve ethical dilemmas methodically and ethically using these models. Researchers may make more ethical and informed decisions using these models, assuring the integrity and credibility of their findings.

A popular paradigm is the Belmont Report, which emphasizes respect for humans, beneficence, and justice (Tsosie et al., 2021; Serpico, 2024). Respecting persons is respecting them as autonomous agents and protecting those with less autonomy, such as children and cognitively impaired people. Beneficence encourages researchers to balance research risks and benefits to maximize benefits and minimize harm. Justice promotes the fair allocation of research gains and burdens to avoid exploitation. The James Rest Four Component Model emphasizes ethical decision-making and is also useful (Warner et al., 2024; Rhim et al., 2021). This approach includes ethical sensitivity, judgment, motivation, and action. Ethical sensitivity recognizes ethical issues, while ethical judgment chooses the most ethical action. Ethics motivation is the determination to act ethically, and ethics action is implementing the selected ethical course (Arar & Saiti, 2022). This model encourages researchers to evaluate ethical challenges from different aspects and promotes holistic ethical decision-making.

Applying these models to real-world research lets researchers make ethical decisions comfortably. When faced with a conflict between participant confidentiality and data disclosure, researchers can use these models to weigh the ethical implications of different actions and choose one that meets ethical and legal standards. Ethical decision-making models urge researchers to document their decisions and rationales, promoting transparency and responsibility (Mökander et al., 2021; Rodgers, 2023). Ethics supervision and reflective practice depend on this record, which helps researchers learn from their mistakes and improve their ethics. In conclusion, ethical decision-making models help social science researchers navigate ethical issues. These models protect participants' rights and research integrity by offering systematic frameworks for ethical issue analysis and resolution.

10.1. The Belmont Report Principles

The Belmont Report, a cornerstone of research ethics, specifies three principles guiding human research: respect for persons, beneficence, and justice (Nagai et al., 2022; Tsosie, 2021). These standards safeguard participation and research integrity. Respect for humans means respecting individual autonomy and protecting those with limited autonomy, such as children or those with cognitive impairments (Peterson et al., 2021; Cohen-Mansfield, 2021) (De Sabbata, 2020). The principle of informed consent requires researchers to provide participants with all study information and to ensure their voluntary participation. Beneficence requires researchers to maximize benefits and minimize harm (Wester et al., 2023; Bester, 2020). It entails carefully weighing the risks and rewards of the research to ensure that the advantages outweigh the hazards to participants. Justice implies fairness in research gains and costs (Davidson, 2021; Bal et al., 2023). Research must not abuse disadvantaged populations and must benefit all groups equally. These principles help researchers better handle ethical issues and protect the rights and welfare of all participants.

10.2. The Four Component Model

Developed by James Rest, the Four Component Model offers a thorough framework for ethical decision-making in social scientific research (Rogers and Breakey, 2023; Savur, 2022 This paradigm comprises four fundamental elements: ethical sensitivity, ethical judgment, ethical motivation, and ethical action, which collectively direct researchers in methodically tackling ethical problems. Ethical sensitivity refers to the ability to identify whether a situation raises an ethical dilemma. Developing an understanding of the impact of research activities on participants and the wider society necessitates the cultivation of empathy and cultural astuteness.

Ethical judgment is the process of identifying the maximum ethically acceptable course of action. In order to make well-informed decisions that maintain ethical standards, researchers must assess several alternatives, taking into account ethical principles, professional requirements, and possible repercussions.

An ethical motive is the dedication to give precedence to ethical principles above personal, professional, or financial concerns. This entails cultivating a robust ethical culture that places high importance on and actively encourages integrity. Ethical action refers to the execution of a selected ethical conclusion. It necessitates not just selecting the correct option but also implementing tangible measures to continually adhere to ethical principles. Incorporating these elements, the Four Component Model assists researchers in managing intricate ethical dilemmas, therefore guaranteeing that their work upholds rigorous ethical principles and makes a constructive contribution to the discipline.

ETHICAL SOLUTIONS AND BEST PRACTICES

Social science research ethical solutions and best practices are essential for resolving complicated ethical issues and conducting respectful research. These solutions combine proactive techniques, ethical rules, and ethical reflection throughout the study process. Community engagement and joint research with study communities are effective. By making research relevant and useful to the community, this strategy promotes ethical research (Lemke et al., 2022; Hoekstra, 2020). It also lets participants voice their concerns, influence the research topic, and ensure fair outcomes sharing. Collaboration promotes trust and respect and minimizes exploitation, creating a more egalitarian research environment.

Ethics training and education are essential. Researchers, especially early careerists, benefit from structured training programs on research ethics principles like informed consent, confidentiality, and data protection (Husband, 2020; Galvão et al., 2020). In training programs, researchers should learn how to recognize and handle ethical issues through case studies, ethical dilemmas, and decision-making frameworks. Research institutes promote ethical awareness and accountability through ongoing education.

IRBs, or ethical review committees, are essential to ethical research. These boards oversee research proposals to ensure ethical conduct and participant well-being (Chou et al., 2024; Xu et al., 2020). IRBs can help researchers navigate difficult ethical issues and discuss ethical dilemmas before they develop in the field.

Researchers should also communicate openly and obtain informed consent. This entails presenting the study's goals, procedures, risks, and benefits to participants and obtaining their consent throughout the trial (Pietrzykowski & Smilowska, 2021; Glaser et al., 2020). Transparency fosters trust and informs people of their rights and responsibilities. In conclusion, ethical solutions and best practices in social science research involve proactive and collaborative community engagement, ongoing education, ethical oversight, and transparent communication. Using these tactics, researchers can respect ethics, protect participant rights, and progress knowledge.

11.1. Community Engagement and Collaborative Research

Community participation and collaborative research are becoming ethical best practices in social science research, especially when investigating marginalized or vulnerable communities (Mertens, 2021; Chesser et al., 2020). This approach actively incorporates community members in research question development, data collection, and analysis to ensure relevance, respect, and benefit to the community. Engaging communities in research builds trust and ownership, lowering the danger of exploitation and ensuring that research outputs match community needs and values (Luger et al., 2020; Herrera et al., 2020). Collaborative research allows community members to influence the research agenda and methods, which can lead to more culturally sensitive and ethical practices.

Community engagement also distributes research gains more evenly by applying findings and ideas to local issues. Community participation and collaborative research improve social science research ethics by increasing transparency, mutual respect, and inclusivity, improving study quality and impact while protecting all parties.

11.2. Ethical Training and Education

Ethics training and education are essential to promoting integrity in social scientific research (Valkenburg et al., 2021; Pizzolato & Dierickx, 2021). Comprehensive ethical training prepares researchers to handle complicated ethical issues. Training programs teach acceptable research ethics, such as informed consent, confidentiality, data security, and damage avoidance.

Researchers learn critical thinking and ethical decision-making through case studies, role-playing, and debates on real-world ethical challenges (Chen & Wu, 2023).

These practical exercises let researchers explore multiple scenarios, evaluate the consequences of their choices, and weigh ethical considerations.

Additionally, ethical education promotes constant reflection and understanding of changing ethical standards. Researchers must remain aware and adapt to new ethical concerns when research contexts and technologies change (Newman et al., 2021; Jongsma & Bredenoord, 2020). Encourage ongoing learning and ethical vigilance to prevent wrongdoing and address ethical issues proactively.

Research institutes can foster ethical behavior by investing in ethical training and teaching. This can increase research quality and credibility while respecting participant rights and well-being.

CONCLUSION AND FUTURE DIRECTIONS IN RESEARCH ETHICS

This chapter concluded that deontological ethics, utilitarianism, and virtue ethics help social science researchers navigate complicated ethical dilemmas. Deontological ethics respects universal values such as consent and privacy, whereas utilitarianism maximizes benefits and minimizes harm. Virtue ethics emphasizes researchers' character and integrity in moral decision-making. These theories handle social science research ethics in a balanced manner.

In the future, practical actions must reinforce ethical standards. To prepare researchers for changing ethical concerns, research institutions should provide continuing ethical education and training. Flexible consent approaches, such as dynamic consent, allow participants to update their consent preferences over time, enhancing autonomy and ethical rigor. The field can promote ethical reflexivity that adapts to new research contexts and technologies by promoting these techniques.

Social settings, technological advances, and new ethical challenges shape social science research ethics. This chapter emphasized the importance of respecting people, practicing beneficence, maintaining fairness, and upholding integrity when navigating real-world research challenges. We must be proactive and agile in ethical decision-making to ensure that research practices evolve with societal and technological advancements. Participant-centered techniques like community participation and collaborative research are potential research ethical trends. These methods build trust and respect, ensuring an equitable distribution of research benefits and a focus on participants' perspectives. Expanding these strategies can reduce power inequalities and improve research ethics.

Developing and improving ethical rules and frameworks is also crucial. New ethical issues arise from big data analytics, artificial intelligence, and internet research, requiring ongoing rule updates. Traditional ethical frameworks should also incorporate data privacy, security, and ethical digital tool use.

Researchers will need ethical training to face future difficulties. Institutions should invest in comprehensive training programs that address fundamental ethical principles, practical applications, critical thinking, and adaptability to new ethical concerns. A culture of ethical reflection and learning will help researchers avoid misbehavior and uphold the highest ethical standards. By improving ethical rules, participant-centered research, and ethical training, the field can ensure that research benefits society while respecting participants' rights and dignity. As social circumstances and technologies change, our ethical research procedures must keep social science research relevant and ethical.

REFERENCES

Akpa-Inyang, F., & Chima, S. C. (2021). … values and beliefs regarding informed consent and limitations of thc principle of respect for autonomy in African communities: A cross-cultural qualitative study. *BMC Medical Ethics, 22*(1), 111. Advance online publication. DOI: 10.1186/s12910-021-00678-4 PMID: 34391415

Andrew, J., & Baker, M. (2021). The general data protection regulation in the age of surveillance capitalism. [REMOVED HYPERLINK FIELD]. *Journal of Business Ethics, 168*(3), 565–578. Advance online publication. DOI: 10.1007/s10551-019-04239-z

Arar, K., & Saiti, A. (2022). Ethical leadership, ethical dilemmas and decision making among school administrators. *Equity in Education & Society, 1*(1), 126–141. Advance online publication. DOI: 10.1177/27526461211070828

Aurey, X. (2020). *From the Martens Clause to Consent to Human Experimentations, the Legal Journey of the Judges during the Nuremberg Doctors' Trial.* Biolaw and International Criminal Law., DOI: 10.1163/9789004364424_015

Bal, M., Stok, M., Bombaerts, G., Huijts, N., Schneider, P., Spahn, A., & Buskens, V. (2023). A fairway to fairness: Toward a richer conceptualization of fairness perceptions for just energy transitions. *Energy Research & Social Science, 103*, 103213. DOI: 10.1016/j.erss.2023.103213

Bester, J. C. (2020). Beneficence, interests, and wellbeing in medicine: What it means to provide benefit to patients. *The American Journal of Bioethics, 20*(3), 53–62. Advance online publication. DOI: 10.1080/15265161.2020.1714793 PMID: 32105204

Bhatt, B. (2022). Ethical complexity of social change: Negotiated actions of a social enterprise. *Journal of Business Ethics, 177*(4), 743–762. Advance online publication. DOI: 10.1007/s10551-022-05100-6

Bishop, D. V. M. (2020). The psychology of experimental psychologists: Overcoming cognitive constraints to improve research: The 47th Sir Frederic Bartlett Lecture. *Quarterly Journal of Experimental Psychology, 73*(1), 1–19. Advance online publication. DOI: 10.1177/1747021819886519 PMID: 31724919

Bitter, C. C., Ngabirano, A. A., Simon, E. L., & Taylor, D. M. (2020). Principles of research ethics: A research primer for low-and middle-income countries. *African Journal of Emergency Medicine : Revue Africaine de la Médecine d'Urgence, 10*, S125–S129. DOI: 10.1016/j.afjem.2020.07.006 PMID: 32837877

Bloemraad, I., & Menjívar, C. (2022). Precarious times, professional tensions: The ethics of migration research and the drive for scientific accountability. *The International Migration Review*, *56*(1), 4–32. Advance online publication. DOI: 10.1177/01979183211014455 PMID: 36467777

Bray, T. (2020). Military Information Technology Certification Training Addressing Implementation Procedures Against Cyber-Attacks: An Exploratory Case Study. [HTML]

Brenner, E., Bleys, R. L., de Caro, R., Catereniuc, I., Chirculescu, A. R., Destrieux, C., & Osorio, M. T. V. (2023). The legal and ethical framework governing body donation in Europe–2nd update on current practice. *Annals of Anatomy-Anatomischer Anzeiger*, *152195*. Advance online publication. DOI: 10.1016/j.aanat.2023.152195 PMID: 38042354

Brown, C., Spiro, J., & Quinton, S. (2020). The role of research ethics committees: Friend or foe in educational research? An exploratory study. *British Educational Research Journal*, *46*(4), 747–769. DOI: 10.1002/berj.3654

Burgess, T., Rennie, S., & Moodley, K. (2023). Key ethical issues encountered during COVID-19 research: A thematic analysis of perspectives from South African research ethics committees. *BMC Medical Ethics*, *24*(1), 11. Advance online publication. DOI: 10.1186/s12910-023-00888-y PMID: 36793067

Chen, H. L., & Wu, C. T. (2023). A digital role-playing game for learning: Effects on critical thinking and motivation. *Interactive Learning Environments*, *31*(5), 3018–3030. Advance online publication. DOI: 10.1080/10494820.2021.1916765

Chesser, S., Porter, M. M., & Tuckett, A. G. (2020). Cultivating citizen science for all: Ethical considerations for research projects involving diverse and marginalized populations. [REMOVED HYPERLINK FIELD]. *International Journal of Social Research Methodology*, *23*(5), 497–508. DOI: 10.1080/13645579.2019.1704355

Chou, C., Lee, I. J., & Fudano, J. (2024). The present situation of and challenges in research ethics and integrity promotion: Experiences in East Asia. [REMOVED HYPERLINK FIELD]. *Accountability in Research*, *31*(6), 576–599. Advance online publication. DOI: 10.1080/08989621.2022.2155144 PMID: 36641764

Cohen-Mansfield, J. (2021). The rights of persons with dementia and their meanings. *Journal of the American Medical Directors Association*, *22*(7), 1381–1385. Advance online publication. DOI: 10.1016/j.jamda.2021.03.007 PMID: 33844989

Corneli, A., Dombeck, C. B., McKenna, K., & Calvert, S. B. (2021). The Single IRB Review Process: Stakeholder Experiences and Recommendations for Guidance by the US Food and Drug Administration. *Ethics & Human Research*, *43*(3), 26. DOI: 10.1002/eahr.500092 PMID: 34019339

Corneli, A., Dombeck, C. B., McKenna, K., & Calvert, S. B. (2021). Stakeholder experiences with the single IRB review process and recommendations for Food and Drug Administration guidance. [wiley.com]. *Ethics & Human Research*, *43*(3), 26–36. DOI: 10.1002/eahr.500092 PMID: 34019339

Cosman, D. (2023). The Nuremberg Code on Human Experimentation. A Historical Overview. Romanian Journal of Artistic Creativity. [HTML]

Darr, R. (2020). Virtues as qualities of character: Alasdair MacIntyre and the situationist critique of virtue ethics. [REMOVED HYPERLINK FIELD]. *The Journal of Religious Ethics*, *48*(1), 7–25. Advance online publication. DOI: 10.1111/jore.12297

Davidson, M. D. (2021). How fairness principles in the climate debate relate to theories of distributive justice. *Sustainability (Basel)*, *13*(13), 7302. Advance online publication. DOI: 10.3390/su13137302

De Sabbata, K. (2020). Dementia, treatment decisions, and the UN convention on the rights of persons with disabilities. A new framework for old problems. [REMOVED HYPERLINK FIELD]. *Frontiers in Psychiatry*, *11*, 571722. Advance online publication. DOI: 10.3389/fpsyt.2020.571722 PMID: 33240127

Di Feliciantonio, C. (2024). 'I guess I really survived many crises': On the benefits of longitudinal ethnographic research. *Area*, *56*(1), e12886. Advance online publication. DOI: 10.1111/area.12886

Drolet, M. J., Rose-Derouin, E., Leblanc, J. C., Ruest, M., & Williams-Jones, B. (2023). Ethical issues in research: Perceptions of researchers, research ethics board members and research ethics experts. *Journal of Academic Ethics*, *21*(2), 269–292. DOI: 10.1007/s10805-022-09455-3 PMID: 35975128

Edmondson, A. (2023). Right Kind of Wrong: Why Learning to Fail Can Teach Us to Thrive. [HTML]

Elaad, E. (2022). Tunnel vision and confirmation bias among police investigators and laypeople in hypothetical criminal contexts. *SAGE Open*, *12*(2), 21582440221095022. Advance online publication. DOI: 10.1177/21582440221095022

Ess, C. M. (2020). Internet research ethics and social media. Handbook of research ethics and scientific integrity. https://doi.org/ [REMOVED HYPERLINK FIELD] DOI: 10.1007/978-3-030-16759-2_12

Fardid, G., Akbari, A., & Ghafoori Nezhad, M. (2023). The Relationship between Dewey's Moral Contextualism and Kant's Deontological Ethics in Abortion. *International Journal of Ethics and Society*, *5*(1), 20–26. DOI: 10.52547/ijethics.5.1.4

Favaretto, M., De Clercq, E., Gaab, J., & Elger, B. S. (2020). First do no harm: An exploration of researchers' ethics of conduct in Big Data behavioral studies. *PLoS One*, *15*(11), e0241865. Advance online publication. DOI: 10.1371/journal.pone.0241865 PMID: 33152039

Favaretto, M., Shaw, D., De Clercq, E., Joda, T., & Elger, B. S. (2020). Big data and digitalization in dentistry: A systematic review of the ethical issues. *International Journal of Environmental Research and Public Health*, *17*(7), 2495. DOI: 10.3390/ijerph17072495 PMID: 32268509

Ferretti, A., Ienca, M., Hurst, S., & Vayena, E. (2020). Big data, biomedical research, and ethics review: New challenges for IRBs. [wiley.com]. *Ethics & Human Research*, *42*(5), 17–28. DOI: 10.1002/eahr.500065 PMID: 32937036

Flynn, S. V. (2021). Research design for the behavioral sciences: An applied approach. [HTML]

Galvão, A., Marques, C., & Ferreira, J. J. (2020). The role of entrepreneurship education and training programmes in advancing entrepreneurial skills and new ventures. [REMOVED HYPERLINK FIELD]. *European Journal of Training and Development*, *44*(6/7), 595–614. DOI: 10.1108/EJTD-10-2019-0174

Glaser, J., Nouri, S., Fernandez, A., Sudore, R. L., Schillinger, D., Klein-Fedyshin, M., & Schenker, Y. (2020). Interventions to improve patient comprehension in informed consent for medical and surgical procedures: An updated systematic review. *Medical Decision Making*, *40*(2), 119–143. https://doi.dox.org/10.1177/0272989X19896348. DOI: 10.1177/0272989X19896348 PMID: 31948345

Hassan, J., Shehzad, D., Habib, U., Aftab, M. U., Ahmad, M., Kuleev, R., & Mazzara, M. (2022). [Retracted] The Rise of Cloud Computing: Data Protection, Privacy, and Open Research Challenges—A Systematic Literature Review (SLR). *Computational Intelligence and Neuroscience*, *2022*(1), 8303504. DOI: 10.1155/2023/9838129 PMID: 35712069

Haven, T., Gopalakrishna, G., Tijdink, J., van der Schot, D., & Bouter, L. (2022). Promoting trust in research and researchers: How open science and research integrity are intertwined. *BMC Research Notes*, *15*(1), 302. DOI: 10.1186/s13104-022-06169-y PMID: 36127719

Head, G. (2020). Ethics in educational research: Review boards, ethical issues and researcher development. *European Educational Research Journal*, *19*(1), 72–83. Advance online publication. DOI: 10.1177/1474904118796315

Henrickson, M., Giwa, S., Hafford-Letchfield, T., Cocker, C., Mulé, N. J., Schaub, J., & Baril, A. (2020). Research ethics with gender and sexually diverse persons. *International Journal of Environmental Research and Public Health*, *17*(18), 6615. DOI: 10.3390/ijerph17186615 PMID: 32932832

Herrera, S. G., Porter, L., & Barko-Alva, K. (2020). Equity in school–parent partnerships: Cultivating community and family trust in culturally diverse classrooms. [HTML]

Ho, M. (2024). Relational tactics and trust in high-risk activism: Anonymity, preexisting ties, and bonding in Hong Kong's 2019–2020 protest. [HTML]. *International Journal of Comparative Sociology*, *65*(4), 499–516. DOI: 10.1177/00207152231220524

Hoekstra, F., Mrklas, K. J., Khan, M., McKay, R. C., Vis-Dunbar, M., Sibley, K. M., Nguyen, T., Graham, I. D., & Gainforth, H. L. (2020). A review of reviews on principles, strategies, outcomes and impacts of research partnerships approaches: A first step in synthesising the research partnership literature. [springer.com]. *Health Research Policy and Systems*, *18*(1), 1–23. DOI: 10.1186/s12961-020-0544-9 PMID: 32450919

Hosseini, M., Wieczorek, M., & Gordijn, B. (2022). Ethical issues in social science research employing big data. Science and Engineering Ethics. springer.com

Hota, P. K., Bhatt, B., & Qureshi, I. (2023). Institutional work to navigate ethical dilemmas: Evidence from a social enterprise. [HTML]. *Journal of Business Venturing*, *38*(1), 106269. DOI: 10.1016/j.jbusvent.2022.106269

Husband, G. (2020). Ethical data collection and recognizing the impact of semi-structured interviews on research respondents. Education Sciences. mdpi.com

Ivey, G. (2023). Interpreting hidden meaning in qualitative research interview data: Opportunities and challenges. [HTML]. *Qualitative Research in Psychology*, *20*(1), 21–51. DOI: 10.1080/14780887.2022.2067509

Jain, A. K., Sahoo, S. R., & Kaubiyal, J. (2021). Online social networks security and privacy: comprehensive review and analysis. Complex & Intelligent Systems. springer.com

Jamrozik, E., Littler, K., Bull, S., Emerson, C., Kang, G., Kapulu, M., Rey, E., Saenz, C., Shah, S., Smith, P. G., Upshur, R., Weijer, C., & Selgelid, M. J. (2021). Key criteria for the ethical acceptability of COVID-19 human challenge studies: Report of a WHO Working Group. [sciencedirect.com]. *Vaccine*, *39*(4), 633–640. DOI: 10.1016/j.vaccine.2020.10.075 PMID: 33341309

Jongsma, K. R. & Bredenoord, A. L. (2020). Ethics parallel research: an approach for (early) ethical guidance of biomedical innovation. BMC Medical Ethics. springer.com

Karimova, G. S., Hoffmann, N. C., Heidbrink, L., & Hoffmann, S. (2020). Virtue ethics between east and west in consumer research: Review, synthesis and directions for future research. [HTML]. *Journal of Business Ethics*, *165*(2), 255–275. DOI: 10.1007/s10551-019-04321-6

Keshta, I. & Odeh, A. (2021). Security and privacy of electronic health records: Concerns and challenges. Egyptian Informatics Journal. sciencedirect.com

Klykken, F. H. (2022). Implementing continuous consent in qualitative research. Qualitative Research. sagepub.com

Knöpfle, P., Haim, M., & Breuer, J. (2024). Key topic or bare necessity? How Research Ethics are Addressed and Discussed in Computational Communication Science. Publizistik. springer.com

Kyngäs, H., Kääriäinen, M., & Elo, S. (2020). The trustworthiness of content analysis. The application of content analysis in nursing science research, 41-48. [HTML]

Labrecque, L. I., Markos, E., Swani, K., & Peña, P. (2021). When data security goes wrong: Examining the impact of stress, social contract violation, and data type on consumer coping responses following a data breach. [HTML]. *Journal of Business Research*, *135*, 559–571. DOI: 10.1016/j.jbusres.2021.06.054

Laurijssen, S. J., van der Graaf, R., van Dijk, W. B., Schuit, E., Groenwold, R. H., Grobbee, D. E., & de Vries, M. C. (2022). When is it impractical to ask informed consent? A systematic review. [sagepub.com]. *Clinical Trials*, *19*(5), 545–560. DOI: 10.1177/17407745221103567 PMID: 35775421

Lemke, A. A., Esplin, E. D., Goldenberg, A. J., Gonzaga-Jauregui, C., Hanchard, N. A., Harris-Wai, J., Ideozu, J. E., Isasi, R., Landstrom, A. P., Prince, A. E. R., Turbitt, E., Sabatello, M., Schrier Vergano, S. A., Taylor, M. R. G., Yu, J.-H., Brothers, K. B., & Garrison, N. A. (2022). Addressing underrepresentation in genomics research through community engagement. [cell.com]. *American Journal of Human Genetics*, *109*(9), 1563–1571. DOI: 10.1016/j.ajhg.2022.08.005 PMID: 36055208

Lowe, K., & Ridgway, A. (2023). The keepers of secrets: Ethics and the emotional labour of working with privileged populations during criminological research. In *Ethical Dilemmas in International Criminological Research* (pp. 113–129). Routledge. [HTML] DOI: 10.4324/9781003241515-9

Luger, T. M., Hamilton, A. B., & True, G. (2020). Measuring community-engaged research contexts, processes, and outcomes: a mapping review. The Milbank Quarterly. wiley.com

Lynch, H. F., Eriksen, W., & Clapp, J. T. (2022). "We measure what we can measure": Struggles in defining and evaluating institutional review board quality. [HTML]. *Social Science & Medicine*, *292*, 114614. DOI: 10.1016/j.socscimed.2021.114614 PMID: 34861569

MacArthur Clark, J. A., & Sun, D. (2020). Guidelines for the ethical review of laboratory animal welfare People's Republic of China National Standard GB/T 35892-2018 [Issued 6 February 2018 Effective from 1 September 2018] [wiley.com]. *Animal Models and Experimental Medicine*, *3*(1), 103–113. DOI: 10.1002/ame2.12111 PMID: 32318667

McGowan, W. (2020). 'If you didn't laugh, you'd cry': Emotional labour, reflexivity and ethics-as-practice in a qualitative fieldwork context. Methodological innovations. sagepub.com

Mertens, D. M. (2021). Transformative research methods to increase social impact for vulnerable groups and cultural minorities. International Journal of Qualitative Methods. sagepub.com

Mökander, J., Morley, J., Taddeo, M., & Floridi, L. (2021). Ethics-based auditing of automated decision-making systems: Nature, scope, and limitations. [springer.com]. *Science and Engineering Ethics*, *27*(4), 44. DOI: 10.1007/s11948-021-00319-4 PMID: 34231029

Moriña, A. (2021). When people matter: The ethics of qualitative research in the health and social sciences. [HTML]. *Health & Social Care in the Community*, *29*(5), 1559–1565. DOI: 10.1111/hsc.13221 PMID: 33170531

Nagai, H., Nakazawa, E., & Akabayashi, A. (2022). The creation of the Belmont Report and its effect on ethical principles: a historical study. Monash bioethics review. springer.com

Newman, P. A., Guta, A., & Black, T. (2021). Ethical considerations for qualitative research methods during the COVID-19 pandemic and other emergency situations: Navigating the virtual field. [sagepub.com]. *International Journal of Qualitative Methods, 20*, 16094069211047823. DOI: 10.1177/16094069211047823

Parker, M. (2023). *Burnout in Information Security: The Case of Healthcare*. Advanced Health Technology. [HTML]

Paunova, M. (2023). Principlism in Practice: Ethics in Survey Research. The Sage Handbook of Survey Development and Application. London ua: Sage, 54-80. [HTML]

Peterson, A., Karlawish, J., & Largent, E. (2021). Supported decision making with people at the margins of autonomy. [tandfonline.com]. *The American Journal of Bioethics, 21*(11), 4–18. DOI: 10.1080/15265161.2020.1863507 PMID: 33372858

Petkov, C. I., Flecknell, P., Murphy, K., Basso, M. A., Mitchell, A. S., Hartig, R., & Thompson-Iritani, S. (2022). Unified ethical principles and an animal research 'Helsinki'declaration as foundations for international collaboration. [sciencedirect.com]. *Current Research in Neurobiology, 3*, 100060. DOI: 10.1016/j.crneur.2022.100060 PMID: 36386861

Pietilä, A. M., Nurmi, S. M., Halkoaho, A., & Kyngäs, H. (2020). Qualitative research: Ethical considerations. The application of content analysis in nursing science research, 49-69. [HTML]

Pietrzykowski, T. & Smilowska, K. (2021). The reality of informed consent: empirical studies on patient comprehension—systematic review. Trials. springer.com

Pizzolato, D. & Dierickx, K. (2021). Stakeholders' perspectives on research integrity training practices: a qualitative study. BMC medical ethics. springer.com

Podschuweit, N. (2021). How ethical challenges of covert observations can be met in practice. Research Ethics. sagepub.com

Poulton, P. (2023). Being a teacher-researcher: Reflections on an insider research project from a virtues-based approach to research ethics. [HTML]. *Educational Action Research, 31*(3), 575–591. DOI: 10.1080/09650792.2021.1962379

Pritchett, M., Ala'i-Rosales, S., Cruz, A. R., & Cihon, T. M. (2022). Social justice is the spirit and aim of an applied science of human behavior: Moving from colonial to participatory research practices. [springer.com]. *Behavior Analysis in Practice, 15*(4), 1074–1092. DOI: 10.1007/s40617-021-00591-7 PMID: 34178290

Putra, B. A. (2023). Complexities of the ethical dilemmas in qualitative international relations research: Research subjects, ethical codes, and constructing qualitative rigor. Social Sciences. mdpi.com

Queloz, M. & van Ackeren, M. (2024). Virtue Ethics and the Morality System. Topoi. springer.com

Rafiq, F., Awan, M. J., Yasin, A., Nobanee, H., Zain, A. M., & Bahaj, S. A. (2022). Privacy prevention of big data applications: A systematic literature review. [sagepub.com]. *SAGE Open*, *12*(2), 21582440221096445. DOI: 10.1177/21582440221096445

Rahman, S. A., Tuckerman, L., Vorley, T., & Gherhes, C. (2021). Resilient research in the field: Insights and lessons from adapting qualitative research projects during the COVID-19 pandemic. [sagepub.com]. *International Journal of Qualitative Methods*, *20*, 16094069211016106. DOI: 10.1177/16094069211016106

Razaque, A., Frej, M. B. H., Alotaibi, B., & Alotaibi, M. (2021). Privacy preservation models for third-party auditor over cloud computing: A survey. Electronics. mdpi.com

Resnik, D., & Hecking, J. (2020). Cultural variation in informed consent for clinical research participation. In *Handbook of Ethnography in Healthcare Research* (pp. 33–46). Routledge. [HTML] DOI: 10.4324/9780429320927-6

Resnik, D. B. (2021). Standards of evidence for institutional review board decision-making. Accountability in research. nih.gov

Rhim, J., Lee, J. H., Chen, M., & Lim, A. (2021). A deeper look at autonomous vehicle ethics: an integrative ethical decision-making framework to explain moral pluralism. Frontiers in Robotics and AI. frontiersin.org

Robinson, K., McKenna, B., & Rooney, D. (2022). The relationship of risk to rules, values, virtues, and moral complexity: What we can learn from the moral struggles of military leaders. Journal of Business Ethics. springer.com

Rodgers, W., Murray, J. M., Stefanidis, A., Degbey, W. Y., & Tarba, S. Y. (2023). An artificial intelligence algorithmic approach to ethical decision-making in human resource management processes. [sciencedirect.com]. *Human Resource Management Review*, *33*(1), 100925. DOI: 10.1016/j.hrmr.2022.100925

Rogers, J., & Breakey, H. (2023). *James Rest's four component model (FCM): A case for its central place in legal ethics*. Leading Works in Legal Ethics. [HTML]

Saleh, S., Sambakunsi, H., Nyirenda, D., Kumwenda, M., Mortimer, K., & Chinouya, M. (2020). Participant compensation in global health research: A case study. [oup.com]. *International Health, 12*(6), 524–532. DOI: 10.1093/inthealth/ihaa064 PMID: 33165559

Sartoretto, P., & Custódio, L. (2020). Dealing with ethical dilemmas in activist research on social movement media. Media Activist Research Ethics: Global Approaches to Negotiating Power in Social Justice Research, 51-66. [HTML]

Savulescu, J., Persson, I., & Wilkinson, D. (2020). Utilitarianism and the pandemic. Bioethics. wiley.com

Savur, S. G. (2022). Ethical Decision-Making—Synthesizing SK Chakraborty's Classification of Ethics with Levels of Moral Judgement and the Four-Component Model. In Global Perspectives on Indian Spirituality and Management: The Legacy of SK Chakraborty (pp. 107-121). Singapore: Springer Nature Singapore. [HTML]

Serpico, K. (2024). Defining institutional review board application quality: critical research gaps and future opportunities. Research Ethics. sagepub.com

Serpico, K. (2024). The Belmont Report doesn't need reform, our moral imagination does. Research Ethics. sagepub.com

Shilton, K., Moss, E., Gilbert, S. A., Bietz, M. J., Fiesler, C., Metcalf, J., Vitak, J., & Zimmer, M. (2021). Excavating awareness and power in data science: A manifesto for trustworthy pervasive data research. [sagepub.com]. *Big Data & Society, 8*(2), 20539517211040759. DOI: 10.1177/20539517211040759

Sperling, D. (2022). "Could You Help Me Die?": On the Ethics of Researcher-Participant Relationship and the Limitations of Qualitative Research. International Journal of Qualitative Methods. sagepub.com

Stilgoe, J., Owen, R., & Macnaghten, P. (2020). Developing a framework for responsible innovation. In The Ethics of Nanotechnology, Geoengineering, and Clean Energy (pp. 347-359). Routledge. sciencedirect.com DOI: 10.4324/9781003075028-22

Tamminen, K. A., Bundon, A., Smith, B., McDonough, M. H., Poucher, Z. A., & Atkinson, M. (2021). Considerations for making informed choices about engaging in open qualitative research. [tandfonline.com]. *Qualitative Research in Sport, Exercise and Health, 13*(5), 864–886. DOI: 10.1080/2159676X.2021.1901138

Taquette, S. R., & Borges da Matta Souza, L. M. (2022). Ethical dilemmas in qualitative research: A critical literature review. [sagepub.com]. *International Journal of Qualitative Methods, 21*, 16094069221078731. DOI: 10.1177/16094069221078731

Thapa, C., & Camtepe, S. (2021). Precision health data: Requirements, challenges and existing techniques for data security and privacy. [PDF]. *Computers in Biology and Medicine, 129*, 104130. DOI: 10.1016/j.compbiomed.2020.104130 PMID: 33271399

Tsosie, K. S., Claw, K. G., & Garrison, N. A. (2021). Considering "Respect for Sovereignty" beyond the Belmont report and the common rule: Ethical and legal implications for American Indian and Alaska native peoples. [tandfonline.com]. *The American Journal of Bioethics, 21*(10), 27–30. DOI: 10.1080/15265161.2021.1968068 PMID: 34554064

Valkenburg, G., Dix, G., Tijdink, J., & de Rijcke, S. (2021). Expanding research integrity: A cultural-practice perspective. [springer.com]. *Science and Engineering Ethics, 27*(1), 10. DOI: 10.1007/s11948-021-00291-z PMID: 33559767

Wallace, S. E. & Miola, J. (2021). Adding dynamic consent to a longitudinal cohort study: a qualitative study of EXCEED participant perspectives. BMC medical ethics. springer.com

Warner, C. H., Fortin, M., & Melkonian, T. (2024). When are we more ethical? A review and categorization of the factors influencing dual-process ethical decision-making. [HTML]. *Journal of Business Ethics, 189*(4), 843–882. DOI: 10.1007/s10551-022-05281-0

West, S. & Schill, C. (2022). Negotiating the ethical-political dimensions of research methods: a key competency in mixed methods, inter-and transdisciplinary, and co-production research. Humanities and Social Sciences Communications. nature.com

Wester, K. L., Deaton, J. D., & Jayamohan, H. (2023). *Ethics in Research.* Reimagining Research. [HTML]

White, T., Blok, E., & Calhoun, V. D. (2022). Data sharing and privacy issues in neuroimaging research: Opportunities, obstacles, challenges, and monsters under the bed. Human Brain Mapping. wiley.com

Wieringa, J., Kannan, P. K., Ma, X., Reutterer, T., Risselada, H., & Skiera, B. (2021). Data analytics in a privacy-concerned world. [sciencedirect.com]. *Journal of Business Research, 122*, 915–925. DOI: 10.1016/j.jbusres.2019.05.005

Williams, O., Sarre, S., Papoulias, S. C., Knowles, S., Robert, G., Beresford, P., Rose, D., Carr, S., Kaur, M., & Palmer, V. J. (2020). Lost in the shadows: Reflections on the dark side of co-production. [springer.com]. *Health Research Policy and Systems, 18*(1), 1–10. DOI: 10.1186/s12961-020-00558-0 PMID: 32380998

Xu, A., Baysari, M. T., Stocker, S. L., Leow, L. J., Day, R. O., & Carland, J. E. (2020). Researchers' views on, and experiences with, the requirement to obtain informed consent in research involving human participants: A qualitative study. [springer.com]. *BMC Medical Ethics, 21*(1), 1–11. DOI: 10.1186/s12910-020-00538-7 PMID: 33008387

ADDITIONAL READING

Drolet, M. J., Rose-Derouin, E., Leblanc, J. C., Ruest, M., & Williams-Jones, B. (2023). Ethical issues in research: Perceptions of researchers, research ethics board members, and research ethics experts. *Journal of Academic Ethics, 21*(2), 269–292. DOI: 10.1007/s10805-022-09455-3 PMID: 35975128

Favaretto, M., Shaw, D., De Clercq, E., Joda, T., & Elger, B. S. (2020/2021). Big data and digitalization in research: Ethical challenges and review practices. *Science and Engineering Ethics, 27*(2), 24. Advance online publication. DOI: 10.1007/s11948-021-00287-9

Hosseini, M., Wieczorek, M., & Gordijn, B. (2022). Ethical issues in social science research employing big data. *Science and Engineering Ethics, 28*(1), 8. Advance online publication. DOI: 10.1007/s11948-022-00380-7 PMID: 35705883

Knöpfle, P., Haim, M., & Breuer, J. (2024). How research ethics are addressed in computational social science: Between formal requirements and ethical reflection. *Publizistik, 69*, 23–44. DOI: 10.1007/s11616-023-00815-9

Mökander, J., Morley, J., Taddeo, M., & Floridi, L. (2021). Ethics-based auditing of automated decision-making systems: Nature, scope, and limitations. *Science and Engineering Ethics, 27*(4), 44. Advance online publication. DOI: 10.1007/s11948-021-00319-4 PMID: 34231029

Newman, P. A., Guta, A., & Black, T. (2021). Ethical considerations for qualitative research methods during emergency situations. *International Journal of Qualitative Methods, 20*, 1–13. DOI: 10.1177/16094069211047823

Shilton, K., Moss, E., Gilbert, S. A., Bietz, M. J., Fiesler, C., & Zimmer, M. (2021). Excavating awareness and power in data science: A manifesto for trustworthy research. *Big Data & Society, 8*(2), 1–15. DOI: 10.1177/20539517211040759

Taquette, S. R., & Borges da Matta Souza, L. M. (2022). Ethical dilemmas in qualitative research: A critical literature review. *International Journal of Qualitative Methods, 21*, 1–14. DOI: 10.1177/16094069221078731

Valkenburg, G., Dix, G., Tijdink, J., & de Rijcke, S. (2021). Expanding research integrity: A cultural-practice perspective. *Science and Engineering Ethics*, *27*(1), 10. Advance online publication. DOI: 10.1007/s11948-021-00291-z PMID: 33559767

KEY TERMS AND DEFINITIONS

Confidentiality: A moral duty necessitating that researchers safeguard participants' personal information against unlawful exposure during data collection, analysis, storage, and dissemination.

Data Privacy: The ethical and legal obligation to preserve participants' personal data, guaranteeing secure management, regulated access, and safeguarding against misuse or unauthorized disclosure.

Deontological Ethics: A moral framework that assesses research activities based on compliance with ethical obligations and principles, including informed consent, integrity, and respect for participant rights, irrespective of results.

Ethical Decision-Making Models: Systematic frameworks employed by researchers to discover, assess, and address ethical challenges by reconciling moral principles, professional obligations, and contextual limitations.

Informed Consent: A practice whereby research volunteers willingly consent to participate in a study after being provided with explicit and thorough information regarding its objectives, methodologies, risks, and advantages.

Institutional Review Board: An official regulatory entity tasked with evaluating research ideas that involve human subjects to guarantee ethical adherence, risk reduction, and safeguarding of participant well-being.

Research Integrity: The dedication to integrity, openness, and responsibility across all phases of the research process, encompassing data collection, analysis, reporting, and authorship.

Utilitarian Ethics: A consequentialist ethical framework evaluates research actions by their outcomes, deeming an action morally permissible if it optimizes total benefits and mitigates harm.

Virtue Ethics: A moral viewpoint that prioritizes the researcher's character, integrity, and intentions as fundamental to ethical research practices and decision-making.

Chapter 10
Research Ethics in Islam:
Principles, Practices, and Contemporary Case Reflections

Naseem Akhter
Shaheed Benazir Bhutto Women University, Peshawar, Pakistan

Arshad Munir
Ghazi University, Dera Ghazi Khan, Pakistan

EXECUTIVE SUMMARY

Research ethics is essential for responsible knowledge production in today's globalized academic environment. While largely shaped by secular Western traditions, Islamic scholarship provides a deep ethical framework grounded in the Qur'an, Hadith, and jurisprudence. Core principles include truthfulness (sidq), justice (adl), intention (niyyah), public good (maslahah), and accountability (amanah). This chapter examines how these values guide research in theory and practice, using case studies from Muslim-majority contexts and global scholarship. It calls for an inclusive model that integrates Islamic ethics with international standards without compromising science. Addressing biomedical research, social sciences, and emerging technologies, it offers practical guidance for researchers and institutions worldwide.

INTRODUCTION

The pursuit of knowledge has been considered a cardinal value in the Islamic tradition. Prophet Muhammad (PBUH) was sent down the first revelation, and the command was Read! The pursuit of knowledge is a religious obligation based on the declaration of the pursuit of knowledge as a religious obligation (Qur'an 96:1).

DOI: 10.4018/979-8-3693-8302-5.ch010

In the history of Islam, scholars constructed sophisticated ways for obtaining, verifying, and transmitting knowledge, and thus elaborated what could be considered the first guidelines of research ethics (Al-Attas, 1993). However, ethical issues facing the modern research environment are new challenges that require a show of etiquette in how they are dissected using both the traditional Islamic standards and the contemporary ethical standards.

The word ethics in research has gained a modern meaning of the concept, including a set of principles and rules of conducting scientific research and academic studies, which are the basis for high integrity, human dignity, and social responsibility. Although Western models such as the Belmont Report (1979) and the Declaration of Helsinki have been the focus of the debate on research ethics around the world, there has been a growing realization that culturally and religiously sensitive methods appealing to diverse communities around the world are required (Padela et al., 2014).

Islamic ethics is an excellent platform to address the research problems of the day. Based on the concept of tawhid (the oneness of Allah (SWT), Islamic ethics is based on the bond of human responsibility as the vicegerent to Allah (SWT) (*khalifah*) on the earth and one who is responsible in the worldly and spiritual aspects. The same is true for the research area where the quest for knowledge is to be within ethical boundaries, which are established in the Islamic law and morality (Iqbal, 2007).

The fact that Islamic ethics have been integrated into modern research paradigms is not only an academic project but an exigency in practice for a number of reasons. First, there are 1.8 billion Muslims worldwide with cultural and religious sensitivities that cannot be adequately addressed by scientific theories of ethics. Second, the Muslim-majority countries are also increasing their research and development activities and, therefore, need an ethical code based on their cultural and religious norms. Third, Islam has a rich ethical culture that provides valuable insights that could lead to the discussion of research ethics in the world and more specifically, research related to the areas of social justice, community welfare, and moral responsibility (Fadel, 2019).

The ethics of Islamic research are based on divine guidance, which balances the spiritual intention and the responsibility to society. These foundations are outlined in this chapter, which expresses the fundamental principles derived on the basis of the Holy Quran, the Hadith, and classical jurisprudence, and then explains how they apply to the modern research methods. Besides a theoretical exposition, the chapter provides practical tools- including ethical checklists and relevant guidelines- that can help scholars, ethics committees and institutional bodies to be ethical in various research situations.

The chapter is organized progressively from basic to functional levels. It starts by providing an overview of Islamic ethical principles (Table 1), moves on to a comparative analysis with Western frameworks to pinpoint areas of convergence and divergence and then situates Islamic ethics in the framework of the discussion of ethics in international research, in particular, artificial intelligence, digital research, and decolonising methodologies. Subsequent sections present direct application tools and examples of case studies based on biomedical, social, and environmental research, thereby demonstrating how theory can inform practice.

A conclusion is drawn from the current issues and suggests strategic ideas for improving Islamic input in international research ethics.

RESEARCH OBJECTIVES

The purpose of this chapter is to address in a systematic manner and explain the convergence of the Islamic ethical principles and the modern research practices, giving both the theoretical background and practical implementation. These research objectives have been properly formulated to cover the gaps in the existing literature, besides providing useful contributions to the intellectual world of Islam, as well as the research ethics discourse in the world.

Foundational Principles

To define the principles of research ethics in Islam, using the primary sources (the Quran and Hadith) and classical jurisprudence, it becomes necessary to create a consistent ethical system based on Islamic tradition.

Comparative Analysis

To compare Islamic ethics with modern world standards (e.g., APA, Belmont Report, Helsinki Declaration), trace areas of overlap, complement, and distinct contributions.

Case Studies

In order to examine case studies that can demonstrate the ethical dilemmas in research on Muslim populations or using the Islamic contexts, it is essential to demonstrate how the Islamic principles of ethics can be applied practically.

Integration

To learn about how Islamic ethics may help illuminate and enrich contemporary research practices, such as data collection, informed consent, intellectual honesty, and authorship in a variety of research fields.

Global Discourse

To contribute to the international debate on decolonizing research ethics by bringing in faith-based research paradigms grounded in the Islamic tradition, and advancing cultural sensitivity in the international research standards.

All these objectives are to develop a universal view of Islamic research ethics that is theory-grounded and practical. Through such goals, this chapter will be an invaluable resource to the researchers, ethics committees, and academic institutions to overcome ethical problems in research involving Muslim communities or Islamic values. In addition, it will have a role to play in the larger objective of designing more inclusive, culturally attentive research ethics on an international scale.

RESEARCH GAP

Despite the ethical framework offered by the Islamic tradition and the increasing literature of Muslim-majority backgrounds, there are significant gaps in how Islamic ethical principles are applied to current research. The literature is fragmented in disciplines, with bioethics being overrepresented, while fields such as social science, humanities, and emerging technologies are underdeveloped (Alahmad & Dierickx, 2012). This fragmentation restricts the development of an overall framework. Moreover, the lack of integration between Islamic principles and the existing international research standards, such as the Belmont Report or the APA Ethics Code, leads to practical problems on the part of the ethics committees and researchers working across cultures (Padela 2013).

Other key gaps include:

Lack of case-based learning: Islamic jurisprudence places a premium on case-based reasoning (*qiyas*), and yet there is a dearth of published case studies that apply Islamic ethics to real-world research (Ramadan, 2009).

Lack of methodological guidance: Although principles are articulated, there is no step-by-step guidance provided for researchers on how to apply them at various points in the research process (Ahsan, 2018).

Ignorance of emerging technologies: Ethical issues raised by artificial intelligence, genomics, and digital surveillance are left almost entirely unaddressed from an Islamic point of view.

The purpose of this chapter is to address these gaps, to offer an interdisciplinary study on the topic of Islamic research ethics, to unite theory and praxis, to present case-based studies across disciplines, and to propose the explicit guidance of research methodology. By doing so, it not only contributes to the current Islamic scholarship on the subject of ethics but it also aids the pursuit of culturally mindful studies more specifically and the research practices at large.

IMPORTANCE OF THE RESEARCH

Islamic research ethics is a field of prime concern with exceptional implications in numerous contexts that range as far as theories in the development of research ethics to potential theories of research ethics implementation into relevant research platforms. In this segment, the shape-shift in the ethics of research in an overall holistic Islamic framework is emphasized as a multidimensional significance encompassed in the building-up process of the entire internationalised academic landscape.

Theoretical Significance

It is because this chapter contributes to the world ethical discussion as it introduces an Islamic paradigm that enriches and opposes the existing Western paradigms, which further reinforces its ontological and epistemological advantages in promoting pluralism, human dignity, and scientific integrity (Padella, 2018). In addition to this, it translates the traditional Islamic concepts to the modern problems of research, thus showing that they remain relevant and versatile, and contribute to the further evolution of Islam ethical reasoning into the framework of contemporary knowledge (Kamali, 2010).

Practical Significance

The chapter supports the practical needs of researchers, institutional ethics committees, and organizations that work with Muslim communities by providing culturally competent ethical advice that would help to reconcile the gaps between Western normative models and the local practices (Alahmad et al., 2017). It forms the basis of developing culturally competent research methods that enhance participant recruitment, retention, as well as the quality of research carried out in communities with deep religious beliefs (Laird et al., 2007). By merging Islamic worldviews with the domain of global research ethics, this study makes the knowledge production more inclusive and fairer to over 1.8 billion Muslims all over the world (Tibi, 2009) and promotes the collaboration between Islamic and Western cultures by identifying common ethical ground and respecting differences (Fadel, 2019).

LITERATURE REVIEW: FOUNDATIONS OF ISLAMIC RESEARCH ETHICS

The academic study of Islamic research ethics has been transformed during the last decades from a general discourse on Islamic bioethics into more detailed applications of Islamic concepts to research situations. This literature review looks at the more important contributions to this new field, explores the development of this field, and identifies some important themes and perspectives.

Basic Islamic Code of Ethics

The academic sources underline a group of Islamic principles of research ethics, including *maslaha* (the interest of the population), la darar wa la dirar (no harm), *urf* (custom), and *karamah* (human dignity). These values confirm the sacred character of the humanistic dignity as the vicegerents of Allah (SWT) (Al-Bar and Chamsi-Pasha, 2015; Sachedina, 2009). Additionally, Ramadan (2009) claims that an ethical guidance should be based on higher goals (*maqasid*) of Islamic law, which puts classical ideals in context to answer the current research questions and provides the constructions in case of reconciliation of traditional ethics with a modern research environment.

Islamic Bioethics Bibliography

The most mature field of Islamic research ethics is the field of bioethics. Ghaly (2016) traces the appearance of Islamic bioethics as a new field, a process that evolved through the connections of religious scholars (*ulama*) with medical practitioners. Aasim Padela's multi-volume work (2013, 2018) has been invaluable for theorising Islamic bioethical frameworks and applying them to today's medical research dilemmas. Daar and al Khitamy (2001) have published an early comparative study of Islamic views on major bioethical topics, and Brockopp and Eich (2008) collect varying scholarly views on Islamic medical and scientific ethics.

Social Sciences Research Ethics Review

Second, Islamic perspectives on the ethics of research in social sciences are less, but are increasingly researched. Akhtar (2014) reviews ethics in the conduct of social research in Muslim communities, underlining the need to respect community values and cultural sensitivity. Fadel (2019) outlines a framework for the application of Islamic legal and ethical guidelines to social research and includes considerations for privacy, gender, and community welfare.

Comparative Ethical Standards

Comparative analysis of Islamic and Western ethics indicates similarities and differences especially in the field of bioethics. Sachedina (2012), Alahmad and Dierickx (2012), and Mustafa (2014) are some researchers who have highlighted the purpose of international and Islamic codes in informing the behavior of research.

The literature is in greater demand for the contextualized ethical approaches to the reflection of the Islamic values and their compatibility with international research standards. Nevertheless, other areas of the application of Islamic ethics to non-bioethical questions like digital research, artificial intelligence, and environmental studies also have a gap, and little empirical evidence of their practical application is observed across Muslim-majority populations (Alahmad et al., 2017).

The chapter is aimed at bridging these gaps by presenting a cross-disciplinary approach to integrate the Islamic ethical theory with real research.

RESEARCH METHODOLOGY

The chapter applies a qualitative methodological approach, which incorporates a textual approach, jurisprudential approach, comparative approach, and case study approach to explore the field of Islamic research ethics. It starts by providing a careful exegesis of the major sources, the Quran and the Hadith, focusing on such themes as veracity, justice, and social responsibility. In this aspect, the inquiry uses classical *tafsir* methodologies and hadith scholarship simultaneously with the modern hermeneutic in addressing the current ethical dilemma.

Besides the major texts, both classical and modern Islamic jurisprudence are also used in the study. It cites such foundational works of scholarship as al-Ghazali, Ihya Ulum al-Din and institutional *fatwas* as issued by organizations such as the Islamic Fiqh Academy. The methodological design also implies the systematic content analysis of research practices that are common in Muslim majority academic societies and a critical investigation of Islamic and secular theories of ethics, in particular, the Belmont Report and the American Psychological Association Ethics Code.

Lastly, the chapter gives a set of case studies across the health sciences, education, sociology and emergent technology domains, thus showing how definite ethical dilemmas can be resolved through the application of Islamic ethical guidelines such as the principle of *qiyas* (analogical reasoning). The chapter as a whole provides a cross-disciplinary, world-relevant schema to mediate between theoretical constructs and practice within the field of Islamic research ethics.

THE FOUNDATIONS OF RESEARCH ETHICS IN ISLAM

Islamic research is rooted in a strong statement of principles based upon the source text of the Quran, the Prophetic (Hadith), and the ancient intellectual tradition in Islamic jurisprudence (*Fiqh*). These are what control the religious business of education, and what the business of education ought to be conducted according to the moral limit.

The (Moral) Responsibility of Striving for Knowledge (*Talab al-Ilm*)

The faith that knowledge-seeking is not only permitted but it is also a religious duty that Islam demands is one of the foundations of the Islamic tradition. According to the Holy Prophet (peace and blessings of Allah be upon him), Allah obliged every Muslim to know something (Sunan Ibn Majah). Because of such a method, the investigation has become a truly spiritual practice when appropriately conducted, given the correct motive. However, the encouragement does not come gratuitously; there are some ethical obligations that are attached to the purpose and the methods of knowledge generation.

The Quran is full of its value to knowledge and the intellectual activity: "Are the men who know and the men who do not know the same? (Qur'an 39:9). Such an epistemology system enhances the pursuit of truth to a guiding value as well as to a morally oriented behavior where an overruling divine supervision is employed (Al-Attas, 1993).

The frequency of citation of the main ethical values from the Qur'an and the authorized corpus of Hadith highlights their seminal value in shaping and sustaining the Islamic ethical system.

Values of the Quran and of the Prophethood, which are more pertinent to the thesis in research practice, are:

Intention (*Niyyah*)

Intention is an important Islamic ethical criterion of moral value. The Prophet Muhammad (peace and blessings of Allah be upon him) said: "Actions are determined by intentions" (Sahih al-Bukhari). In research, this principle stressed that knowledge should be pursued for beneficial purposes and not for acquisitiveness, material gain, or harming others. Research that is driven by the public good (*maslahah*) is morally worthy, whereas research that is carried out with bad intentions or with disregard for consequences is unethical regardless of the study's methodological soundness (Kamali, 2008).

Truthfulness (Sidq)

Truthfulness is one of the cardinal virtues in Islamic ethics, and there are numerous injunctions against falsehood in the Qur'an: "O you who believe! Qur'an 9:119: Be fearful of Allah and be companions of those who are truthful. In the context of research, this principle requires honesty in every part of the research process, from data collection and analysis to reporting of results. It categorically forbids fabrication, falsification, and plagiarism as not only a violation of academic integrity, but also of religious ethics (Ramadan, 2009).

Trust (*Amanah*)

The notion of trust (*amanah*) is a main theme in Islamic ethics: "Indeed, Allah commands you to render trusts to whom they are due" (Qur'an 4:58). In research contexts, this principle sets out responsibilities to research participants, colleagues, sponsoring institutions, and society more generally. Researchers have obligations to fulfil, to keep confidentiality, to reward contributions, and to use research resources (Fadel, 2019).

Harm Avoidance (*La Darar*)

That edict that was set by Prophet Muhammad (peace and blessings be upon him) is, there shall be no harm, nor returning of harm (Sunan Ibn Majah), and is a principle of Islamic research ethics. This maxim requires researchers to reduce the harm done towards subjects, communities and the environment, and ensure that any benefits that may arise are more than the harm that may be caused (Al-Bar and Chamsi-Pasha, 2015). It highlights the fact that the production of knowledge in Islam is an ethically motivated one that is guided by values that protect human dignity, truthfulness and wilful investigation. These are the ethical values that form the foundation of every research process such as recruiting the participants, informed consent, analyzing data and the presentation of the findings.

These are the main principles that describe the Islamic moral responsibilities in the research field. The following chapters of the chapter then transfer the concepts to different research contexts so that they do not appear redundant, but rather contribute to concept clarity (Rattani and Hyder, 2019).

Table 1. Core Islamic Ethical Principles for Research

Principle	Qur'anic/Prophetic Basis	Research Application
Adl **(Justice)**	Qur'an 16:90	Fair participant selection; equitable distribution of benefits
Amānah **(Trust)**	Qur'an 8:27	Data accuracy, confidentiality, and transparency
Karāmah **(Human Dignity)**	Qur'an 17:70	Respectful treatment of participants
Ma la ah **(Public Good)**	Qur'an 2:201	Promoting community benefit and preventing harm
Ni ām **(Order/Balance)**	Qur'an 55:7–9	Maintaining harmony in scientific and social practices

ISLAMIC PRINCIPLES OF ETHICAL RESEARCH PRACTICE

According to the aforementioned core values, Islamic ethics prescribes certain specific guidelines for the various aspects of the research process. These guidelines are intended to provide concrete advice to researchers who want to carry out their activities in accordance with Islamic ethics and are operating within the framework of contemporary research approaches.

Informed Consent and Dignity of Participants

The Islamic concept of human dignity (*karamah*) extracted from the Qur'anic term of humans as Allah's (SWT) vicegerents on earth provides a good basis for research on subject rights. The Qur'an says: "We have honoured the procreation of Adam" (17:70), which creates an innate dignity that is due to us in all our human relations, such as research subjects (Sachedina, 2009).

Informed consent in Islamic ethics is not just about getting permission but about a genuine understanding and voluntary participation without coercion. The principle, enshrined by the Prophet Muhammad (pbuh) as "The property of a Muslim is not lawful [for another] except by his willing consent" (Musnad Ahmad), applies to taking part in research. This requires clear information to individuals about the purposes, procedures, risks, and benefits of research in language that is comprehensible to study participants (Alahmad and Dierickx, 2012).

Islamic ethics also acknowledges community consent as complementary to individual consent, especially in research on identifiable communities. This is an expression of the values of *maslaha*, that is, the welfare of the community, and the Islamic principle of *shura*, that is, mutual consultation. In practice, this may encompass the inclusion of community leaders or representatives in research planning, whilst also respecting individual autonomy (Fadel, 2019).

Confidentiality, Privacy, and Respect for Community Values

The Islamic right to privacy is based on the bar against intrusion into the affairs of others as summarized in the injunction of not spying into the business of another (Qur'an 49:12). This principle puts a heavy responsibility on local researchers in ensuring the security of the participants and other sensitive data. In addition, the jurisprudential principle of eliminating harm (al-darr al-yuzal) highlights this necessity where investigators are encouraged to pre-empt harm that occurs due to privacy encroachments (Kamali, 2008).

Research projects should be designed and conducted in a culturally sensitive manner thus ensuring that they are in line with the beliefs of the host community. The sensitivity of this matter requires adherence to gender norms, religious practices, and cultural taboos that may form the process of recruiting the participants, as well as acquiring the data. In the Islamic normative theory, the norms that exist in the culture are validated by customary practice (*urf*); they must not violate the essence of the core ethical values; hence, culturally sensitive research processes are acceptable (Ramadan, 2009).

The ethics of consent of informed consent in Islamic research is based on three main principles, which include transparency, dignity, and informed understanding.

Research Integrity: Plagiarism, Fabrication, and Falsification

Islamic ethics offers a holistic model of research ethics rooted in the central virtue of truthfulness (*sidq*) and trust (*amanah*). The Qur'an condemns clearly falsehood: "Do not clothe the truth with falsehood and do not conceal the truth while you know [it]" (2:42). This sets the highest standards for data integrity, honest reporting, and appropriate attribution (Iqbal, 2007).

Plagiarism *(Al-Sirqah Al-'Ilmiyyah)*

Islamic ethics disapproves of plagiarism as a kind of theft and falsehood. The Prophet Muhammad (peace and blessings of Allah be upon him) highlighted the importance of giving thanks by saying, “Whoever does not thank people does not thank Allah” (Sunan Abu Dawud). This is the means of providing adequate citation and acknowledgment as both scholarly and religious commandments. As has been described by Islamic scholars themselves, there were in the past elaborate chains of transmission (*isnad*) for knowledge, which laid the groundwork for strict forms of attributional practice (Al-Attas, 1993).

Fabrication (*Ikhtilaq*)

In Islamic ethics, the creation of data or outcomes is a grave transgression against honesty. The Qur'an cautions: “Do not say what you do not know of.” (17:36) This verse provides that claims cannot be fabricated, but must be founded on authentic evidence. The Prophet Muhammad (peace and blessings be upon him) further warned: “Whoever lies about me knowingly, then let him take his seat in Hell” (Sahih al-Bukhari) and so accentuates the importance of the avoidance of fabrication in the transmission of knowledge (Ramadan, 2009).

Falsification (*Tahrif*)

Falsification or reporting selectively to fit the objective of reaching preconceived conclusions is contrary to the Islamic injunction on truthfulness. The Qur'an denounces those who “twist the Scripture with their tongues” (3:78), thereby creating a rule that applies to all kinds of misrepresentation of knowledge. In the research context, this precludes selective reporting, p-hacking, and other forms of data manipulation (Akhtar, 2014).

Equity and Justice in who participates and how data is used

Justice (*'adl*) is one of Islam's cardinal values, and is enjoined in the Qur'an numerous times to “stand firmly for justice” (4:135). In research settings, this principle involves fair selection of research participants, not exploiting vulnerable populations, equitable sharing of research benefits and burdens, and attention to groups of people who have historically been more marginalized (Padela, 2013).

The Islamic principle of averting harm overrides the principle of acquiring benefit (*dar' al-mufasid muqaddam 'ala jalb al-masalih*) and creates a hierarchy of ethical considerations in research. This places research participant protection above any possible knowledge gains, especially if research is conducted with vulnerable populations or sensitive to learn (Al-Bar & Chamsi-Pasha, 2015).

The Use of Public Funds and Accountability in Research

Islamic ethics provides a stringent code of financial ethics rooted in the belief that all material resources are God's and man is a mere trustee. The Qur'an warns against misappropriation: "Do not eat of each other's wealth unjustly" (2:188). In research settings, this involves open and accountable use of research funds, absence of conflicts of interest, and accountability to stakeholders (Kamali 2008).

These principles together provide a systematic ethical code of practice for the conduct of research that calls for the need to balance the pursuit of knowledge with moral integrity. By incorporating these principles into research design, implementation, and dissemination, researchers can engage in research that not only meets scientific goals but also aligns with Islamic values.

PRACTICAL IMPLEMENTATION TOOLS FOR RESEARCHERS AND ETHICS COMMITTEES

The Islamic research ethics goes beyond the theoretical discussions by offering practical guidelines to ethical decision-making based on Quranic values, namely, *adl* (justice), *amanah* (trust), and *karamah* (human dignity). These value propositions guide the design and review of research with quantifiable benchmarks, putting the welfare of the people (*maslahah*) first, informed consent, data confidentiality, fairness and accountability. The compliance with the principles makes sure that scholarly inquiry is aligned with the procedural and general needs of spiritual stewardship and social well-being (Rattani and Hyder, 2019; Fadel, 2024).

COMPARATIVE INSIGHTS: ISLAMIC AND WESTERN RESEARCH ETHICS

A comparative study of Islamic research ethics and the dominant paradigms in the West shows areas of overlap, complementarity, and unique contribution. This comparative understanding promotes cross-cultural exchange and contributes to the development of more inclusive ethical frameworks that draw on a diversity of intellectual traditions.

Philosophical Principles: Convergence and Divergence

The ethics of Islamic and Western research have similar values, like the truth, non-maleficence and human dignity, but also differ in terms of philosophical background. Western ethics are usually the product of secular tradition, where personal autonomy and individual rights are prominently displayed, and Islamic ethics stand on the foundation of the divine revelation provided through the Quran and the Sunnah and combine the spiritual and secular aspects (Sachedina, 2012). Such divine grounding brings about unique motivations such as responsibility to God (taqwa), conscience of the repercussion in this world and the next world, obedience to Allah's (SWT) wisdom, whereby ethical research behavior becomes a spiritually oriented and who listic activity (Padela, 2013).

Connections with the Belmont Report

The three central values of the Belmont Report-respect for persons, beneficence, and justice-have clear and similar counterparts in Islamic moral tradition, albeit with definite differences:

Respect for Persons vs *Karamah* (Dignity)

While the Belmont Report's basis for respect for persons is found in individual autonomy, Islamic ethics stresses human dignity *(karamah)* bestowed by God. This broader conception acknowledges autonomy, but locates it in the context of relationships and values of community. Both traditions recognise informed consent, although Islamic approaches might emphasize greater importance of family in decision-making and community wellbeing in addition to individual choice (Alahmad et al., 2017).

Maslahah vs. Beneficence (Public Welfare)

The Islamic notion of *maslahah* (public welfare) is similar to the Belmont principle of beneficence (maximization of benefit while minimizing harm). Islamic ethics, on the other hand, has a more sophisticated formulation of the benefit and harm calculus in the form of *maqasid al-shari'ah* (higher objectives of Islamic law), which gives priority to the protection of faith, life, intellect, lineage, and property. The ethical dilemma resolution framework offers systematic directions for solving situations in which benefits and harms have to be weighed against each other (Al-Bar & Chamsi-Pasha, 2015).

Justice vs. '*Adl*

Both traditions stress equality in the distribution of benefits and burdens of research. The Islamic notion of justice ('adl) has distributive, procedural, and social aspects with special focus on safeguarding vulnerable groups and rectifying structural unfairness. Islamic ethics puts particular emphasis on the preferential option for the vulnerable in light of many Qur'anic injunctions to take care of the disadvantaged (Ramadan, 2009).

Comparative Insights on Specific Research Practices

Beyond broad principles, comparing specific research practices reveals both similarities and distinctive approaches:

Table 2. Research Practices in Western and Islamic Approaches

Research Practice	Western Approach	Islamic Approach
Informed Consent	Individual-focused, written documentation emphasized	Values individual consent while recognizing family/ community contexts; verbal processes may be culturally appropriate
Privacy & Confidentiality	Primarily framed as an individual right	Protected based on the prohibition of spying and exposing others' faults; community reputation is also considered
Risk Assessment	Typically utilitarian calculation of the risk/benefit ratio	Employs legal maxims like "harm must be eliminated" and hierarchical assessment of necessity
Vulnerable Populations	Protected based on autonomy limitations and exploitation risk	Special protection based on religious obligation to care for the disadvantaged; community responsibility emphasized
Publication Ethics	Professional standards focused on integrity and transparency	Religious duty to disseminate beneficial knowledge while avoiding harm; intention in publication matters

Potential for Integration and Mutual Enrichment

The comparative considerations suggest that there is a considerable space for integration and mutual fertilization between Islamic and Western traditions of research ethics. Islamic ethics can inform in unique ways our understanding of intention, theologically motivated justifications for ethics, community content of ethics, and decisional hierarchies with respect to ethics. On the other side are the Western frameworks, which introduce complex procedural solutions, institutional mechanisms, and a rich case literature that can act as a precedent for application (Mustafa 2014).

Rather than dissolve them or homogenize them, a comprehensive approach would neither discard nor dissolve these traditions, but would recognize their distinctive contributions while identifying points of convergence in their commitment to common ethical ideals that foster research practice in global contexts. In this respect, the practice can act towards unifying more inclusive ethical traditions, which would be sensitive to cultural and religious diversity, while being anchored to fundamental commitments to human dignity, veracity, and social responsibility in research (Fadel 2019).

ISLAMIC ETHICS AND GLOBAL RESEARCH ETHICS DISCOURSE

Islamic research ethics has a number of core principles in common with international ethical codes like the Declaration of Helsinki, the CIOMS Guidelines and, in particular, its focus on justice, respecting persons, and beneficence. Still, it provides a unique structure and is based on the divine responsibility and social good.

In modern discussions of artificial intelligence, genomics, and decolonisation of ethics, Islamic ethics offers another epistemic paradigm, which is based on moral intention (*niyyah*) and the responsibility of a group (*takaful*) (Ghaly, 2021).

This paradigm goes in line with the suggestions of decolonial theorists that more comprehensive, non-Western moral logic should be included in the paradigm (Smith, 2021; Mignolo, 2020).

Based on the *maqāsid al-Shariah* which are the aims of the Islamic law, Islamic research ethics is a combination of faith, reason and social justice. Accordingly, it brings out an equal and more global discourse of research ethics.

CONTEMPORARY CASE REFLECTIONS

The application of Islamic ethics to modern research situations also exposes the vitality of the ethics and the problems of its application. This section looks at four case studies that present ethical dilemmas in a variety of research settings and discusses the ways in which Islamic ethics could guide how to address them.

Case Study 1: Biomedical Research in IDP Camps

Background: European scientists have suggested that a vaccine efficacy study should be performed in a large Muslim refugee camp in Jordan. The trial aims to test a new vaccine against a disease that is prevalent in the camp, which could help the residents of the camp. However, issues concerning the safety of respondents, the suitability of informed consent procedures, and the cultural suitability of the study procedures are a matter of concern.

Ethical Analysis: The case raises a number of Islamic ethical principles, such as defending vulnerable populations, the validity of informed consent, and benefits versus harm. Human dignity (*karamah*) is the principle that states that refugees are not to be considered only as research subjects but as moral beings that deserve respect. The principle of harm prevention (*la darar*) thus creates problems of sufficient protection in the light of the vulnerability of the population (Alahmad et al., 2017).

Islamic Ethical Approach: An Islamic ethics would indicate:

- Community consultation by camp religious leaders and representatives to make the culture appropriate.
- Improved circumvention of limited literacy and power inequalities in informed consent.
- Guarantee of physical goods must come to the camp people even after the research.
- Bearing in mind gender norms when interacting with the research and ensuring equal access to possible benefits.
- Open communication about the financial support of research, possible commercialisation and ownership of data.

This model highlights the importance of procedural justice and community involvement and respects the autonomy of individuals, which explains how Islamic ethics may provide a feasible path to follow in complicated cases of research (Padela, 2013).

Case Study 2: Gender and Religion Study among Conservative Communities.

Context: A social science researcher tries to explore religious interpretations and religious practices of women in a conservative Muslim community in South Asia. Through the study, the researcher seeks to chain different interpretations of religious texts on gender roles to establish the reason as to why some gender roles are practiced by Muslims. The issues of female accessibility, sensitivity of the subject, possible community disapproval and positionality of the researcher as an outsider are ethical issues.

Ethical Dilemma: Optimizing the goals of research with consideration to the community standards and guarding and protecting the participants.

Islamic Principles: right to privacy (*sitr*), gender relationship regulations, community welfare, and honesty.

Ethical Approach:

- A participatory approach, which includes female research assistants.
- Community involvement in the course of research.
- Careful presentation of results to maintain integrity.

Research implications: The research must be respectful of the community, record voices and experiences.

Ethical Analysis: This case brings out the conflicts existing between the aims of research and cultural sensitivity. Islamic morality enhances intimacy of privacy (*sitr*), proper gender relations, and life in society. The importance of the truth, which needs to be communicated, has to be weighed against the possible distortion or damage in case the findings have not been explained in the context (Akhtar, 2014).

Islamic Ethical Approach: The Islamic ethical approach would propose:

- Use of female researchers of the same cultural background when dealing with female participants.
- Building community-based research designs and interpretation.
- Privacy and confidentiality assurance by using proper data collection strategies.
- Reporting results cautiously without supporting stereotyping and being truthful at the same time.
- Giving back research results to society to be validated and beneficial.

This method shows how Islamic ethics can help conduct culturally sensitive studies and maintain scientific integrity and the values of the community (Fadel, 2019).

Case Study 3: Online Ethics and Social Media Studies.

Context: A researcher will analyze the social media posts of Muslim youths that are open to the public and are related to religious identity and practice. The analysis aims to shed some light on the way the religious interpretations of the digital natives develop. Even though the data might seem publicly available, there are questions concerning the expectations of privacy, study approval, personal identification, and integrity.

Ethical Analysis: The case raises new issues on the border of digital ethics and Islamic values. Although the technical legality of working with public data can be determined, the Islamic ethics focus on intent and privacy respect and contextual harm. The Quranic prohibition of spying (49:12) is an indication that care should be taken even when the information seems open, especially when the users did not envisage the use of research (Mustafa, 2014).

Islamic Ethical Approach:

- Remember the initial background and purpose of social media users repurposing their content.
- Utilised greater anonymization than usual to conceal identity in religious settings.
- Be open with research purposes where face-to-face interaction is possible.
- Reason the results (Be sensitive to the stage of religious identity youths are in)
- Provide the research advantages to the populations being studied.

This strategy shows that traditional Islamic values can be used to address new ethical issues in digital research situations, which guides in areas where the traditional paradigm can prove to be insufficient (Ghaly, 2016).

Case Study 4: Cross-cultural Partnership Research

Background: The practice of environmental sustainability is researched in terms of a joint work between the institutions of the Western world and the countries dominated by Muslims. There is a problem of power asymmetry in terms of funding and publication, research priorities, data ownership and cultural differences in the research practice and ethical review processes.

Ethical Analysis: The case brings up issues of impartiality, dignity, and reciprocity of cross-cultural alliances. Justice (*adl*), consultation (*shura*) and reciprocity are the main elements of ethical collaboration in Islam. Being based on the principle of useful knowledge (*ilm nafi*), research must benefit the community instead of outside interests (Ramadan, 2009).

Islamic Ethical Approach:

- Have a clear understanding of who owns the data, who is the author and who has the intellectual rights.

- Make sure that the project design is based on the local research priorities under the basis on the real consultation.
- Capacity-building among the local researchers and not extractive models.
- Recognise each other in respect of institutional ethical review processes and find a common ethical understanding.
- Promote fair sharing of research findings, such as publication and application.

In addition, they illustrate how Islamic ethical values can be used as a source of guidance in a variety of research settings to complement international ethical standards. They speak both about the general nature of Islamic morals and the contextualization of this general nature to the specific challenges that come up with research (Alahmad et al., 2017).

EMERGING FIELDS: DIGITAL, SOCIAL, AND ENVIRONMENTAL RESEARCH

Biomedical ethics as an academic discipline among the Islamic world has been recognized as a serious area of study since ancient times, but with the advent of artificial intelligence, digital research, and environmental studies, there are new ethical issues that require special academic interest.

Amanah (trust), *adl* (justice), and *niyyah* (intent) are some of the Islamic values that are mentioned in the spheres of artificial intelligence and data research to stress the principles of being transparent, just, and morally intended (Hayat, 2024). Equally, such values as *karamah* (human dignity), khilafah (stewardship) and *mizan* (balance) are encouraged within the social and environmental studies and thus encourage respect, sustainability, and well-being of the communities (Waqas, 2025).

When combined with each other, the principles provide evidence that Islamic ethics is a highly integrated system that transcends the boundaries of disciplines and integrates different moral, social, and ecological aspects of knowledge.

CHALLENGES AND THE WAY FORWARD

Although Islamic research ethics has relevant principles and guidance, it has various challenges when applied in the current research settings. To counteract these issues, there must be approaches to strike a balance between loyalty to the Islamic codes of ethics and active involvement in the world of research. This part explores several important issues and suggests directions in the Muslim research ethics on the theory and practice.

Institutionalization of Islamic Research Ethics

The fact that Islamic research ethics is not institutionalized in academic/research institutions is considered one of the greatest threats. Although there are partially developed Islamic structures in bioethics, like the Islamic Organization of Medical Sciences (IOMS), in most fields, there are no organizational arrangements to implement Islamic ethics in research. There is also the approach whereby the same ethics review committees in Muslim-majority countries are modelled after the Western approach, with little emphasis on religious and cultural localities (Alahmad and Dierickx, 2012).

Challenges

- Limited representation of Islamic perspectives in global research ethics discourse
- Insufficient integration of Islamic ethics in institutional review processes
- Gaps between theoretical principles and practical implementation guidelines
- Diversity of interpretations within Islamic tradition
- Balancing universal ethical principles with contextual applications

Opportunities

- There has been an increasing concern in decolonized and culturally responsive ethical frameworks.
- The research capacity is growing in the Muslim-majority countries.
- The Islamic ethical viewpoints can be spread more widely by the means of digital platforms.
- Interreligious dialogue between Islamic thinkers and research practitioners is being encouraged.
- It is possible to establish integrated ethical models that do not ignore diversity and still have the main standards.

The possible remedy to this issue is to set up institutional frameworks that can transform Islamic ethics into viable guidelines and procedures. Possible solutions would be:

- Establishing specialized Islamic research ethics committees that include both religious scholars and research methodologists
- Developing institutional review board (IRB) guidelines that incorporate Islamic ethical considerations alongside international standards

- Creating networks of ethics consultants with expertise in Islamic ethics who can advise researchers and institutions
- Advocating for inclusion of Islamic perspectives in international research ethics guidelines and governance structures

These institutional methods would facilitate the transition from theory to practice and would give the researchers tangible guidelines on how Islamic ethics can be applied in their work (Ghaly, 2016).

Training and Awareness for Researchers and Ethics Committees

The other problem is that researchers and ethics review committees are ignorant and not trained in Islamic research ethics. Most Muslim scholars are also taught solely the western standards of research ethics, but scholars taught within the Islamic world are rarely exposed to modern research methodology and ethical issues. Such a discrepancy forms a knowledge gap that inhibits the application of the Islamic ethical principles to the context of research (Alahmad et al., 2017).

To solve this problem, it is required to adopt extensive educational campaigns:

- Preparation of curriculum material about Islamic research ethics as part of research methods courses;
- Development of special training for the members of the ethics committee on Islamic principles to use in reviewing research;
- Implementing mentorship initiatives that match accomplished researchers with familiarity in Islamic ethics with novice investigators at the beginning of their careers;
- Publication of available materials, including cases, decision models, and ethics textbooks based on Islamic values;
- Holding conferences and workshops where Islamic scholars, as well as research practitioners, can learn together.

The long-term goals of these educational programs include developing the ability to ethically conduct research based on the Islamic principles and also developing a community of practice that will be able to solve the emerging ethical challenges (Padeda, 2013).

Potential for Integrated Ethical Models

The most promising way seems to be the development of combined ethical models which combine the universal research standards with the Islamic values. They are supposed to demonstrate Islamic ethics as an alternative to the international norms, but as their extension that can help expand the ethical analysis with new insights and foundations. The combination is an integrative form, but still open to the unique contributions of the Islamic tradition, yet engaged in the universal ethical conversation (Sachedina, 2012).

Potential approaches to integration include:

- Identifying areas of convergence between Islamic principles and international standards as a foundation for integrated frameworks
- Developing "ethical enhancement" approaches that supplement basic ethical requirements with Islamic ethical considerations
- Creating decision matrices that incorporate both universal ethical principles and Islamic-specific considerations
- Fostering cross-cultural dialogue on research ethics that respects diverse religious and cultural perspectives
- Documenting case studies where Islamic ethics has successfully informed research in diverse contexts

These collectivises of frameworks are an act of respect for the thesis that ethical traditions are not fixed, but always in movement, and run through communication between traditions. These institutions of higher learning can provide conduits for Islamic ethics, contributing to this global conversation and increasing awareness among researchers who work across cultures, in order to empower real-life views (Fadel, 2019).

Institutionalisation, training, and integration can overcome these obstacles and contribute to the practical application of Islamic research ethics from theory to practice, so that it may serve to enrich Islamic scholarship tradition and contribute to the research ethics literature globally. This development has been accomplished through an alliance between the custodians of religious knowledge, the researchers, and the ethical commission in a bid to create an ethical framework that is appropriate to Islamic cultural values but is also capable of addressing the demands of contemporary research.

CONCLUSION

This chapter has presented briefly the rich history of the research ethics of the Islamic world and has examined its components, direct consequences, and application in the contemporary world. In this examination, the Islamic ethical tradition provides an extensive ethical scheme for the research that meets and supplements the global ethical requirements, and also establishes the path for the Muslim researchers' and subjects' needs.

Foundational Principles

It is now disclosed that the Islamic research ethics is based on the significant pillars of the Quran and the Prophetic tradition, that is, the *biti's* (*sidq*: honesty), *madh-humiyya* (*adl:* justice), *niyyah* (intention), *amanah* (trust), and awf (not causing harm). These are values expressing an extended moral code, not only in the context of research but also in the motivations of the researcher and social consequences of knowledge-making in the broadest sense of the collective.

Practical Applications

Such guiding principles at a range of points can be woven together, forming a kind of professional toolkit of ethical advice at various points of the research process, on the following issues but not at all limited to: from informed consent and choice of participants, data basing, for instance, to publication ethics. Case studies would have been presented with regard to the limitations of Islam for the solutions of intrinsic multi-dimensional ethical issues in biomedical studies, social sciences, the digital world, and inter-cultural practices in research studies.

Comparative Insights

Our comparative study with the Western ethical traditions helped us to describe the areas where they overlap and complement, and what value was added that Islamic ethics contributed. This analogy suggests that Islamic ethics can substantiate some contribution to the discourse on intention, community well-being, and hierarchical ethical argument and, thus, enhances the international debate on ethical research.

Implementation Challenges

Some challenges that the chapter has recognized in the implementation of the Islamic research ethics into practice include this: there is no institutionalization as well as training as well as there is a need to have integrated models that may be reflective as far as Islamic values are concerned as well as international standards. We recommended institutional building, educational programmes, and integrative solutions, amongst others, as ways of coping with them.

There are quite important findings in this examination. Firstly, Islamic research ethics presents a spiritually based approach to ethics that connects ethical behavior with religious concepts that providing additional inspirational resources to ethical conduct that may not rely on professional standards or legal requirements of ethical action. The second factor that may help the researcher comply with ethical practice by being accountable to the signs of God *(taqwa)* is the spirituality of the incorporation of the Quran by the researcher.

Second is the Islamic ethics, which is highly social and communal-based based in which personal research activities have a wide range of settings the welfare and social responsibility of the community. It is a methodology that attempts to bypass that overly individualistic approach to the problem of ethics so common in many Western moral traditions, and that is concerned with the implications that the research will have on societies, cultures, and social arrangements.

Third, Islamic moral tradition provides advanced categories of moral thinking through the categories of *maqasid al-shari'ah* or the higher end (s) of Islamic law, and through rules of law that are then used as tools to resolve ethical dilemmas. These frameworks provide the means through which competing ethical variables are weighed and a decision taken on which aspects of the common situations are to dominate when it tends to be the case that ethical principles are at odds with each other.

Finally, the analysis shows that Islamic research ethics is not a strict historical tradition, but a resource that could be used in managing the problems of ethics today. In the field, Islamic ethics has been and continues to evolve through a sustained conversation between its classical foundations and 21st-century reality, and can provide the relevant outlook to scholars operating in complex ethical environments in various contexts.

These facts demonstrate that the ethics of Islamic research can play a significant role, which was once referred to as the ethics of research at the World Conference, and that such perspectives would render the improvement of ethical frameworks across cultures on the one hand and influence the minds of Muslim researchers and the Muslim population work respectively. Imagine that through the absence of a single form of cultural and other inconsistent contributions of our own contributions to ethical discourse, other various ethical traditions, we can come up with the more moderating, self-awareness-directed conceptions of research ethics, somewhat more acquiescent of alternative values, and unswervingly committed to the principles of human dignity, veracity, and social accountability.

RECOMMENDATIONS AND FUTURE DIRECTIONS

Based on the analysis that has been drawn in this chapter, it follows that there are a few recommendations that can be made on the formulation of Islamic research ethics in both theory and practice. These are recommendations to different stakeholders like researchers, universities, ethical review boards, and policy makers to entrench stronger and contextually relevant ethical frameworks that integrate the Islamic standards with international research standards.

Practice Implications for Research

To individual researchers and research teams, we suggest:

Guided Ethical Thinking

Include reflective planning (*niyyah*) as part of planning the research, and consider not only methodological rigor but also purpose and implications of the research. This deliberation is to be based on the issues of social benefit, constraint from evil, and the compatibility with the principles of Islam.

Better Community Connectivity

Continue to develop best practices for research consultations with Muslim communities, such as involving religious as well as community leaders and other stakeholders during the research design, implementation, and dissemination process. This must go beyond a tokenistic consultation process and be an explicit partnership where community knowledge and priorities are explored.

Culturally Relevant and Culturally Proficient Strategies of Teaching

Adapt research design to Islamic cultural and religious sensitivities, especially issues related to gender relations, privacy, and community sensitivities. This accommodation needs to be medically reasonable, but it also needs to be religiously reasonable for the participants.

Ethical Data Management

Data management practices founded on Islamic principles of trust (*amanah*), confidentiality, and privacy (including improved communication on data use, data security, and deferential treatment to participants as the owners of their information).

Recommendations for Institutional Development

To the academic institutions, research organizations, and funding agencies, we suggest:

Specialized Ethics Special Committees: create Special Ethics Committees that encompass expertise in Islamic ethics with conventional research ethics, which give researchers a framework of multiple ethical perspectives.

- Establish Institutional Principles that are specifically designed to provide strong guidance on the role of Islamic ethics in research and yet allow themselves to be interpreted broadly by the Islamic tradition.
- Establish Islamic research ethics training programs among researchers, ethics committee members, and research administrators, and develop the capacity to conduct ethical research practice based on Islamic principles.
- Foster Interdisciplinary Discussion among Islamic scholars, research methodologists, and specialists in ethics to come up with practical frameworks to promote the alignment between theoretical knowledge and the practice of research.

Fund Research on Islamic research ethics, and empirical research on how ethical principles are implemented in practice, and how to develop empirical case-based learning materials.

Recommendations for Policy Development

To policy makers, regulatory agencies, and international organizations, we propose:

The inclusion of Islamic Perspectives concerning the development of national and international research ethics policy in a way that will help to have a wider scope of religious and cultural viewpoints in the policy documents.
Institute Cross-cultural Ethical Review Mechanism(s) in deference to the predictability of ethical norms and contextual suaveness of mores and values to facilitate ethical research working across a variety of environments.

Preservation The International Exchange in the field of integration of religious Ethics and research practice to encourage clarity in dialogue between other ethical customs and paradigms.

Understand Alternative Models of Informed Consent and Participant Engagement that Might Better Fit Islamic Cultural Contexts While Retaining Core Ethical Commitments to Participant Welfare and Autonomy

Allow for Ethical Pluralism that recognizes that multiple valid models of research ethics exist and should not be replaced by a single standardized model in every situation.

Future Research Directions

To spread the awareness and practice of the ethics in conducting Islamic research, several promising avenues of further investigation emerge:

Emerging Technologies

Explore Islamic moral wisdom about new technologies in research, including artificial intelligence, genomics, neuro-technology, and digital surveillance, and establish ethical frameworks that address emerging challenges in the context of their Islamic insights.

Diverse Muslim Contexts

Examine meanings and applications of Islamic research ethics in the context of Muslims, and how they are understood, with respect to geographic distinctions, legal schools (madhabs), and cultural differences, to come up with a more nuanced sense of local practices;

Empirical Studies

Prepare empirical studies on the conceptualisation and operationalisation of Islamic ethical principles by researchers, by ethics committees, and participants, and trace gaps in the theory and practice of research.

Integrated Models

Build and experiment with integrated frames of ethics that connect the Islamic values with international research standards, and evaluate them in a range of research situations and in order to refine strategies on the basis of real-world experience.

REFERENCES

Akhtar, S. (2014). The ethics of conducting research with Muslim participants. *Research Ethics, 10*(3), 138–149.

Al-Attas, S. M. N. (1993). *Islam and secularism.* International Institute of Islamic Thought and Civilization.

Al-Bar, M. A., & Chamsi-Pasha, H. (2015). *Contemporary bioethics: Islamic perspective.* Springer. DOI: 10.1007/978-3-319-18428-9

Alahmad, G., Al-Jumah, M., & Dierickx, K. (2017). Confidentiality, informed consent and children's participation in research involving stored tissue samples: Interviews with medical professionals from the Middle East. *Narrative Inquiry in Bioethics, 7*(1), 71–79. PMID: 28713147

Alahmad, G., & Dierickx, K. (2012). What do Islamic institutional fatwas say about medical and research confidentiality and breach of confidentiality? *Developing World Bioethics, 12*(2), 104–112. DOI: 10.1111/j.1471-8847.2012.00329.x PMID: 22702346

Brockopp, J. E., & Eich, T. (2008). *Muslim medical ethics: From theory to practice.* University of South Carolina Press.

Daar, A. S., & Al Khitamy, A. B. (2001). Bioethics for clinicians: 21. Islamic bioethics. *Canadian Medical Association Journal, 164*(1), 60–63. PMID: 11202669

Fadel, H. E. (2024). Ethics of clinical research: An Islamic perspective. *Journal of the Islamic Medical Association of North America, 56*(2), 45–53.

Fadel, M. (2019). The Islamic ethical framework for religious minority rights in research contexts. *The Journal of Religious Ethics, 47*(3), 430–458.

Ghaly, M. (2016). Islamic bioethics in the twenty-first century. *Zygon, 51*(3), 732–751.

Ghaly, M. (2021). Islamic perspectives on genomics and global bioethics. *Developing World Bioethics, 21*(3), 105–113.

Hayat, I. (2024). Islamic ethical perspectives on artificial intelligence and digital surveillance: Principles, challenges, and policy recommendations. *International Research Journal of Arts. Humanities and Social Sciences, 2*(1), 12–24.

Iqbal, M. (2007). *Science and Islam.* Greenwood Press. DOI: 10.5040/9798216011477

Kamali, M. H. (2008). *Principles of Islamic jurisprudence.* Islamic Texts Society.

Kamali, M. H. (2010). *Islamic law in the modern world: Malaysian perspectives*. Islamic Book Trust.

Laird, L. D., de Marrais, J., & Barnes, L. L. (2007). Portraying Islam and Muslims in MEDLINE: A content analysis. *Social Science & Medicine*, *65*(12), 2425–2439. DOI: 10.1016/j.socscimed.2007.07.029 PMID: 17767988

Mignolo, W. D. (2020). *On decoloniality: Concepts, analytics, praxis*. Duke University Press. DOI: 10.1515/9780822371779

Mustafa, Y. (2014). Islam and the four principles of medical ethics. *Journal of Medical Ethics*, *40*(7), 479–483. DOI: 10.1136/medethics-2012-101309 PMID: 23975951

Padela, A. I. (2013). Islamic bioethics: Between sacred law, lived experiences, and state authority. *Theoretical Medicine and Bioethics*, *34*(2), 65–80. DOI: 10.1007/s11017-013-9249-1 PMID: 23588855

Padela, A. I. (2018). Islamic bioethics and animal research: Principles, attitudes, and practices. *Cambridge Quarterly of Healthcare Ethics*, *27*(1), 121–132.

Padela, A. I., Malik, A. Y., Curlin, F., & De Vries, R. (2014). [Re]considering respect for persons in a globalizing world. *Developing World Bioethics*, *15*(2), 98–106. DOI: 10.1111/dewb.12045 PMID: 24720355

Ramadan, T. (2009). *Radical reform: Islamic ethics and liberation*. Oxford University Press.

Rattani, A., & Hyder, A. A. (2019). Developing an Islamic research ethics framework. *Journal of Religion and Health, 58*(1), 3–17.* https://doi.org/DOI: 10.1007/s10943-017-0508-8

Sachedina, A. (2009). *Islamic biomedical ethics: Principles and application*. Oxford University Press. DOI: 10.1093/acprof:oso/9780195378504.001.0001

Sachedina, A. (2012). Islam and biomedical ethics. In Kuhse, H., & Singer, P. (Eds.), *A companion to bioethics* (pp. 106–114). Wiley-Blackwell.

Smith, L. T. (2021). *Decolonizing methodologies: Research and Indigenous peoples* (3rd ed.). Zed Books. DOI: 10.5040/9781350225282

Tibi, B. (2009). *Islam's predicament with modernity: Religious reform and cultural change*. Routledge. DOI: 10.4324/9780203881552

Waqas, M. (2025). Exploring the ethical framework of science: A critical study from the Islamic perspective. *Pakistan Journal of Islamic Philosophy*, *7*(1), 18–31.

Yin, R. K. (2018). *Case study research and applications: Design and methods* (6th ed.). Sage Publications.

KEY TERMS AND DEFINITIONS

Maslahah (public good):: stipulates the fact that research activities must foster societal good and make favorable contributions towards communal good.

***Niyyah* (intention):** concept is the one that commits the scholars to conduct research not because it benefits themselves but because it aims at the promotion of human good and the pursuit of truth, and not because it involves self-enrichment or evil motives.

Sidq **(truthfulness):** requires that investigators are completely honest and truthful in the whole research process, and this implies that no fabricated, falsified, or plagiaristic data should be produced.

The concept of *adl* **(justice):** commits research to follow the principles of fairness and equity in the selection of the participants, allocation of benefits and protecting vulnerable groups.

The term *Amanah:*: (trust) requires researchers to be custodians of knowledge, data and resources, to maintain confidentiality, attribute and be responsible custodians of knowledge.

Chapter 11
Ethics of Using Artificial Intelligence and Digital Tools in Research

Debajit Karmakar
http://orcid.org/0000-0001-9272-0627
Lakshmibai National Institute of Physical Education, India

Anindita Das
http://orcid.org/0000-0002-6502-9619
Lakshmibai National Institute of Physical Education, India

Tashi Ongmu Bhutia
http://orcid.org/0009-0000-0127-5709
Government B.Ed. College, Soreng, India

Karuppasamy Govindasamy
http://orcid.org/0000-0002-3019-5545
Symbiosis International University, India

Abhishek Kumar
http://orcid.org/0009-0009-6607-5375
Indira Gandhi Institute of Physical Education and Sports Sciences, India

EXECUTIVE SUMMARY

The rapid advancement of artificial intelligence and digital technologies has transformed the research landscape, offering powerful tools for data collection, analysis, and interpretation. Alongside these opportunities, new ethical challenges have emerged, extending beyond traditional research ethics. Issues such as algorithmic bias, data privacy, ownership, transparency, and accountability have become cen-

DOI: 10.4018/979-8-3693-8302-5.ch011

tral to maintaining integrity in scientific inquiry. This chapter explores key ethical principles, including respect for persons, beneficence, non maleficence, and justice, in the context of AI assisted research. It examines risks of overreliance on automated decision making, threats to authenticity in digital environments, and complexities of intellectual property. Emphasis is placed on responsible innovation, human oversight, global guidelines, and frameworks to balance technological potential with ethical responsibility.

INTRODUCTION

The rapid advancement of artificial intelligence and digital technologies has fundamentally transformed the landscape of academic and scientific research. From sophisticated machine learning algorithms that can analyze vast datasets in minutes to automated research platforms that streamline data collection processes, these technological innovations have opened unprecedented opportunities for discovery and knowledge creation. However, with these powerful capabilities comes an equally significant responsibility to consider the ethical implications of their implementation in research contexts (Walker, 2007).

The integration of artificial intelligence and digital tools in research presents a complex web of ethical considerations that extend far beyond traditional research ethics frameworks. While conventional research ethics has long focused on protecting human subjects, ensuring informed consent, and maintaining data confidentiality, the emergence of AI-driven research methodologies introduces new dimensions of ethical complexity. Questions about algorithmic bias, data ownership, transparency in automated decision-making, and the potential for technology to inadvertently perpetuate social inequalities have become central concerns for researchers, institutions, and policymakers alike (Powell, 2020).

This chapter provides a comprehensive examination of the ethical landscape surrounding the use of artificial intelligence and digital tools in research. It explores the fundamental principles that should guide researchers in making ethical decisions about technology adoption while addressing the specific challenges that arise when human intelligence intersects with artificial intelligence in the pursuit of knowledge. The discussion encompasses various research domains, from biomedical studies that rely on AI-powered diagnostic tools to social science research that utilizes digital platforms for data collection and analysis (Walker, 2007). The significance of addressing these ethical considerations cannot be overstated. As research increasingly depends on AI and digital technologies, the decisions made today about their ethical implementation will shape the future of scientific inquiry and its impact on society. Researchers must navigate the delicate balance between leveraging technological capabilities to advance knowledge and ensuring that such advancement occurs within ethical boundaries that respect human dignity, promote social good, and maintain the integrity of the scientific enterprise (Pennock, 2018).

UNDERSTANDING ARTIFICIAL INTELLIGENCE AND DIGITAL TOOLS IN RESEARCH

- **Defining Artificial Intelligence (AI)**

Artificial Intelligence, in the context of research, refers to computational systems that can perform tasks typically requiring human intelligence, including pattern recognition, decision-making, natural language processing, and predictive analysis. Within research environments, AI encompasses various technologies ranging from simple rule-based systems to sophisticated neural networks capable of deep learning. The scope of AI in research includes machine learning algorithms that can identify patterns in complex datasets, natural language processing tools that can analyze textual data, computer vision systems that can interpret visual information, and predictive models that can forecast outcomes based on historical data (Webster et al., 2013).

The evolution of AI in research has moved from basic statistical analysis tools to comprehensive systems capable of autonomous hypothesis generation and testing. Modern AI research tools can process multimodal data, integrate information from diverse sources, and provide insights that might not be readily apparent through traditional analytical methods. This transformation has made AI an indispensable component of contemporary research methodology, particularly in fields dealing with large-scale data analysis, complex modeling, and pattern recognition (Rani & Sharma, 2012). Understanding AI in research also requires recognizing its limitations and potential for error. AI systems operate based on the data they are trained on and the algorithms that govern their processing. This dependence means that AI tools can perpetuate biases present in training data, make errors in novel situations not represented in their training sets, and produce results that may lack the contextual understanding that human researchers bring to their work. These characteristics have significant ethical implications that researchers must consider when incorporating AI into their methodologies (McFee, 2014).

- **Digital tools in research (data management, simulation, analytics, etc.)**

Digital tools in research encompass a broad spectrum of technologies that facilitate various aspects of the research process, from initial data collection to final dissemination of results. Data management systems have revolutionized how researchers collect, store, organize, and share research data. These platforms enable real-time collaboration among research teams, automated data validation and cleaning processes, and sophisticated version control systems that maintain the integrity and traceability of research datasets (Rani & Sharma, 2012).

Simulation tools represent another critical category of digital research instruments, allowing researchers to model complex systems and test hypotheses in virtual environments. These tools are particularly valuable in fields where direct experimentation may be costly, dangerous, or ethically problematic. Digital simulations can model everything from molecular interactions in drug development to social dynamics in behavioural research, providing researchers with powerful means to explore scenarios that would otherwise be impossible to investigate (Walker, 2007). Analytics platforms powered by digital technologies have transformed how researchers extract insights from data. These tools can handle vast quantities of information, perform complex statistical analyses, and generate visualizations that help researchers understand patterns and relationships in their data. Advanced analytics platforms increasingly incorporate machine learning capabilities, enabling them to identify non-obvious correlations and generate predictive models that enhance research outcomes (Webster et al., 2013).

- **Growth of AI integration in academic and scientific research**

The integration of AI in academic and scientific research has experienced exponential growth over the past decade, driven by advances in computational power, algorithm development, and data availability. This growth is evident across virtually all research disciplines, from the natural sciences where AI assists in drug discovery and climate modeling, to the humanities where machine learning tools analyze literary texts and historical documents (Rani & Sharma, 2012).

In biomedical research, AI integration has reached particularly sophisticated levels, with machine learning algorithms now capable of diagnosing diseases from medical images, predicting treatment outcomes, and identifying potential drug compounds. Research institutions have invested heavily in AI infrastructure, creating specialized computing centers and hiring interdisciplinary teams that combine domain expertise with AI capabilities (Walker, 2007). This investment has accelerated the pace of discovery in many fields and enabled research questions that were previously computationally intractable. The social sciences have similarly embraced AI integration, using machine learning techniques to analyze social media data, predict behavioural outcomes, and model complex social phenomena. Digital humanities projects now routinely employ AI tools for text analysis, image recognition, and pattern identification in large cultural datasets. This widespread adoption has democratized access to sophisticated analytical capabilities while also raising questions about the changing nature of scholarly expertise and the role of human interpretation in research (Webster et al., 2013).

The growth of AI integration has also been facilitated by the development of user-friendly platforms and tools that make advanced AI capabilities accessible to researchers without extensive technical backgrounds. Cloud-based AI services, automated machine learning platforms, and drag-and-drop analytics tools have lowered the barriers to AI adoption, enabling researchers across disciplines to incorporate these technologies into their work (McFee, 2014).

CORE ETHICAL PRINCIPLES IN RESEARCH

- **Respect for Persons and Autonomy**

The principle of respect for persons, fundamental to research ethics since the Belmont Report, takes on new dimensions in the context of AI and digital tools. Traditional interpretations of this principle emphasized treating individuals as autonomous agents capable of making informed decisions about their participation in research. In the digital age, this principle must be extended to consider how AI systems interact with human autonomy and decision-making processes (Haines, 2017).

Respect for persons in AI-assisted research requires ensuring that participants understand not only the nature of the research but also how AI tools will be used to collect, analyze, and interpret their data. This understanding becomes particularly complex when AI systems operate in ways that are not easily explainable to research participants. Black box algorithms, for instance, may make decisions or classifications about participants' data without providing clear reasoning that can be communicated during the informed consent process (Walker, 2007).

The autonomy aspect of this principle is challenged by AI systems that can influence human behaviour and decision-making. Adaptive research platforms that modify their approach based on participant responses, or AI systems that provide personalized feedback or recommendations, potentially impact participants' autonomous choices throughout the research process. Researchers must consider how these dynamic interactions affect participants' ability to make informed decisions about their continued participation (Munhall, 1988).

- **Beneficence and Non-Maleficence**

The principles of beneficence and non-maleficence, which require researchers to maximize benefits while minimizing risks and harm, become complex in AI-assisted research due to the multifaceted nature of potential benefits and risks associated with these technologies. Beneficence in this context extends beyond immediate research benefits to consider the broader societal impacts of AI-driven research outcomes and the potential for these tools to advance knowledge in ways that benefit humanity (Haines, 2017).

AI and digital tools offer significant potential benefits, including the ability to process large datasets that can reveal important patterns relevant to public health, social policy, and scientific understanding. These tools can accelerate research timelines, reduce costs, and enable studies of previously impossible scope and scale (Munhall, 1988). However, realizing these benefits while adhering to the principle of beneficence requires careful consideration of how the benefits are distributed and who has access to the insights generated through AI-assisted research. Non-maleficence in the AI research context involves identifying and mitigating both direct and indirect harms that may result from the use of these technologies. Direct harms may include privacy breaches, data misuse, or psychological harm from participation in AI-mediated research processes. Indirect harms may be more subtle and long-term, such as the perpetuation of social biases through biased algorithms or the contribution to societal inequalities through research that reinforces existing power structures (Haines, 2017).

The temporal dimension of non-maleficence in AI research is particularly important, as the consequences of AI-assisted research may not be immediately apparent. Algorithms trained on research data may be used in future applications that could impact research participants or similar populations in unforeseen ways. This requires researchers to consider not only immediate harm but also potential future applications of their AI-driven research methods and findings (Moore, 2003).

- **Justice and Fairness**

Justice and fairness in AI-assisted research encompass both procedural justice in how research is conducted and distributive justice in how benefits and burdens are allocated. The principle of justice traditionally focused on fair subject selection and equitable distribution of research benefits and risks. In the AI context, this principle must address algorithmic fairness, representation in training data, and equitable access to AI-enhanced research opportunities (Munhall, 1988).

Algorithmic fairness presents unique challenges for research ethics, as AI systems may inadvertently perpetuate or amplify existing social inequalities. Training data used to develop AI research tools may underrepresent certain populations or contain biases that affect how these tools perform across different demographic groups. This can result in research findings that are less accurate or applicable for marginalized populations, violating the principle of justice by creating unequal benefits from research participation (Connelly, 2014). The global nature of many

AI research projects raises additional justice concerns related to data extraction and benefit sharing. Researchers in well-resourced institutions may use AI tools to analyze data collected from populations in less developed regions without ensuring that the benefits of the research flow back to those communities. This dynamic can perpetuate global inequalities and exploit vulnerable populations for the benefit of those with greater technological capabilities (Munhall, 1988).

Fairness in AI research also requires attention to the digital divide and ensuring that research participation does not depend on access to advanced technology or high levels of digital literacy. Researchers must consider how their choice of digital tools and AI platforms may inadvertently exclude certain populations or create barriers to participation that compromise the representativeness and fairness of their studies (Webster et al., 2013).

- **Transparency and Accountability**

Transparency and accountability in AI-assisted research present unique challenges due to the complexity and often opaque nature of AI systems. Traditional transparency in research involved clear documentation of methods, open sharing of data and results, and reproducible analytical procedures. AI introduces complications in each of these areas, as machine learning algorithms may be proprietary, their decision-making processes may not be fully interpretable, and their performance may vary across different contexts and datasets (Etzel & Skvarla, 2017).

Methodological transparency in AI research requires researchers to provide clear documentation not only of their research procedures but also of the AI tools used, including their training data, performance characteristics, and known limitations. This documentation must be accessible to other researchers and to research participants, requiring translation of technical information into understandable terms. The challenge is particularly acute with proprietary AI tools where the underlying algorithms may not be fully disclosed. Accountability in AI research involves establishing clear responsibility for the decisions made by AI systems and their consequences (Ketefian, 2015). When AI tools are used to collect data, analyze results, or make recommendations, researchers must maintain responsibility for these outcomes while also acknowledging the role of the AI system in the research process. This requires developing governance structures that can address situations where AI systems produce unexpected or potentially harmful results. The reproducibility of AI-assisted research presents additional transparency challenges (Moore, 2003). AI systems may produce different results when run on different hardware, with different software versions, or at different times due to random elements in their algorithms. Ensuring that AI-assisted research meets standards of scientific reproducibility requires new approaches to documentation and method sharing that account for these technical complexities (Webster et al., 2013).

ETHICAL CHALLENGES OF USING AI IN RESEARCH

- **Data Privacy and Confidentiality**

Data privacy and confidentiality in AI-assisted research present unprecedented challenges that extend far beyond traditional privacy protection measures. The vast data processing capabilities of AI systems, combined with their ability to identify patterns and make inferences from seemingly anonymized data, have fundamentally altered the landscape of privacy protection in research contexts. Traditional anonymization techniques that were considered adequate for protecting participant privacy may be insufficient when applied to datasets that will be processed by sophisticated AI algorithms (Ketefian, 2015).

The concept of differential privacy has emerged as a more robust approach to protecting individual privacy in AI research contexts. This technique involves adding carefully calibrated noise to datasets to prevent the identification of individual records while preserving the overall statistical properties of the data for research purposes. However, implementing differential privacy requires technical expertise and may reduce the accuracy of research findings, creating a tension between privacy protection and research utility that researchers must carefully navigate (Heilferty, 2011). AI systems often require large amounts of data for training and validation, leading researchers to combine datasets from multiple sources or to use data for purposes beyond the original consent scope. This practice raises significant ethical concerns about secondary use of data and the limits of informed consent. Participants who consented to specific research uses of their data may not have anticipated that their information would be used to train AI algorithms or combined with other datasets in ways that could reveal new insights about them (Etzel & Skvarla, 2017).

The global nature of many AI research projects compounds privacy challenges, as data may be transferred across jurisdictions with different privacy laws and regulations. Cloud-based AI platforms may store and process data in multiple countries, making it difficult to ensure consistent privacy protection and to provide participants with clear information about how their data will be handled. Researchers must navigate this complex regulatory landscape while ensuring that privacy protections meet the highest standards regardless of where data processing occurs (Andersen, 1994).

- **Intellectual Property and Ownership Issues**

The integration of AI tools in research has created complex intellectual property and ownership issues that challenge traditional notions of authorship, invention, and data ownership. When AI systems contribute to research discoveries or generate novel insights, questions arise about who owns the intellectual property rights to these contributions and how credit should be attributed. These issues are particularly complex when AI tools are developed by third parties and used by researchers who may not fully understand the underlying algorithms or training data (Andersen, 1994).

Data ownership becomes particularly complex in AI research contexts where multiple datasets may be combined or where AI systems generate new data through synthesis or simulation. Researchers must navigate questions about whether they own the outputs of AI systems trained on their data, whether data generated by AI tools can be owned at all, and how ownership rights are affected when AI systems combine data from multiple sources. These questions become even more complex when considering the rights of research participants whose data contributes to AI training datasets (Stapleton et al., 2010). The use of proprietary AI tools in research raises additional intellectual property concerns related to dependency on commercial platforms and potential restrictions on publication and data sharing. Researchers who rely on commercial AI services may face limitations on their ability to share their methods openly or to reproduce their findings using alternative tools. This dependency can compromise the principles of open science and may create barriers to research transparency and reproducibility (Madak, 1994).

Collaborative AI research projects involving multiple institutions or international partnerships present additional ownership challenges. When AI systems are developed jointly by multiple parties or when research involves sharing data and algorithms across institutional boundaries, determining ownership rights and responsibilities becomes increasingly complex. Clear agreements about intellectual property rights must be established early in collaborative projects to avoid conflicts and ensure that all parties understand their rights and obligations (Andersen, 1994).

- **Overreliance on Automated Decision-Making**

The sophistication and convenience of AI tools create risks of overreliance on automated decision-making that can compromise research quality and ethical standards. When researchers become overly dependent on AI systems for critical research decisions, they may lose the ability to critically evaluate these decisions or to recognize when AI tools are producing inappropriate or biased results. This overreliance can lead to a degradation of research skills and judgment that are essential for maintaining research integrity (Madak, 1994).

Automation bias occurs when researchers accept AI-generated results without adequate critical evaluation, particularly when these results align with their expectations or when the AI system is perceived as highly accurate. This bias can be particularly problematic when AI tools produce results that appear sophisticated or authoritative but may contain errors or reflect biases in training data. Researchers must maintain healthy skepticism about AI-generated results and develop skills for critically evaluating automated outputs (Watson II et al., 2020). The black box nature of many AI systems can contribute to overreliance by making it difficult for researchers to understand how decisions are made or to identify potential problems with AI-generated results. When researchers cannot understand or explain how AI tools arrive at their conclusions, they may be less likely to question these results or to recognize when additional human oversight is needed. This lack of transparency can compromise research integrity and make it difficult to identify and correct errors (Harriss et al., 2022).

Overreliance on AI can also lead to a loss of domain expertise and analytical skills among researchers. When AI tools automate complex analytical processes, researchers may become less skilled at performing these analyses manually or at understanding the underlying principles that govern these processes. This skill degradation can make researchers more dependent on AI tools and less capable of recognizing when these tools are inappropriate or when human judgment is needed (Oliver, 2010).

Preventing overreliance requires maintaining appropriate human oversight of AI-assisted research processes, developing skills for critically evaluating AI outputs, and establishing clear guidelines for when human intervention is required. Researchers must also maintain their domain expertise and analytical skills to ensure they can provide appropriate oversight of AI systems and make informed decisions about when to trust or question automated results (Watson II et al., 2020).

DIGITAL TOOLS AND RESEARCH INTEGRITY

- **Risks of Plagiarism and Automated Content Generation**

The proliferation of sophisticated text generation AI tools has introduced new dimensions to plagiarism and academic misconduct that challenge traditional notions of original authorship and intellectual honesty. Large language models capable of generating human-like text across various domains have made it possible to produce research content that may appear original but is actually generated by AI systems. This capability raises fundamental questions about what constitutes original scholarly work and how academic institutions should address the use of AI-generated content in research and publication (Harriss et al., 2022).

Traditional plagiarism detection systems, designed to identify copied text from existing sources, may be inadequate for detecting AI-generated content that does not directly copy from specific sources but instead synthesizes information from numerous training examples. This limitation creates challenges for maintaining research integrity and for ensuring that published research represents genuine scholarly contribution. New detection methods and policies are needed to address AI-generated content while distinguishing between appropriate and inappropriate uses of AI writing tools. The use of AI for literature reviews and background research presents particular ethical challenges. While AI tools can efficiently synthesize large amounts of literature and identify relevant sources, relying on AI-generated summaries without adequate human verification can lead to inaccuracies, misrepresentations, or the perpetuation of errors present in training data. Researchers must develop strategies for using AI tools to enhance rather than replace careful literature review processes (Oliver, 2010).

Code generation AI tools present similar challenges in computational research fields where software development is a significant component of the research process. AI systems that can generate functional code based on natural language descriptions may produce work that appears original but incorporates patterns or solutions from training data without proper attribution. Researchers using these tools must understand their obligations for attribution and ensure that AI-generated code meets standards for originality and proper citation (Harriss et al., 2022).

- **Manipulation of Data through Digital Means**

Digital tools have created unprecedented capabilities for data manipulation that can be used for legitimate research purposes but also present new opportunities for research misconduct. Advanced image editing software, statistical manipulation tools, and data generation algorithms can alter research data in sophisticated ways that may be difficult to detect using traditional quality control measures. These capabilities require new approaches to data integrity verification and new policies for addressing digital data manipulation.

Deepfake technology and advanced image synthesis tools present particular challenges for research involving visual data. These tools can create realistic but fabricated images, videos, or audio recordings that could be used to falsify research evidence or to create misleading documentation of research procedures. The sophistication of these tools makes it increasingly difficult to distinguish between authentic and fabricated multimedia content, requiring new verification methods and protocols (Oliver, 2010).

Statistical software and data analysis tools, while essential for legitimate research, can also be misused to manipulate datasets or analysis results in ways that support desired conclusions while appearing methodologically sound. The complexity of modern statistical methods and the availability of automated analysis tools can make it easier to engage in questionable research practices such as p-hacking or selective reporting while maintaining a veneer of methodological rigor. Data synthesis and simulation tools present additional integrity challenges when the boundary between real and synthetic data becomes blurred. While synthetic data can serve legitimate research purposes, such as protecting privacy or augmenting small datasets, the use of synthetic data must be clearly disclosed and appropriately validated. Failure to distinguish between real and synthetic data or inappropriate use of synthetic data to support research conclusions can compromise research integrity (Harriss et al., 2022).

- **Maintaining Authenticity in Virtual and Online Research Environments**

The shift toward virtual and online research environments, accelerated by digital technologies and global circumstances, has created new challenges for maintaining research authenticity and participant verification. Online research platforms may make it difficult to verify participant identities, ensure that participants meet inclusion criteria, or prevent multiple enrollments by the same individual. These challenges can compromise research validity and raise ethical concerns about data quality and participant rights (Webster et al., 2013).

Bot detection and automated participation represent significant threats to research integrity in online environments. Sophisticated bots or automated systems may be able to complete surveys, participate in online experiments, or engage with research platforms in ways that mimic human behaviour while providing invalid data. Researchers must implement verification methods and monitoring systems to detect and prevent automated participation while balancing security measures with participant accessibility and privacy. Virtual research environments may also alter participant behaviour in ways that affect research authenticity. Online interactions may differ significantly from face-to-face interactions, potentially affecting the validity of research findings (Oliver, 2010). Participants may feel less accountable in virtual environments or may behave differently when they know they are interacting with digital systems rather than human researchers. These behavioural changes must be considered when designing online research and interpreting results. The digital divide and varying levels of technological access among potential participants can compromise the representativeness and authenticity of online research. Participants with limited internet access, older devices, or lower digital literacy may be systematically excluded from online studies or may have different experiences that affect their responses. Researchers must consider these factors when designing online research and when interpreting and generalizing their findings (Resnik, 2014).

RESPONSIBLE USE OF AI IN DATA COLLECTION AND ANALYSIS

- **Ethical Considerations in Big Data Analytics**

Big data analytics powered by AI systems presents both unprecedented research opportunities and significant ethical challenges that require careful consideration and proactive management. The scale and scope of big data often exceed traditional research paradigms, involving datasets that may contain millions or billions of data points representing diverse populations and activities. This scale introduces ethical considerations related to consent, privacy, representation, and the responsible use of data that may not have been originally collected for research purposes (Devine et al., 2021).

The retrospective nature of much big data analytics raises important questions about informed consent and the limits of secondary data use. Many large datasets used in AI research were originally collected for commercial, administrative, or other non-research purposes, often without explicit consent for research use. While these datasets may offer valuable insights for advancing scientific knowledge and addressing societal challenges, their use in research must be carefully evaluated to ensure that it aligns with ethical principles and respects the rights and expectations of the individuals represented in the data. Data quality and representativeness present additional ethical challenges in big data analytics (Milanović et al., 2021). Large datasets may contain systematic biases, missing data, or quality issues that could affect research conclusions. The scale of big data may make it difficult to thoroughly validate data quality or to understand the processes that generated the data. These issues can lead to research findings that appear robust due to large sample sizes but may actually reflect biases or errors in the underlying data. The power dynamics inherent in big data analytics raise concerns about equity and justice in research. Organizations with access to large datasets and sophisticated AI analytics capabilities may have significant advantages over researchers and communities with fewer resources. This disparity can lead to research that primarily serves the interests of well-resourced institutions while potentially exploiting the data of less powerful populations without providing corresponding benefits (Resnik, 2014).

- **Validity and Reliability of AI-driven Research Findings**

Ensuring the validity and reliability of AI-driven research findings requires new approaches to research design, analysis, and interpretation that account for the unique characteristics of AI systems. Traditional concepts of validity and reliability must be reconsidered in light of AI capabilities and limitations, including their sensitivity to training data, potential for overfitting, and challenges in interpretability. Researchers must develop new standards and practices for evaluating the quality and trustworthiness of AI-assisted research (Harriss et al., 2022).

Internal validity in AI research is complicated by the black box nature of many machine learning algorithms and the potential for these systems to identify spurious correlations that do not reflect genuine causal relationships. AI systems may achieve high predictive accuracy while relying on features or patterns that are not scientifically meaningful or that may not generalize to new contexts. Researchers must carefully evaluate whether AI-identified patterns represent genuine phenomena or artifacts of the training data and analytical process. External validity presents additional challenges in AI research due to the dependence of AI systems on their training data and the potential for distribution shift when these systems are applied to new populations or contexts (Oliver, 2010). AI models trained on specific datasets may not perform as well when applied to different populations, time periods, or settings. This limitation requires careful consideration of the generalizability of AI-driven research findings and transparent reporting of the conditions under which these findings are valid. Reliability of AI systems can be affected by various technical factors including random initialization, hardware differences, software versions, and data pre-processing choices. These factors can lead to different results when the same analysis is repeated, challenging traditional notions of reproducibility in research. Establishing reliability in AI research requires developing new protocols for testing system stability and documenting the technical conditions that affect results (Resnik, 2014).

- **Protecting Participant Rights in Automated Surveys and Experiments**

Automated surveys and experiments powered by AI systems offer significant advantages in terms of scale, efficiency, and personalization, but they also present unique challenges for protecting participant rights and maintaining ethical standards. AI-driven research platforms can adapt their behavior based on participant responses, provide personalized feedback, and make real-time decisions about participant eligibility or study procedures. These capabilities require new approaches to informed consent, ongoing consent validation, and participant protection (Devine et al., 2021).

Dynamic consent presents a particular challenge in AI-driven research where study procedures may change based on algorithm decisions or participant responses. Traditional informed consent processes assume that participants can be provided with complete information about study procedures before participation begins. However, adaptive AI systems may make decisions during the research process that could not be fully anticipated at the time of initial consent. This requires developing new models for ongoing consent that allow participants to understand and approve changes to study procedures as they occur (Panter, 2011).

Algorithmic personalization in automated research raises concerns about manipulation and undue influence. AI systems that adapt their approach based on participant characteristics or responses may inadvertently use persuasive techniques that compromise participant autonomy. For example, an AI system might identify that certain types of questions or incentives are more effective for particular participants, potentially creating situations where participants are subjected to individually tailored influence that they may not recognize or consent to. The automated nature of AI-driven research can create challenges for providing adequate human oversight and support for participants. While automation can improve efficiency and reduce costs, it may also create situations where participants have questions, concerns, or problems that cannot be adequately addressed by automated systems. Maintaining appropriate human oversight and providing accessible support channels becomes crucial for ensuring participant rights and wellbeing in automated research environments (Rao et al., 2024).

Data portability and participant control present additional challenges in AI-driven research where participant data may be processed in complex ways that make it difficult to provide participants with meaningful access to or control over their information. AI systems may transform participant data through various processing steps, combine it with other data sources, or use it to train models that cannot be easily separated from other data. Providing participants with meaningful control over their data in these contexts requires developing new technical and procedural approaches (Milanović et al., 2021).

ETHICAL FRAMEWORKS AND GUIDELINES

- **International Guidelines (e.g., UNESCO, OECD, EU AI Act)**

The development of international guidelines for AI ethics in research represents a critical step toward establishing global standards for responsible AI use in scientific and academic contexts. UNESCO's Recommendation on the Ethics of Artificial Intelligence, adopted in 2021, provides a comprehensive framework that addresses AI applications across various domains, including research and scientific inquiry. This recommendation emphasizes the importance of human rights, transparency, accountability, and cultural diversity in AI development and deployment, principles that have direct relevance for research applications (Panter, 2011).

The UNESCO framework specifically addresses the use of AI in scientific research, highlighting the need for researchers to maintain scientific integrity while leveraging AI capabilities. The recommendation emphasizes that AI should enhance rather than replace human judgment in research and that researchers must maintain responsibility for the ethical implications of their AI-assisted work. This framework also addresses issues of open science and data sharing, advocating for approaches that balance the benefits of open research with privacy protection and equitable access to AI tools and insights (Watson II & Etzel, 2019).

The OECD AI Principles, established in 2019 and regularly updated, provide another influential international framework that emphasizes inclusive growth, sustainable development, and well-being in AI applications. For research contexts, these principles highlight the importance of ensuring that AI benefits are broadly shared and that AI development is guided by respect for human rights and democratic values. The OECD framework also emphasizes the need for transparency and explainability in AI systems, which has particular relevance for research applications where understanding and reproducing results is essential (Zyphur & Pierides, 2017). The European Union's AI Act represents the most comprehensive legal framework for AI governance to date, with specific provisions that affect research applications. The Act establishes a risk-based approach to AI regulation, with different requirements depending on the potential risks associated with AI applications. Research uses of AI may be subject to various provisions depending on the specific application and context, with particular attention to high-risk applications that could affect individual rights or safety (Panter, 2011).

These international frameworks share common themes that are particularly relevant for research applications: the importance of human oversight and responsibility, the need for transparency and explainability, the requirement to address bias and discrimination, and the necessity of protecting individual rights and privacy. However, translating these high-level principles into practical guidance for researchers requires additional work to develop specific protocols and procedures that can be implemented in diverse research contexts (Rao et al., 2024).

- **Institutional Review Boards (IRBs) and AI-based Research Protocols**

Institutional Review Boards and ethics committees face significant challenges in reviewing and overseeing AI-based research protocols due to the technical complexity of AI systems and the novel ethical issues they present. Traditional IRB frameworks were developed for conventional research methods and may not adequately address the unique risks and considerations associated with AI-assisted research. This has led to calls for updating IRB procedures and training to better handle AI-related research proposals (Isbell et al., 2022). IRBs must develop new expertise and procedures for evaluating AI research protocols, including understanding the technical characteristics of AI systems, assessing the risks associated with algorithmic bias and discrimination, and evaluating the adequacy of privacy protections in AI-assisted research. This requires either training existing IRB members in AI ethics and technology or recruiting new members with relevant technical expertise. Many institutions are struggling with this capacity building while trying to keep pace with the rapid adoption of AI tools in research (Watson II & Etzel, 2019).

The review of AI research protocols requires attention to several unique considerations that may not be present in traditional research reviews. These include evaluating the training data used for AI systems, assessing the potential for algorithmic bias, reviewing the transparency and explainability of AI tools, and ensuring that informed consent processes adequately address AI-related risks and limitations. IRBs must also consider the dynamic nature of AI systems that may learn and change during the research process, requiring ongoing monitoring and potentially updated consent procedures (Rao et al., 2024). Risk assessment for AI research presents particular challenges due to the difficulty of predicting how AI systems will behave in new contexts and the potential for emergent behaviours that were not apparent during initial system development. IRBs must develop frameworks for assessing both direct risks to research participants and broader societal risks that could result from AI research findings or applications. This requires considering not only immediate research impacts but also potential future uses of AI tools and findings (Zyphur & Pierides, 2017).

- **Codes of Conduct for Digital Research Tools**

Professional codes of conduct for digital research tools are emerging as essential frameworks for guiding researchers in the responsible use of AI and digital technologies. These codes address the specific ethical challenges associated with digital research tools while providing practical guidance for researchers navigating complex ethical decisions. Different professional organizations and research communities are developing specialized codes that reflect the unique characteristics and challenges of their fields. General principles common to many digital research codes include requirements for transparency in method reporting, respect for participant privacy and autonomy, attention to algorithmic fairness and bias, and maintenance of human oversight in automated processes. These codes typically emphasize the importance of understanding the tools being used, including their limitations and potential biases, and taking responsibility for the ethical implications of tool selection and use (Bulger & Heitman, 2007).

Disciplinary codes of conduct for digital research often address field-specific considerations and challenges. For example, codes for social science research may emphasize issues related to behavioural manipulation and informed consent in online environments, while codes for biomedical research may focus on data security and the clinical implications of AI-assisted diagnosis or treatment recommendations. These specialized codes provide researchers with guidance that is tailored to their specific contexts and challenges. Implementation and enforcement of digital research codes present ongoing challenges for professional organizations and institutions. Unlike traditional research ethics violations, which may be relatively straightforward to identify and address, digital research ethics violations may involve complex technical issues that require specialized expertise to evaluate. This has led to calls for developing new mechanisms for ethics education, monitoring, and enforcement that are appropriate for the digital research environment (Isbell et al., 2022).

BALANCING INNOVATION AND ETHICS

- **Encouraging Responsible Innovation**

Responsible innovation in AI-assisted research requires creating environments and incentive structures that promote both scientific advancement and ethical excellence. This approach recognizes that ethical considerations should not be viewed as barriers to innovation but rather as essential components that ensure technological progress serves societal good and maintains public trust in scientific research. Responsible innovation frameworks emphasize the importance of embedding ethical considerations throughout the research and development process rather than treating ethics as an afterthought. The concept of responsible innovation involves anticipatory governance, which requires researchers and institutions to consider potential future impacts of their AI-assisted research and to develop mechanisms for addressing these impacts proactively. This includes conducting ethical impact assessments before implementing new AI tools, establishing monitoring systems to track the effects of AI-assisted research, and creating feedback mechanisms that allow for course correction when problems are identified. Such approaches help ensure that innovation proceeds in directions that are socially beneficial and ethically sound (Watson II & Etzel, 2019).

Stakeholder engagement represents another crucial component of responsible innovation in AI research. This involves including diverse voices in the development and evaluation of AI research tools and protocols, including community representatives, ethicists, social scientists, and affected populations. Meaningful stakeholder engagement can help identify potential risks and benefits that researchers might not consider independently and can provide valuable perspectives on how AI-assisted research should be designed and implemented. Institutional support for responsible innovation requires developing reward systems that recognize and incentivize ethical excellence alongside scientific achievement. This may involve modifying promotion and tenure criteria to include ethical considerations, providing resources for ethics training and consultation, and creating institutional cultures that value responsible conduct. Funding agencies and publishers also play important roles in promoting responsible innovation by incorporating ethical criteria into their evaluation processes and requiring evidence of ethical consideration in research proposals and publications (Cortina, 2020).

- **Ethical Use of AI for Enhancing Research Productivity**

AI tools offer tremendous potential for enhancing research productivity through automation of routine tasks, acceleration of data analysis, and identification of patterns that might not be apparent through traditional methods. However, realizing these productivity gains while maintaining ethical standards requires careful consideration of how AI tools are implemented and used. The goal should be to use AI to augment human capabilities rather than to replace human judgment and oversight in critical research decisions (Bulger & Heitman, 2007).

Appropriate applications of AI for research productivity enhancement include automating data preprocessing and cleaning tasks, conducting preliminary analyses to identify promising research directions, and assisting with literature reviews and information synthesis. These applications can free researchers to focus on higher-level tasks that require human creativity, judgment, and domain expertise. However, even in these applications, human oversight remains essential to ensure quality and to catch potential errors or biases. Quality assurance in AI-enhanced research productivity requires establishing validation procedures for AI-generated outputs and maintaining human verification of critical results. This includes developing protocols for spot-checking AI analyses, establishing thresholds for when human review is required, and creating audit trails that document how AI tools contributed to research outcomes. Such measures help ensure that productivity gains do not come at the expense of research quality or integrity (Laryeafio & Ogbewe, 2023).

Training and skill development are essential for enabling researchers to use AI productivity tools effectively and ethically. This includes technical training on how to use AI tools appropriately, education about the limitations and potential biases of these tools, and development of skills for critically evaluating AI outputs. Researchers must understand both the capabilities and limitations of AI tools to use them responsibly and to maintain appropriate skepticism about automated results (Cortina, 2020).

- **Limits of Automation in Human-Centered Research**

While AI and automation offer significant benefits for research productivity and capability, there are important limits to automation that must be recognized and respected, particularly in human-centered research contexts. These limits arise from the inherently human nature of many research questions, the importance of contextual understanding and interpretation, and the ethical requirements for human oversight and responsibility in research involving human participants.

Human judgment remains irreplaceable in research design, where decisions about research questions, methodological approaches, and ethical considerations require understanding of human values, social contexts, and cultural nuances that AI systems cannot fully capture. While AI tools can assist with various aspects of research design, such as power calculations or literature synthesis, the fundamental decisions about what questions to ask and how to ask them require human insight and responsibility. Interpretation of research findings presents another area where human expertise cannot be fully automated (Cunningham & McCrum-Gardner, 2007). While AI systems can identify patterns and correlations in data, understanding the meaning and implications of these patterns requires contextual knowledge, theoretical understanding, and consideration of alternative explanations that AI systems may not possess. Human researchers bring domain expertise, cultural understanding, and ethical sensitivity that are essential for meaningful interpretation of research results. Ethical decision-making throughout the research process requires human judgment and cannot be fully automated. While AI tools can assist with identifying potential ethical issues or flagging concerning patterns, the resolution of ethical dilemmas requires human values, moral reasoning, and consideration of multiple stakeholder perspectives. Automated systems may be able to apply rule-based ethical guidelines but cannot handle the complex value judgments and contextual considerations that characterize many ethical decisions in research (Watson II & Etzel, 2019).

Communication and dissemination of research findings to diverse audiences require human skills in translation, contextualization, and responsiveness to audience needs and concerns. While AI tools can assist with certain aspects of communication, such as translation or formatting, the core tasks of explaining research significance, addressing stakeholder concerns, and facilitating meaningful dialogue about research implications require human engagement and empathy (Cortina, 2020).

CASE STUDIES AND REAL-WORLD EXAMPLES

- **Ethical Dilemmas in AI-driven Biomedical Research**

The integration of AI in biomedical research has created numerous ethical dilemmas that illustrate the complex challenges researchers face when using these powerful technologies. One prominent example involves the use of AI diagnostic tools in clinical research studies. A major medical center implemented an AI system for analyzing medical images as part of a research study comparing AI diagnosis with human physician diagnosis. However, the AI system showed systematic performance differences across racial groups, performing less accurately for underrepresented minorities due to training data that predominantly included images from white patients.

This situation created an ethical dilemma about how to proceed with the research. Continuing the study as designed would likely produce results that reflected these racial disparities, potentially reinforcing healthcare inequities. However, stopping or modifying the study would delay research that could benefit patient care and would waste resources already invested in the project. The research team ultimately decided to expand their dataset to include more diverse training examples and to explicitly study algorithmic fairness as part of their research objectives, but this decision required additional time, funding, and institutional support (Laryeafio & Ogbewe, 2023).

Another significant ethical challenge emerged in a large-scale genomics research project that used machine learning algorithms to identify genetic variants associated with disease risk. The AI system identified patterns that appeared to predict disease susceptibility, but the algorithms were too complex to provide clear explanations for their predictions. This created a dilemma about whether and how to share these findings with research participants. While the predictions might provide valuable health information, the lack of interpretability made it difficult to provide participants with meaningful explanations of what the results meant for their health. The research team faced competing ethical obligations: the principle of beneficence suggested they should share potentially valuable health information with participants, but the principle of respect for persons required providing participants with understandable information. The team ultimately developed a protocol for sharing results that included extensive counseling about the limitations and uncertainties of the AI predictions, but this approach was resource-intensive and may not be scalable to larger studies (Watson II & Etzel, 2019).

- **AI in Social Science and Behavioural Research**

Social science research using AI tools has generated numerous ethical challenges related to privacy, consent, and the potential for research to influence or manipulate participant behaviour. A notable case involved a large-scale study of social media behaviour that used machine learning algorithms to analyze posts and interactions from millions of users. The research aimed to understand how social networks influence political attitudes and behaviour, but the scale and automated nature of the data collection raised significant ethical concerns. The researchers initially planned to collect and analyze social media data without explicit consent from users, arguing that the posts were publicly available and that seeking individual consent from millions of users would be impractical. However, ethical review revealed that many users may not have understood that their posts could be used for research purposes and that the AI analysis could reveal sensitive information about political beliefs, mental health, and personal relationships that users did not intend to share publicly (Bulger & Heitman, 2007).

The research team faced a fundamental tension between the scientific value of large-scale behavioural data analysis and the ethical requirements for informed consent and privacy protection. They ultimately developed a multi-layered approach that included obtaining platform-level agreements for research use, implementing privacy-preserving techniques to reduce the identifiability of individual users, and providing opt-out mechanisms for users who did not want their data included in research studies. Another challenging case involved an online experiment that used AI-powered recommendation algorithms to study how exposure to different types of content affects user attitudes and behaviour. The AI system dynamically adjusted the content shown to participants based on their responses and behaviour patterns, creating personalized experimental conditions that could potentially influence participants' political views, emotional states, or purchasing decisions (Cunningham & McCrum-Gardner, 2007).

This raised significant ethical concerns about the potential for the research itself to cause harm by manipulating participant behaviour in ways that could have lasting effects beyond the research context. The institutional review board struggled with how to evaluate the risks and benefits of this type of dynamic, AI-driven experimentation, ultimately requiring the researchers to implement extensive monitoring procedures and to provide participants with detailed debriefing about how the AI system had influenced their experience (Laryeafio & Ogbewe, 2023).

- **Digital Platforms and Online Learning Analytics**

Educational technology research using AI-powered learning analytics has created numerous ethical dilemmas related to student privacy, algorithmic bias in educational assessments, and the potential for AI systems to influence educational opportunities and outcomes. A significant case involved a university that implemented an AI system for predicting student success and providing early intervention recommendations. The system analyzed data from learning management systems, campus card transactions, library usage, and other digital footprints to identify students at risk of academic failure (Bulger & Heitman, 2007).

While the system showed promise for improving student retention and success rates, it also raised concerns about surveillance, privacy, and the potential for algorithmic bias. The AI system relied heavily on behavioural data that might reflect socioeconomic status, cultural differences, or learning styles rather than academic ability, potentially creating systematic disadvantages for certain student populations. Students from lower-income backgrounds, for example, might be flagged as at-risk based on patterns of campus resource usage that reflected financial constraints rather than academic struggles (Laryeafio & Ogbewe, 2023).

The university faced difficult decisions about how to use the AI predictions ethically. Providing interventions based on AI recommendations could help students who were genuinely struggling, but it could also stigmatize students who were flagged by the system and potentially become self-fulfilling prophecies that negatively affected student performance. The institution ultimately implemented the system with extensive human oversight, transparency measures that allowed students to understand why they were flagged, and opt-out provisions for students who did not want to be included in the predictive modelling (Milanović et al., 2021). A related case involved an online learning platform that used AI to personalize educational content and pacing for individual learners. The AI system tracked detailed learning behaviours, including how long students spent on different topics, their error patterns, and their emotional responses as measured through various digital signals. This data was used to adapt the learning experience in real-time, potentially improving learning outcomes but also raising concerns about the collection and use of sensitive behavioural and emotional data (Panter & Sterba, 2011).

The ethical challenges were compounded by the fact that many users of the platform were minors, requiring additional protections for vulnerable populations. Parents and educators raised concerns about the long-term implications of collecting detailed behavioural and emotional data about children's learning processes, particularly given the potential for this data to be used in ways that could affect educational opportunities or outcomes later in life. The platform developers had to balance the potential benefits of personalized learning with privacy protection and parental rights, ultimately implementing strict data governance policies and providing extensive controls for parents and educators (Laryeafio & Ogbewe, 2023).

CONCLUSION

The integration of artificial intelligence and digital tools in research marks one of the most significant transformations in the history of scientific inquiry. These technologies provide advanced capabilities for data collection, analysis, and interpretation, accelerating discovery, revealing hidden patterns, and addressing complex research questions that traditional methods cannot manage. At the same time, they introduce ethical challenges that require careful attention and ongoing responsibility from the research community. Traditional research ethics principles such as respect for persons, beneficence, non-maleficence, and justice remain essential, yet their application in AI-assisted research requires new interpretations. Respect for persons now includes algorithmic transparency, meaningful consent, and autonomy protection in contexts influenced by AI systems. Justice requires attention to algorithmic bias, equitable access, and fair distribution of benefits and burdens. Data privacy concerns extend beyond anonymization to differential privacy, data synthesis, and the implications of machine learning. Issues of authorship, ownership, and intellectual property are also more complex in AI-driven research.

Research integrity is shaped by both opportunities and risks. AI tools can enhance quality control and detect misconduct, but they can also enable misuse and weaken oversight. Responsible data practices require fairness, transparency, and protection of participant rights, while the validity of AI-driven findings depends on ethical study design and interpretation. Global frameworks from organizations such as UNESCO, OECD, and the European Union provide principles that institutions must adapt into practice. Review boards and professional codes are evolving to meet these new demands. The way forward lies in education, governance, ethical tool design, and collaborative dialogue. Researchers must learn to use AI responsibly, institutions must support ethical innovation, and stakeholders must work together to ensure equity. The future of research depends on balancing technological potential with ethical responsibility to advance knowledge while safeguarding human values and social justice.

REFERENCES

Andersen, M. B. (1994). Ethical considerations in the supervision of applied sport psychology graduate students. *Journal of Applied Sport Psychology*, *6*(2), 152–167. DOI: 10.1080/10413209408406291

Bulger, R. E., & Heitman, E. (2007). Expanding Responsible Conduct of Research Instruction across the University. *Academic Medicine*, *82*(9), 876–878. DOI: 10.1097/ACM.0b013e31812f7909 PMID: 17726396

Connelly, L. M. (2014). Ethical considerations in research studies. *Medsurg Nursing*, *23*(1), 54–56. https://go.gale.com/ps/i.do?id=GALE%7CA360608990&sid=googleScholar&v=2.1&it=r&linkaccess=abs&iss PMID: 24707669

Cortina, J. M. (2020). On the whys and hows of quantitative research. *Journal of Business Ethics*, *167*(1), 19–29. https://idp.springer.com/authorize/casa?redirect_uri=https://link.springer.com/article/10.1007/s10551-019-04195-8&casa_token=pFPN_pEjrIEAAAAA:a7jsAzcP8Hm4W6wK6-3f6xraknWh930FRFSHrnaqzlPat7_SpNuyxuBIRici2x7ZxIJ86nCVCjJCbUdRvw. DOI: 10.1007/s10551-019-04195-8

Cunningham, J. B., & McCrum-Gardner, E. (2007). Power, effect and sample size using GPower: Practical issues for researchers and members of research ethics committees. *Evidence-Based Midwifery*, *5*(4), 132–137. https://go.gale.com/ps/i.do?p=HRCA&sw=w&issn=14794489&v=2.1&it=r&id=GALE%7CA172050741&sid=googleScholar&linkaccess=abs

Devine, E. G., Pingitore, A. M., Margiotta, K. N., Hadaway, N. A., Reid, K., Peebles, K., & Hyun, J. W. (2021). Frequency of concealment, fabrication and falsification of study data by deceptive subjects. *Contemporary Clinical Trials Communications*, *21*, 100713. DOI: 10.1016/j.conctc.2021.100713 PMID: 33604482

Etzel, E. F., & Skvarla, L. A. (2017). Ethical considerations in sport and performance psychology. In *Oxford Research Encyclopedia of Psychology*. https://oxfordre.com/psychology/display/10.1093/acrefore/9780190236557.001.0001/acrefore-9780190236557-e-141

Haines, D. (2017). Ethical considerations in qualitative case study research recruiting participants with profound intellectual disabilities. *Research Ethics*, *13*(3–4), 219–232. DOI: 10.1177/1747016117711971

Harriss, D. J., Jones, C., & MacSween, A. (2022). Ethical Standards in Sport and Exercise Science Research: 2022 Update. *International Journal of Sports Medicine*, *43*(13), 1065–1070. DOI: 10.1055/a-1957-2356 PMID: 36495253

Heilferty, C. M. (2011). Ethical considerations in the study of online illness narratives: A qualitative review. *Journal of Advanced Nursing*, *67*(5), 945–953. DOI: 10.1111/j.1365-2648.2010.05563.x PMID: 21231955

Isbell, D. R., Brown, D., Chen, M., Derrick, D. J., Ghanem, R., Arvizu, M. N. G., Schnur, E., Zhang, M., & Plonsky, L. (2022). Misconduct and Questionable Research Practices: The Ethics of Quantitative Data Handling and Reporting in Applied Linguistics. *Modern Language Journal*, *106*(1), 172–195. DOI: 10.1111/modl.12760

Ketefian, S. (2015). Ethical considerations in research. Focus on vulnerable groups. *Investigacion y Educacion en Enfermeria*, *33*(1), 164–172. http://www.scielo.org.co/scielo.php?pid=S0120-53072015000100019&script=sci_arttext. DOI: 10.17533/udea.iee.v33n1a19 PMID: 26148168

Laryeafio, M. N., & Ogbewe, O. C. (2023). Ethical consideration dilemma: Systematic review of ethics in qualitative data collection through interviews. *Journal of Ethics in Entrepreneurship and Technology*, *3*(2), 94–110. https://www.emerald.com/insight/content/doi/10.1108/JEET-09-2022-0014/. DOI: 10.1108/JEET-09-2022-0014

Madak, P. R. (1994). Ethical Considerations When Using Qualitative Methods in Evaluation. *The Canadian Journal of Program Evaluation*, *9*(2), 1–13. DOI: 10.3138/cjpe.9.001

McFee, G. (2014). Ethical considerations. In *Research methods in sports coaching* (pp. 98–108). Routledge., https://www.taylorfrancis.com/chapters/edit/10.4324/9780203797549-12/ethical-considerations-graham-mcfee DOI: 10.4324/9780203797549-10

Milanović, L. Ran\djelović, N., Živković, D., & Savić, Z. (2021). Ethics and sports, ethics in sports, sports ethics: Aspects of consideration of different authors. *Physical Education and Sport through the Centuries, 8*(2), 16–32. https://scindeks.ceon.rs/Article.aspx?artid=2335-05982102016M

Moore, Z. E. (2003). Ethical dilemmas in sport psychology: Discussion and recommendations for practice. *Professional Psychology, Research and Practice*, *34*(6), 601–610. https://psycnet.apa.org/fulltext/2003-10053-004.html. DOI: 10.1037/0735-7028.34.6.601

Munhall, P. L. (1988). Ethical Considerations in Qualitative Research. *Western Journal of Nursing Research*, *10*(2), 150–162. DOI: 10.1177/019394598801000204 PMID: 3394317

Oliver, J. (2010). Ethical practice in sport psychology: Challenges in the real world. In *Routledge handbook of applied sport psychology* (pp. 60–68). Routledge., https://www.taylorfrancis.com/chapters/edit/10.4324/9780203851043-8/ethical-practice-sport-psychology-jonah-oliver

Panter, A. T. (2011). *Handbook of ethics in quantitative methodology*. Taylor & Francis., https://books.google.com/books?hl=en&lr=&id=DKgYxsYQDqsC&oi=fnd&pg=PP1&dq=ethics+in+quantitative+research+&ots=UNEt8NsWNC&sig=enO34T-SNnNmf4HHjOqK4__Iox4 DOI: 10.4324/9780203840023

Panter, A. T., & Sterba, S. K. (2011). Ethics in quantitative methodology: An introduction. In *Handbook of ethics in quantitative methodology* (pp. 1–11). Routledge., https://api.taylorfrancis.com/content/chapters/edit/download?identifierName=doi&identifierValue=10.4324/9780203840023-1&type=chapterpdf DOI: 10.4324/9780203840023

Pennock, R. T. (2018). Beyond research ethics: How scientific virtue theory reframes and extends responsible conduct of research. In *Cultivating Moral Character and Virtue in Professional Practice*. Routledge. DOI: 10.4324/9781315182414-13

Powell, T. C. (2020). Can Quantitative Research Solve Social Problems? Pragmatism and the Ethics of Social Research. *Journal of Business Ethics*, *167*(1), 41–48. DOI: 10.1007/s10551-019-04196-7

Rani, R., & Sharma, R. K. (2012). Ethical Consideration in Research. *International Journal of Nursing Education*, *4*(1). http://ijone.org/scripts/IJONE%20Jan-June%202012.pdf#page=49

Rao, K. N., Mair, M., Arora, R. D., Dange, P., & Nagarkar, N. M. (2024). Misconducts in research and methods to uphold research integrity. *Indian Journal of Cancer*, *61*(2), 354–359. DOI: 10.4103/ijc.ijc_4_23 PMID: 39016268

Resnik, D. B. (2014). Data Fabrication and Falsification and Empiricist Philosophy of Science. *Science and Engineering Ethics*, *20*(2), 423–431. DOI: 10.1007/s11948-013-9466-z PMID: 23982326

Stapleton, A. B., Hankes, D. M., Hays, K. F., & Parham, W. D. (2010). Ethical dilemmas in sport psychology: A dialogue on the unique aspects impacting practice. *Professional Psychology, Research and Practice*, *41*(2), 143a. https://psycnet.apa.org/record/2010-06890-009. DOI: 10.1037/a0017976

Walker, W. (2007). Ethical considerations in phenomenological research. *Nurse Researcher*, *14*(3), 36–45. DOI: 10.7748/nr2007.04.14.3.36.c6031 PMID: 17494467

Watson, J. C., II, & Etzel, E. F. (2019). *Ethical and legal issues in sport and performance psychology.* https://psycnet.apa.org/record/2019-05400-041

Watson, J. C.II, Harris, B. S., & Baillie, P. (2020). Ethical Issues Impacting the Profession of Sport Psychology. In *Handbook of Sport Psychology* (pp. 751–772). John Wiley & Sons, Ltd., DOI: 10.1002/9781119568124.ch36

Webster, S., Lewis, J., & Brown, A. (2013). Considerations in qualitative research. *Qualitative Research Practice: A Guide for Social Science Students and Researchers, 1*, 77. https://www.torrossa.com/gs/resourceProxy?an=5017584&publisher=FZ7200#page=102

Zyphur, M. J., & Pierides, D. C. (2017). Is quantitative research ethical? Tools for ethically practicing, evaluating, and using quantitative research. *Journal of Business Ethics, 143*(1), 1–16. https://idp.springer.com/authorize/casa?redirect_uri=https://link.springer.com/article/10.1007/s10551-017-3549-8&casa_token=RBVoBlCDYocAAAAA:FloGK-8NKEx3MjMP8zzMpiDbaCAqRAoqA_SS0dsFqGAT1vl3sNAUpqvZWK-nqIVq6nv5KrbD8wgbwEtuWA. DOI: 10.1007/s10551-017-3549-8

Chapter 12
Ethical Dimensions of AI-Powered Business Decisions Ensuring Transparency and Responsibility

Kiran Ari
http://orcid.org/0009-0008-1203-7915
Dayananda Sagar College of Engineering, India

U. Roopa
http://orcid.org/0000-0002-1813-7203
Dayananda Sagar College of Engineering, India

EXECUTIVE SUMMARY

Artificial Intelligence (AI) is reshaping business decision-making by enabling predictive analytics, automation, and real-time insights across domains like finance, marketing, HR, and operations. While enhancing efficiency, AI also raises ethical challenges of transparency, fairness, privacy, and responsibility—especially when opaque algorithms affect vulnerable groups. This chapter explores the ethical dimensions of AI adoption in business through case-based learning, applying principles of justice, autonomy, beneficence, and non-maleficence. Real-world cases—such as biased loan approvals and facial recognition in retail—illustrate dilemmas of bias, consent, and surveillance. Students engage in reflection exercises, dilemma maps, and role-based analysis, guided by ethical frameworks including OECD, IEEE, the EU AI Act, and a proposed model, Ethical Reflexivity in AI (ER-AI). By combining

DOI: 10.4018/979-8-3693-8302-5.ch012

global perspectives, emerging challenges like generative AI, and the intersection of research and business ethics, the chapter equips future scholars and practitioners to navigate AI responsibly.

INTRODUCTION

Ethical considerations have always been the backbone of responsible research and business practice, but the advent of Artificial Intelligence (AI) has elevated these concerns to new dimensions. In the contemporary era, where organizations and researchers increasingly rely on data-driven algorithms for critical decision-making, the conversation around ethics is no longer confined to theoretical discussions- it has become a matter of immediate and practical urgency. Whether in business, healthcare, academia, or governance, decisions powered by AI are reshaping how information is processed, interpreted, and acted upon. This transformation raises pressing questions about accountability, fairness, and transparency, which traditional ethical frameworks are often ill-equipped to handle. For emerging researchers and practitioners, engaging with these dilemmas is essential not only for maintaining academic integrity but also for ensuring that technological innovations serve society responsibly.

Research ethics, as a discipline, has long been guided by principles of integrity, honesty, and respect for human subjects. Classic frameworks such as the Nuremberg Code, the Belmont Report, and the establishment of Institutional Review Boards (IRBs) laid the foundation for protecting participants and ensuring responsible research conduct. These guidelines were developed primarily in response to historical injustices and unethical practices in biomedical and social research. However, as research methodologies have expanded and technologies like AI have become embedded in scholarly inquiry and business applications, ethical dilemmas have grown more complex. For instance, the ethical issues in digital research do not always revolve around direct harm to human participants but instead around subtler concerns such as algorithmic bias, surveillance, and the misuse of personal data. Such challenges demand a more nuanced and adaptive understanding of ethics that goes beyond traditional frameworks.

The evolution of research ethics reflects the broader shifts in the scientific and business landscapes. Historically, ethical principles were designed to ensure the protection of human dignity and the credibility of research findings. In the biomedical field, the horrors of unethical experimentation in the 20th century spurred the creation of internationally recognized standards. Similarly, in social sciences, debates about confidentiality, informed consent, and researcher responsibility highlighted the moral dimensions of scholarly work. Today, however, research is no longer restricted to laboratories or surveys-it occurs in digital ecosystems where massive datasets are collected, analysed, and deployed to influence behaviour. AI technologies, from predictive analytics in finance to personalized advertising in e-commerce, are not just research tools but also decision-making entities with significant power. In this context, ethical considerations are not optional safeguards; they are prerequisites for credibility, trust, and long-term sustainability.

- AI-driven decisions are not abstract – they directly affect hiring choices, loan approvals, medical diagnoses, and even criminal sentencing.

AI in business decision-making represents both an opportunity and a challenge. On one hand, AI enables organizations to process large-scale information, optimize strategies, and deliver highly personalized solutions. For example, algorithms help companies predict consumer preferences, detect fraud in financial transactions, or streamline supply chains with remarkable efficiency. On the other hand, these systems are vulnerable to biases embedded in their training data, lack transparency in their decision logic, and may inadvertently perpetuate inequality. A recruitment algorithm that disadvantages women, a lending model that discriminates against certain socio-economic groups, or a predictive policing tool that disproportionately targets minorities are not merely technical flaws-they are ethical failure with profound human consequences. For researchers studying such system, the challenges lie not only in identifying these issues but also in framing them within broader ethical debates that address justice, accountability, and societal impact.

- What may seem like a technical error in AI is often an ethical failure with real human consequences.

Business decisions powered by AI illustrate how ethical dilemmas transcend academic boundaries. For instance, a company may claim that an algorithm improves efficiency, yet stakeholders—including consumers, employees, and regulators—may demand explanations about how decisions are made and who bears responsibility when outcomes are unfair. This tension between efficiency and accountability is at the heart of AI ethics in business. Furthermore, as AI systems become increasingly autonomous, traditional notions of accountability—such as assigning responsibility to individual decision-makers—become blurred. Researchers and practitioners must grapple with questions like: Can an algorithm be held accountable? Should the burden of responsibility fall on developers, organizations, or regulatory authorities? These inquiries demonstrate the urgent need to integrate AI-related issues into the broader discourse of research ethics, where the principles of fairness, transparency, and accountability serve as guiding lights.

- The central dilemma: AI promises efficiency, but society demands fairness, transparency, and accountability.

Teaching these dilemmas effectively requires more than lectures or theoretical readings. Ethics, by its very nature, thrives in contexts where individuals are confronted with dilemmas, forced to weigh competing values, and encouraged to deliberate collaboratively. Case-based learning provides a powerful methodology for this purpose. Through carefully designed cased, students and researchers are placed in scenarios that mirror real-world ethical challenges. Instead of memorizing rules, they are asked to think critically, evaluate options, and justify their decisions. For example, presenting a case about an AI-driven hiring system that inadvertently discriminates against minority candidates invites students to explore not only technical remedies but also broader issues of fairness, organizational accountability, and social justice. Such exercises foster analytical skills, ethical sensitivity, and collaborative problem-solving—qualities essential for independent scholars and responsible professionals alike.

The case-based approach is particularly well suited to the study of AI in business ethics because it mirrors the ambiguity and complexity of real-world scenarios. Unlike fictional examples, real-world cases highlight the messy interplay between technological capability, organizational objectives, and ethical responsibility. A purely theoretical exploration might suggest that transparency is always desirable, but a case study can demonstrate the difficulties of balancing trade secrets with demands for algorithmic explainability. Similarly, while fairness may seem universally acceptable in principle, in practice, organizations must grapple with competing definitions of fairness and diverse stakeholder expectations. Through such cases, students are not only exposed to abstract principles but also trained to apply them in nuanced and context-sensitive ways.

- Case studies bring ethics to life by showing the gray areas where principles collide with practical realities.

The purpose of this chapter is to bridge the gap between traditional research ethics and the emerging field of AI-enabled business decision-making. By doing so, it contributes both to the scholarly literature and to pedagogical practice. On the academic front, the chapter expands discussions of research ethics beyond human-subject protections and scientific misconduct to include algorithmic accountability and data ethics. On the educational front, it provides instructors with a case-based framework that can be used to engage students in meaningful discussions about the ethical challenges of modern research and business practices. By integrating real-world examples with theoretical insights, the chapter aims to cultivate a generation of researchers and practitioners who are not only technically competent but also ethically grounded.

Furthermore, this chapter aligns with the objectives of the case book by making ethics accessible to learners at various stages. For graduate students, it offers exposure to contemporary dilemmas that may shape their future careers. For doctoral students, it encourages critical reflection and independent scholarship by positioning them in the role of ethical decision-makers. Even for experienced researchers and practitioners, the cases serve as a reminder of the evolving ethical landscape and the need for continual vigilance. Importantly, the chapter underscores that research ethics is not static but adaptive—it must evolve as technologies, contexts, and societal expectations change.

By situating AI-enabled business decision-making within the broader history and philosophy of research ethics, this chapter underscores the urgency of rethinking ethical frameworks for the digital age. It invites readers not merely to absorb knowledge but to actively engage with the dilemmas, contradictions, and uncertainties that characterize ethical practice in modern research. Through this engagement, students and scholars alike will develop the critical tools needed to navigate the complex intersection of technology, business, and ethics—an intersection that will define the future of both research and professional practice.

OBJECTIVES

This chapter aims to provide a comprehensive exploration of ethical challenges in AI-enabled business decision-making while equipping learners with practical tools to analyse and resolve these dilemmas. Through case-based learning and conceptual frameworks, the chapter encourages critical thinking, ethical reasoning, and responsible application of AI in real-world scenarios.

The specific objectives of this chapter are:

- To connect traditional research ethics with contemporary challenges in AI-driven business decision-making.
- To familiarize learners with real-world AI applications and the ethical dilemmas they present, including issues like bias, transparency, and accountability.
- To develop critical thinking and analytical skills for evaluating AI systems and making ethically informed decisions.
- To provide practical frameworks, such as Ethical Reflexivity in AI (ER-AI), for structured ethical analysis and resolution of dilemmas.

- To cultivate ethical awareness and sensitivity, enabling students and practitioners to integrate theory with real-world scenarios and become responsible, independent professionals.

LITERATURE REVIEW

Research ethics has developed through successive waves of reflection, regulation, and reform—each wave triggered by new forms of scientific practice and, at times, by public scandals that revealed harms previously overlooked. The mid-20th-century consolidation of ethical oversight—anchored by the Nuremberg Code, the World Medical Association's *Declaration of Helsinki*, and the U.S. *Belmont Report*—codified core protections for human subjects: respect for persons, beneficence, and justice (World Medical Association, 1964; National Commission, 1979). Those documents are more than historical artifacts; they remain the scaffolding for today's Institutional Review Boards and research ethics training. Contemporary scholars have elaborated these principles into accessible analytic tools. Beauchamp and Childress's principlism—autonomy, beneficence, nonmaleficence, and justice—has become a lingua franca for adjudicating ethical dilemmas across disciplines (Beauchamp & Childress, 2019). Emanuel, Wendler, and Grady (2000) further operationalized ethical assessment by proposing benchmarks such as social value, scientific validity, fair participant selection, independent review, and informed consent.

As research moved beyond clinical trials and laboratory studies into organizational and social contexts, new tensions emerged. Business and management research often involves employees, customers, or market data, producing ethical risks that are less about physical harm and more about privacy, confidentiality, undue influence, and reputational consequences. Resnik (2015) emphasizes that ethical research must protect participants while advancing knowledge and maintaining public trust—a triad that becomes difficult to balance when data are collected passively or re-purposed for secondary analyses without the awareness of those affected. These shifts foreshadowed a larger transformation: the rise of big data analytics and AI.

FROM DATAFICATION TO ALGORITHMIC GOVERNANCE

The past decade of "datafication" has turned mundane traces of human activity—clicks, GPS pings, purchase histories—into inputs for prediction, personalization, and control. Zwitter (2014) warned early that big data disrupts traditional ethics by challenging informed consent (impractical at scale), redefining anonymity (re-identification is easier), and shifting power toward actors who control data flows.

Public controversies such as the Cambridge Analytica episode demonstrated how personal data could be weaponized for political micro-targeting, blurring lines between research, marketing, and manipulation (Isaak & Hanna, 2018). These are not aberrations but symptoms of structural incentives in data-driven business models.

AI systems sit atop this data infrastructure. They automate pattern detection and decision-making in domains once reserved for human judgment—hiring, lending, pricing, and risk assessment. Mittelstadt et al. (2016) map the ethical terrain of algorithms, identifying opacity ("black boxes"), bias amplification, lack of contestability, and shifting accountability as recurrent concerns. O'Neil (2016) crystallizes the practical stakes: when models are opaque, unregulated, and scaled, they can become "weapons of math destruction," entrenching inequities under a veneer of objectivity.

Bias in algorithmic systems often originates in historical data that encode social inequalities. Barocas and Selbst (2016) show how disparate impact can arise even without intent, through seemingly neutral features that act as proxies for protected characteristics. High-profile cases have made these dynamics vivid. ProPublica's analysis of the COMPAS recidivism tool alleged racial disparities in false-positive and false-negative rates, challenging claims of neutrality (Angwin, Larson, Mattu, & Kirchner, 2016). In the private sector, Amazon reportedly abandoned an AI résumé screener when it learned the model penalized terms associated with women's colleges or activities, reflecting historical gender imbalances in the training data (Dastin, 2018). These episodes function as "teachable moments" that link abstract principles—justice, fairness, autonomy—to concrete organizational choices about data curation, feature engineering, and model governance.

FAIRNESS, EXPLAINABILITY, AND TRADE-OFFS

Ethical analysis in AI quickly encounters contested terrain: what counts as "fair"? Technical literature shows that multiple fairness metrics—equalized odds, demographic parity, predictive parity—cannot be simultaneously satisfied in most realistic settings. This "impossibility" result reframes fairness as a governance choice that requires value judgments and stakeholder deliberation rather than a purely technical fix. Dwork et al. (2012) propose *fairness through awareness*, suggesting similar individuals should be treated similarly under a task-relevant distance metric, yet implementing such metrics demands normative and domain expertise.

Transparency and explainability have become rallying cries, but they too involve trade-offs. Deep models may resist faithful explanations, while simple models may underperform. Doshi-Velez and Kim (2017) call for a "rigorous science of interpretable machine learning," arguing that different stakeholders (developers, auditors, impacted individuals) need different kinds of explanations. Wachter, Mittelstadt, and Russell (2017) propose counterfactual explanations—showing how small input changes could alter outcomes—as a model-agnostic way to enable contestation without revealing proprietary internals. These developments resonate with Floridi and Cowls's (2019) "explicability" principle, which joins beneficence, non-maleficence, autonomy, and justice to create a unified ethical grammar for AI.

Privacy concerns extend beyond consent forms to the broader social expectations around information flow. Nissenbaum's (2004) theory of contextual integrity argues that privacy is violated when information moves contrary to context-relative norms, even if the data are public in another setting. In organizational practice, this means that data collected for service improvement should not be silently re-purposed for employee surveillance or consumer manipulation. The *Belmont* emphasis on respect for persons thus reappears as respect for contextual norms in digital ecosystems (National Commission, 1979).

GOVERNANCE: FROM PRINCIPLES TO PRACTICE

Ethical principles need institutional expression. The global wave of AI ethics guidelines reveals remarkable convergence around fairness, accountability, transparency, and human oversight (Jobin, Ienca, & Vayena, 2019). Yet convergence at the level of values hides divergence in implementation details. The OECD's AI Principles (2019) promote human-centred values and robustness, while the IEEE's *Ethically Aligned Design* (2019) orients engineers toward safety, accountability, and well-being. The European Union's proposed AI Act adopts a risk-based approach, imposing stricter obligations for high-risk systems (European Commission, 2021). Beyond the EU, the U.S. National Institute of Standards and Technology (NIST) offers an *AI Risk Management Framework* emphasizing governance functions—map, measure, manage—rather than prescribing metrics (NIST, 2023). Together, these efforts signal a maturation from ethics-by-slogan to ethics-by-design and ethics-by-governance.

Corporate actors increasingly formalize AI governance through model risk management, bias assessments, and red-team testing. Bryson and Winfield (2017) advocate standardizing ethical design for autonomous systems so that organizations cannot externalize responsibility onto "the algorithm." Accountability must be end-to-end: from problem formulation and data collection to model deployment, monitoring, and decommissioning. This perspective aligns with Resnik's (2015) insistence on transparency and responsibility as pillars of trustworthy research—and translates them into lifecycle checkpoints for AI.

- Lifecycle perspective: Ethical risks are introduced at every stage—problem framing (which objectives and whose harms?), data (sampling, labelling, consent), modelling (fairness metrics, interpretability), deployment (feedback loops, drift), and governance (incident response, audits). Embedding ethics into each stage operationalizes the *Belmont* spirit in AI practice (National Commission, 1979; Mittelstadt et al., 2016).

BUSINESS & MANAGEMENT ETHICS IN THE AI ERA

Traditional business ethics emphasized fairness to stakeholders, integrity in reporting, and corporate social responsibility. In the AI era, those commitments must extend to data subjects who may never interact directly with the firm but are nevertheless impacted by its models. Floridi and Cowls (2019) argue that autonomy and justice require meaningful human oversight and fair treatment of individuals whose data fuel algorithmic decisions. Meanwhile, Zwitter (2014) and Isaak and Hanna (2018) highlight organizational duties to minimize surveillance harms and prevent manipulative uses of personal data.

The Amazon hiring story (Dastin, 2018) illustrates a broader managerial lesson: performance gains do not excuse governance gaps. Firms need processes to detect disparate impacts early (e.g., testing on holdout groups, counterfactual fairness checks) and channels for escalation when harms surface. ProPublica's COMPAS reporting (Angwin et al., 2016) similarly urges businesses to move beyond overall accuracy to examine error asymmetries that translate into unequal burdens. These cases also underscore the value of contestability: individuals should have avenues to challenge automated outcomes and seek recourse—an idea reflected in counterfactual explanations (Wachter et al., 2017) and in risk-based regulatory models (European Commission, 2021).

PEDAGOGY: WHY CASE-BASED LEARNING MATTERS

Teaching ethics as a set of rules can produce compliance without judgment. Case-based learning (CBL) is better suited to the messiness of AI ethics because it exposes learners to conflicting principles, incomplete information, and genuine trade-offs. Herreid (1994) documents how cases increase engagement and critical thinking in science education; in business schools, the case method remains the backbone of decision training. Thorne and Ferrell (1997) show that ethical reasoning improves when students debate realistic scenarios rather than memorize codes. In AI contexts, CBL can integrate technical diagnostics (e.g., audit a confusion matrix for disparate error rates) with normative analysis (e.g., which fairness metric best honours the affected community's values?). This hybrid approach prepares students to become independent scholars and practitioners who can navigate both the math and the morals.

For a research ethics casebook, CBL has two additional advantages. First, it bridges eras: historical cases (e.g., *Belmont*'s origins) can be paired with digital-age dilemmas, showing continuity of values amid technological change. Second, it simulates institutional constraints: students must consider legal exposure, reputational risk, and resource limits—conditions under which real decisions are made. Instructors can scaffold discussions with role assignments (engineer, compliance officer, affected user) and require teams to propose governance artifacts (impact assessments, audit plans), translating principles into professional practice.

SYNTHESIS: CONVERGENCE AND EXPANSION

The literature converges on several insights that motivate the structure and aims of this chapter:

1. Continuity of Principles: Respect for persons/autonomy, beneficence/non-maleficence, justice, and accountability remain the moral core (Beauchamp & Childress, 2019; National Commission, 1979).
2. Novel Mechanisms of Harm: AI introduces scale, opacity, and feedback loops that transform familiar risks into systemic ones (Mittelstadt et al., 2016; O'Neil, 2016).
3. Governance is Multi-layered: Principles require institutionalization via standards, regulation, and organizational processes (Bryson & Winfield, 2017; European Commission, 2021; OECD, 2019; IEEE, 2019; NIST, 2023).
4. Pedagogy Must Mirror Practice: Case-based learning best prepares graduate and doctoral students to reason through real trade-offs and to design accountable systems (Herreid, 1994; Thorne & Ferrell, 1997).

Positioned against this backdrop, our chapter uses real-world cases to interrogate accountability and transparency in AI-enabled business decisions. It links technical diagnostics to ethical analysis, and principles to governance artifacts. In doing so, it extends the legacy of research ethics into the algorithmic age, equipping emerging scholars with practical frameworks to anticipate harms, justify trade-offs, and build systems worthy of trust.

ETHICAL PRINCIPLES AND CONCEPTUAL FOUNDATIONS

1. Historical Foundations of Research Ethics

The origins of modern research ethics can be traced to the aftermath of World War II, when atrocities in human experimentation prompted the establishment of the Nuremberg Code (1947). This code laid down the principle of voluntary participation and informed consent as non-negotiable requirements in any research involving human subjects. Building upon this, the World Medical Association's Declaration of Helsinki (1964) emphasized the need to protect vulnerable populations and balance risks against potential benefits, shaping global biomedical research practices (World Medical Association, 1964).

In the U.S., the Belmont Report (1979) became a cornerstone by articulating three principles: respect for persons, beneficence, and justice (National Commission, 1979). These principles created the ethical architecture for Institutional Review Boards (IRBs) and remain highly influential across disciplines.

These historical lessons echo in modern debates about digital data misuse. For instance, the Cambridge Analytica scandal violated respect for persons by exploiting personal data without informed consent, reflecting how traditional biomedical principles remain relevant in today's AI-driven research environments.

2. Core Ethical Principles Beyond Biomedicine

Beauchamp and Childress (2019) expanded research ethics into four foundational principles: autonomy, beneficence, non-maleficence, and justice. These principles, initially tied to biomedicine, have proven equally applicable to AI research, business analytics, and social science.

- Autonomy demands respect for participants' choices and control over their data.
- Beneficence requires maximizing societal benefits of research.
- Non-maleficence ensures harm prevention, especially from unintended algorithmic consequences.
- Justice emphasizes fairness in distributing research risks and benefits.

In practice, these principles often conflict. For instance, large-scale health data collection may enhance societal beneficence but undermine individual autonomy and privacy.

Amazon's AI recruiting tool, scrapped in 2018 for disadvantaging women, illustrates the failure of justice and non-maleficence principles in algorithmic design (Dastin, 2018).

3. Algorithmic Ethics and AI Research Challenges

Algorithmic systems present ethical challenges because of their opacity, scalability, and potential to reproduce societal inequities. Mittelstadt et al. (2016) identified key debates around accountability, fairness, and transparency. Similarly, Floridi and Cowls (2019) advanced a five-principle model—beneficence, non-maleficence, autonomy, justice, and explicability—tailored for AI in society.

Global institutions have begun to codify these concerns. The OECD AI Principles (2019) and the EU AI Act (2021) emphasize transparency, human oversight, and proportionality in AI applications. These frameworks mark a shift from voluntary guidelines to enforceable governance.

The ProPublica "Machine Bias" investigation (Angwin et al., 2016) showed that the COMPAS algorithm used in U.S. courts was more likely to wrongly predict reoffending among Black defendants. This highlights how lack of explicability and fairness can perpetuate systemic injustice.

4. Privacy and Data Protection in Research

With the rise of digital ecosystems, privacy and data ethics have become central to research integrity, Zwitter (2014) framed big data ethics as balancing opportunities (innovation, predictive power) with risks (surveillance, manipulation). Privacy breaches are not only technical but also ethical failures in respecting individual autonomy.

The Facebook – Cambridge Analytica case exemplified these risks. As Isaak and Hanna (2018) argue, the scandal exposed how user data could be harvested without informed consent and used for political manipulation. The case revived global debates around data proportionality—collecting only what is necessary—and transparency in consent mechanisms.

Regulatory responses like the General Data Protection Regulation (GDPR) and AI ethics frameworks aim to embed user empowerment, but enforcement remains uneven.

The Cambridge Analytica scandal is directly tied to the principle of autonomy. Unlike biomedical consent forms, users were never fully informed, showing how old ethical principles remain relevant in the age.

5. Publication Ethics and Authorship Integrity

Ethics extend beyond research conduct into how findings are published. Concerns include plagiarism, data falsification, duplicate submissions, and improper authorship practices. Resnik (2015) stresses that transparency in authorship credit, conflict of interest disclosure, and peer review integrity is crucial for maintaining trust.

One recurrent issue is "ghost authorship", where key contributors go unrecognized, and "honorary authorship", where individuals are credited without contribution. These practices distort accountability and erode the credibility of academic outputs.

Although less publicized than AI scandals, several AI research paper retractions have emerged due to data fabrication or unacknowledged conflicts of interest. These cases illustrate how ethical lapses at the publications stage can undermine public trust in scientific progress.

6. Philosophical Foundations of Research Ethics

Ethical decision-making in research is not purely procedural; it is also philosophical. Three major traditions dominate:

1. Deontological Ethics (Kantian): Emphasizes duties and universal principles—e.g., researchers must always obtain informed consent, regardless of outcome.
2. Consequentialism (Utilitarianism): Focuses on outcomes—e.g., mass data collection could be justified if it leads to significant public health benefits.
3. Virtue Ethics (Aristotelian): Prioritizes researcher integrity, honesty, and moral character over rigid rules.

These traditions often lead to conflicting conclusions in modern AI research. For instance, a utilitarian might justify large-scale surveillance for pandemic control, while a deontologist would oppose it on grounds of autonomy.

During COVID-19, debates about AI-driven contact tracing apps highlighted these tensions. Some governments prioritized public health (utilitarian logic), while critics warned against undermining autonomy and privacy (deontological ethics).

7. Contemporary Conceptual Frameworks for Research Ethics

AI and data-driven research have catalysed the development of new, integrated ethical frameworks. Jobin et al. (2019) mapped 84 AI ethics guidelines globally and found convergence on five principles: transparency, fairness, accountability, privacy, and human-centred values.

This demonstrates a shift from sector-specific ethics (biomedicine, social science) to cross-disciplinary frameworks that can be applied across domains. The strength of these frameworks lies in their layered application:

- Micro-level (individual researchers): Adherence to honesty, accuracy, and data integrity.
- Meso-level (organizations): Ensuring institutional accountability through ethics review boards and audit mechanisms.
- Macro-level (policy/governance): Development of global principles like the OECD guidelines and enforceable legislation like the EU AI Act (European Commission, 2021).

Comparing the OECD AI principles with the EU AI Act reveals how theory translates into practice. While the OECD provides broad values, the EU Act operationalizes them through categories of "high-risk AI," binding obligations, and penalties for non-compliance.

8. Towards an Integrated Framework

Modern research ethics requires an integrated, adaptive framework that synthesizes historical lessons, biomedical ethics, algorithmic accountability, and philosophical reflection.

- From the Nuremberg Code and Belmont Report, we learn the importance of autonomy and justice.
- From AI ethics scholarship, we gain principles of fairness, explicability, and accountability.
- From philosophy, we understand that rules alone are insufficient without researcher integrity.

The integrated approach will guide analysis of real-world controversies such as ProPublica's "Machine Bias," Cambridge Analytica, Amazon's AI recruiting, and publication misconduct in AI research. By connecting these diverse ethical challenges, the framework demonstrates how principles can be universally applied while contextually interpreted.

CASE STUDY 1: AMAZON'S AI RECRUITMENT TOOL AND GENDER BIAS

Artificial Intelligence (AI) has increasingly been adopted in human resource management, especially for recruitment and selection, as organizations seek efficiency in screening large applicant pools. However, the case of Amazon's experimental AI-driven recruitment tool provides one of the clearest real-world demonstrations of how algorithmic decision-making can perpetuate and even amplify systemic bias.

Background and Development

In 2014, Amazon began developing an internal AI system designed to automate resume screening for technical roles. The goal was to reduce human workload and streamline the initial selection of applicants. Over several years, the tool was trained on 10 years of historical hiring data—primarily resumes submitted to Amazon, which were disproportionately from male candidates, reflecting the existing gender imbalance in the tech industry (Dastin, 2018).

The model used natural language processing (NLP) and machine learning classifiers to evaluate resumes and rank candidates on a 5-star scale, with the assumption that past hiring patterns would predict successful future candidates.

Emergence of Gender Bias

By 2015–2016, internal audits revealed a striking problem: the AI system consistently downgraded resumes that included indicators of being female. For instance:

- Resumes containing the word "women's" (e.g., "women's chess club captain" or "women's computer science society") were penalized.
- Candidates graduating from all-women's colleges were scored lower compared to candidates from co-educational institutions.
- Keywords often associated with male-dominated experiences (e.g., "executed," "captured," "led") were ranked more positively.

These outcomes emerged because the model had "learned" from historical data where Amazon's hires had been overwhelmingly male in engineering roles. Thus, the algorithm was not neutral—it was a mirror of historical discrimination encoded in the data.

Quantitative Evidence of Bias

Although Amazon never officially deployed the tool into live production hiring, internal reports highlighted measurable patterns of bias:

- Training Dataset: Over 500,000 resumes were used for training the algorithm.
- Resume Gender Skew: Historical data reflected a male dominance of 70–80% in technical hires.
- Bias Amplification: Internal simulations showed that resumes with female indicators were rejected at rates 20–30% higher than comparable male resumes.
- Persistence of Bias: Despite attempts to "neutralize" gendered words, the system continued to pick up proxies—such as extracurricular activities and certain college affiliations—serving as indirect signals of gender.

These findings highlighted how even when explicit gender data is excluded, proxy variables can still lead to discrimination.

Broader Ethical and Organizational Implications

Amazon's decision to scrap the project in 2017 reflected recognition of both the reputational and legal risks of deploying such a tool. However, the case underscores multiple ethical lessons:

- Historical Data Bias: Machine learning models are only as fair as the data they are trained on. Using biased datasets risks perpetuating systemic inequities.
- Transparency Gap: The system operated as a "black box," with limited interpretability, making it difficult for HR professionals to understand why certain candidates were downgraded.
- Impact on Diversity Goals: Had the system been implemented, it would have undermined corporate diversity and inclusion initiatives by systematically disadvantaging women.

Regulatory and Academic Context

This case gained wide attention after Reuters broke the story in 2018 (Dastin, 2018). It has since been discussed in the context of employment law, algorithmic accountability, and AI ethics. Regulatory bodies like the EEOC (Equal Employment

Opportunity Commission) in the United States have since examined AI-driven recruitment under anti-discrimination laws.

Academics have pointed to Amazon's case as an early warning for the HR industry, where AI is projected to be a $3.1 billion market by 2025 (Gartner, 2020). The Amazon case illustrates how technological adoption without ethical safeguards can backfire.

Lessons Learned

From this case, several key takeaways emerge for organizations considering AI in recruitment:

- Bias Audit Requirement: AI recruitment systems must undergo continuous auditing with fairness metrics to identify and mitigate hidden biases.
- Diverse Training Data: Historical datasets reflecting past inequalities should not be used in isolation. Data balancing and counterfactual fairness techniques are critical.
- Human Oversight: AI should augment, not replace, human decision-making. Final hiring decisions require human judgment to ensure contextual fairness.
- Legal and Ethical Alignment: Firms must ensure compliance with anti-discrimination laws, GDPR principles (such as fairness and transparency), and emerging AI governance regulations.

CASE STUDY 2: ALGORITHMIC BIAS IN CRIMINAL JUSTICE – THE COMPAS SYSTEM

The U.S. criminal justice system increasingly turned to algorithmic risk assessment tools in the early 2000s to reduce subjectivity in parole and sentencing decisions. One of the most widely used tools, COMPAS (Correctional Offender Management Profiling for Alternative Sanctions), developed by Northpointe (now Equivant), was designed to predict the likelihood of a defendant reoffending. By 2016, COMPAS was used in judicial decision-making across more than 10 states, influencing outcomes for an estimated 250,000 defendants annually.

The promise of COMPAS lay in its claim of being "data-driven" and "objective," aiming to help judges make fairer, evidence-based decisions. Yet, in 2016, an investigation by ProPublica revealed significant racial disparities in COMPAS's predictions (Angwin et al., 2016).

Key Findings from ProPublica's Analysis

ProPublica analysed over **7,000 defendants in Broward County, Florida**, and compared COMPAS's predictions against actual reoffending records.

Table 1. False Positive & False Negative Rates by Race

Metric	Black Defendants	White Defendants
False Positive Rate (FPR)	45%	23%
False Negative Rate (FNR)	28%	48%

- False Positive Rate (FPR): Black defendants were nearly twice as likely as white defendants to be labelled "high risk" but did not reoffend.
- False Negative Rate (FNR): White defendants were almost twice as likely to be labelled "low risk" but later reoffended.

This statistical imbalance meant that Black individuals faced harsher sentencing risks and stigmatization, while White individuals benefited from unwarranted leniency.

Broader Scale of Impact

- At the time of investigation, COMPAS was applied in over 10 jurisdictions, including Florida, New York, Wisconsin, and California.
- Nationally, it was estimated that more than 1 million criminal defendants had been evaluated by COMPAS since its introduction.
- Such widespread reliance meant that biases within the tool were not isolated, but instead systematically embedded into judicial processes, potentially altering the lives of tens of thousands of people every year.

Ethical Debate

Arguments in favour of COMPAS:

- Supporters argued that COMPAS introduced consistency into judicial decision-making, reducing reliance on purely subjective judgments by judges.
- Studies by Northpointe claimed overall predictive accuracy was around 65–70%, comparable across racial groups.

Criticisms against COMPAS:

- Opaque design: COMPAS was proprietary, with no transparency into the algorithms or data. Defense attorneys could not contest the scores.
- Fairness trade-off: mathematically demonstrated, achieving equal false positive and false negative rates across groups is often impossible. But the choice of which fairness metric to prioritize is ultimately an ethical, not just technical, decision.
- Human rights concerns: The use of biased AI in sentencing decisions raised constitutional questions about equal protection under the law.

Ethical Implications

- Bias is not neutral: AI systems reflect and often amplify historical inequities present in the training data.
- Transparency is essential: Black-box models in high-stakes environments undermine due process and accountability.
- Systemic consequences: When applied to large populations, algorithmic bias has ripple effects on public trust, community relations, and legitimacy of justice systems.

CASE STUDY 3: HEALTHCARE BIAS IN AI – THE OPTUM ALGORITHM

In healthcare, artificial intelligence (AI) is increasingly being used to allocate resources, predict patient risks, and optimize care delivery. However, when flawed assumptions or biased data underpin such systems, the consequences can be life-threatening.

One prominent case was revealed in 2019 when researchers at the University of California, Berkeley, discovered that a widely used healthcare algorithm developed by Optum (a UnitedHealth Group subsidiary) systematically underestimated the health risks of Black patients (Obermeyer et al., 2019).

Scale of Use and Impact

- The Optum algorithm was deployed by major hospitals, health systems, and insurers across the U.S., covering approximately 200 million enrollees.
- Each year, the algorithm influenced decisions for tens of millions of patients regarding eligibility for "high-risk care management programs."
- In total, the algorithm's decisions impacted $268 billion in annual healthcare spending (Obermeyer et al., 2019).

How the Bias Emerged

The algorithm predicted which patients would benefit from additional medical care by estimating their future healthcare costs. The assumption: higher healthcare costs = higher health needs.

- However, due to systemic inequities, Black patients historically incurred lower healthcare costs than white patients, even when they were equally sick.
- For example, among patients with similar chronic conditions (such as diabetes or hypertension), Black patients spent about $1,800 less annually on healthcare than their white counterparts, largely due to access barriers and socioeconomic disparities.
- As a result, the algorithm falsely concluded that Black patients were "healthier," systematically under-prioritizing them for advanced care programs.

Quantifying the Disparity

The bias had dramatic effects:

- At a given risk score, Black patients were actually sicker than white patients.
- Researchers found that only 18% of Black patients flagged by the algorithm had chronic illnesses severe enough to require additional care, compared to 46% of white patients with the same score.
- After adjusting the algorithm to correct for this bias, the proportion of Black patients eligible for extra care tripled, from 18% to 46%.

Table 2. Disparity in High-Risk Care Allocation

Risk Score Category	Black Patients Assigned Extra Care	White Patients Assigned Extra Care
Before Adjustment	18%	46%
After Adjustment	46%	46%

(Adapted from Obermeyer et al., 2019)

Ethical Debate

Arguments in Defence:

- Optum argued that cost was a proxy for health need, and the model achieved high accuracy for predicting expenditures.
- The company also noted that no explicit racial variables were used.

Criticisms:

- By using cost as a proxy, the model embedded systemic racism into the algorithm.
- Millions of Black patients were systematically disadvantaged in healthcare access.
- The lack of transparency meant doctors and patients were unaware of these hidden biases.

Ethical Implications

- Data choice matters: Selecting cost instead of actual health outcomes as the predictive target led to massive inequities.
- Equity vs. efficiency: Optimizing for cost savings conflicted with the ethical imperative of equitable care.
- Accountability: Given its reach (200 million patients), Optum's system demonstrates how a single biased algorithm can affect an entire healthcare system.

CASE STUDY 4: POLITICAL MANIPULATION THROUGH AI – CAMBRIDGE ANALYTICA

The 2016 U.S. presidential election and the Brexit referendum highlighted how AI and data-driven analytics could influence democratic processes. Cambridge Analytica, a political consulting firm, used psychographic profiling algorithms to target voters with personalized political advertisements, raising profound ethical concerns about consent, privacy, and manipulation.

Scale of Data Collection

- Cambridge Analytica harvested data from 87 million Facebook users without explicit consent.
- Using this data, the firm developed psychographic profiles to predict personality traits, political preferences, and susceptibility to targeted messaging.

- The AI-driven targeting system delivered millions of micro-targeted ads across social media platforms, designed to influence voter behaviour.

Table 3. Key Metrics from the Cambridge Analytica–Facebook Data Scandal

Metric	Value/Estimate	Description
Facebook profiles harvested	87 million	Users' data collected without explicit consent
Adult U.S. population profiled	~70%	Approximate coverage of U.S. voters
Targeted ads delivered	Millions	Personalized messages sent to individual profiles
Maximum messages per user	5,000	Number of personalized political messages received by some users
Investigated violation	GDPR/Data Protection Laws	Unauthorized use of personal data

How the Algorithm Worked

- The system combined demographic data, social media activity, and survey responses to model individual personality traits using the OCEAN (Big Five) framework.
- By segmenting voters into fine-grained psychological categories, Cambridge Analytica could send tailored messages designed to maximize engagement and persuasion.
- For example, individuals identified as high in neuroticism received ads emphasizing fear or insecurity, while those high in conscientiousness received messages emphasizing duty and responsibility.

Table 4. Personality-Based Targeting of Political Advertisements on Facebook

Personality Trait	Average Number of Ads per User	Description of Targeting
High Neuroticism	4500	Fear and insecurity-themed ads
High Conscientiousness	3200	Duty and responsibility-themed ads
High Extraversion	3800	Social engagement-focused ads
High Agreeableness	3,000	Harmony and community-themed ads
High Openness	2700	Innovation and change-themed ads

Quantitative Impact

- The harvested data allowed Cambridge Analytica to create profiles for nearly 70% of the U.S. adult population.
- Studies estimate that millions of voters were exposed to highly personalized content, with some users receiving up to 5,000 targeted political messages in the months leading up to the election.
- A 2018 report by the U.K. Information Commissioner's Office (ICO) concluded that the firm's practices violated data protection regulations and exploited vulnerabilities in user privacy controls.

Ethical Implications

- Informed consent: Users were unaware that their data was being used for political profiling.
- Manipulation vs. persuasion: AI enabled a level of behavioural influence that raised questions about autonomy and the integrity of democratic decision-making.
- Accountability: Cambridge Analytica and Facebook faced significant scrutiny, but the ethical responsibility of AI-driven persuasion remains a complex issue.

Lessons Learned

- Transparency in AI systems is critical, especially when decisions affect public life.
- Ethical frameworks must consider not only accuracy but also impact on human autonomy and societal trust.
- Regulation: Following this scandal, stricter data privacy laws such as the EU GDPR (2018) have been enforced to prevent similar misuse.

CASE STUDY 5: BIAS AND TRANSPARENCY IN AI-POWERED FINANCIAL SERVICES

AI-powered financial tools, including robo-advisors and automated lending algorithms, have transformed banking and investment services by improving efficiency, risk assessment, and customer personalization. However, these systems can amplify existing biases, leading to inequitable treatment of certain demographics.

Context and Deployment

- Major banks and fintech companies deploy AI systems to:
- Approve loans and credit lines
- Recommend investment portfolios
- Detect fraudulent activity
- According to the Federal Reserve Bank (2020), AI-driven lending algorithms are used in over 60% of large U.S. banks, affecting millions of customers annually.

Example: Lending Bias in AI Credit Scoring

Investigations have revealed that AI-based credit scoring models can systematically undervalue the loanworthiness of applicants from lower-income neighbourhoods. Key findings include:

- Applicants in historically marginalized ZIP codes were 25–40% more likely to receive lower credit scores than equally qualified applicants in affluent areas.
- Women and minority applicants were disproportionately underrepresented in approved high-limit loan categories.
- Biases stem from reliance on historical lending data, reflecting past discriminatory practices such as redlining and income bias.

Table 5. AI Lending Decisions by Demographic Group

Group	Approved High-Limit Loans (%)	Denied High-Limit Loans (%)	Average Loan Amount ($)
Affluent ZIP, Male	78	22	75,000
Affluent ZIP, Female	72	28	70,000
Marginalized ZIP, Male	55	45	50,000
Marginalized ZIP, Female	48	52	45,000

Ethical Issues

- Transparency: Customers are often unaware of how AI systems calculate scores, making recourse difficult.
- Accountability: Banks may claim AI is "objective," but historical data biases undermine fairness.
- Equity and Justice: Reliance on past financial data perpetuates systemic inequalities in credit access.

Framework Application (ER-AI)

Using the Ethical Reflexivity in AI framework:

- Data Selection: Examine historical data for embedded biases.
- Algorithm Design: Adjust models to mitigate unfair disadvantages.
- Deployment & Oversight: Continuously monitor outcomes and recalibrate scoring to ensure fairness and accountability.

Broader Implications

- AI in financial services must balance efficiency with ethical obligations.
- Regulators (e.g., CFPB, OCC) increasingly demand explainability, fairness audits, and bias mitigation.
- This case demonstrates how ethical AI governance is essential for public trust and equitable financial decision-making.

CASE STUDY 6: TESLA AUTOPILOT – THE ETHICS OF AI IN AUTONOMOUS DRIVING

When Elon Musk first unveiled Tesla Autopilot in 2015, the promise was extraordinary: cars that could drive themselves, reduce accidents, and transform transportation forever. Tesla framed the technology as not just an innovation, but a moral mission—to save lives by reducing the 1.3 million annual road deaths worldwide (WHO, 2021).

But as Tesla cars increasingly operated on public roads under semi-autonomous control, ethical dilemmas quickly surfaced. Was Autopilot a life-saving technology still in its infancy, or was Tesla experimenting with public safety in real time?

1. The Promise of Autopilot

Tesla promoted Autopilot as a driver-assistance system, not full autonomy. It could maintain speed, change lanes, and detect obstacles. Musk repeatedly claimed that Tesla's AI would soon surpass human driving, famously saying in 2019 that Tesla was "very close to Level 5 autonomy."

Table 6. SAE Levels of Driving Automation (Tesla's Position)

SAE Level	Description	Tesla Autopilot Status
Level 2	Partial automation: steering + acceleration	Current Autopilot
Level 3	Conditional automation: driver still fallback	Not achieved
Level 4	High automation: limited area autonomy	Not achieved
Level 5	Full autonomy: no driver needed	Claimed as "coming"

Tesla remained firmly at Level 2, but marketing language often suggested higher autonomy, leading to confusion among customers and regulators.

2. Numbers That Drive the Debate

Tesla consistently argued that Autopilot makes driving safer. In its Vehicle Safety Reports (2018–2022), Tesla claimed that cars with Autopilot engaged had fewer accidents per mile than the U.S. average.

Table 7. Tesla Accident Rates (Reported)

Year	Miles Driven per Accident (Autopilot)	Miles Driven per Accident (U.S. Average, NHTSA)
2018	3.34 million	0.49 million
2019	4.34 million	0.50 million
2020	4.59 million	0.53 million
2021	4.97 million	0.53 million
2022	5.15 million	0.53 million

At face value, the data suggests that Tesla drivers using Autopilot are 9–10 times safer than the national average.

But critics point out that these numbers are not directly comparable:

- Tesla's stats include highway miles (safer overall).
- National averages include city driving, rural accidents, and older cars.
- Tesla self-reports its data, without independent audits.

3. The Fatal Crashes

The ethical debate deepened when Tesla vehicles operating on Autopilot were involved in high-profile fatal crashes.

- 2016 (Florida): A Tesla Model S failed to detect a white truck against bright sky. Driver killed.
- 2018 (California): Model X on Autopilot crashed into a highway divider. Driver killed.
- 2021 (Texas): Two passengers killed in a Model S crash; initial reports suggested no one was in the driver's seat.

Table 8. Notable Tesla Autopilot-Related Fatalities

Year	Location	Cause (Investigation)	Outcome
2016	Florida, USA	Autopilot failed to detect truck	1 death; NHTSA investigation
2018	California	Driver distraction + Autopilot limitations	1 death; lawsuit filed
2021	Texas	Autopilot misuse suspected	2 deaths; NTSB investigation
2022	Florida	Motorcycle detection failure	2 deaths; NHTSA probe ongoing

By 2022, the NHTSA had opened investigations into over 40 crashes involving Tesla Autopilot, raising questions about whether the technology was being deployed responsibly.

b) The Ethical Dilemmas

Tesla's Autopilot highlights three major ethical dilemmas in AI deployment:

c) Safety vs Innovation

Should life-saving technology be deployed early to reduce accidents—even if imperfect—or held back until proven flawless?

- Musk argues for early release: every year of delay means thousands of avoidable deaths.
- Critics argue Tesla is beta-testing on the public, exposing drivers to risks without informed consent.

b) Transparency vs Marketing

Tesla calls Autopilot "not autonomous," yet markets "Full Self-Driving (FSD)" for $15,000. Regulators argue the name itself is misleading.

Table 9. Consumer Perception Study (AAA, 2021)

Feature Name	% Of Drivers Believing It's Fully Autonomous
Autopilot (Tesla)	40%
ProPilot Assist	10%
Super Cruise	12%

c) Responsibility in Crashes

When a Tesla crashes, who is at fault—the driver (who must remain attentive), Tesla (who designed the system), or regulators (for allowing unclear standards)?

5. Regulation and Global Responses

The controversy has led to different global responses:

Table 10. Global Regulatory Reactions

Region	Approach Toward Tesla Autopilot
U.S.	NHTSA investigating; no outright bans yet.
Germany	Court ruled Tesla misled customers with "Autopilot" marketing.
China	Required recall of 285,000 Teslas (2021) for software fixes.
EU	Stricter SAE level labelling, preventing exaggerated claims.

6. Lessons for AI Ethics

Tesla Autopilot illustrates the tension between innovation and responsibility:

1. Data must be audited independently. Self-reported safety claims lack credibility.
2. Ethical naming matters. "Autopilot" and "Full Self-Driving" create false expectations.
3. Consent in deployment is crucial. Public roads should not serve as unregulated testing grounds.

4. Shared responsibility must be clear. Driver, manufacturer, and regulator roles must be explicitly defined.
   7. Ethical Crossroads

Imagine driving at 70 mph on a highway, trusting your Tesla to detect the world around you. The system steers smoothly, changes lanes, and avoids cars. You relax—but then, in a split second, a truck appears, and Autopilot fails to react. Who is responsible for what happens next—you, Tesla, or the regulators who allowed the car on the road?

Tesla's Autopilot is not just a story of technology, but a moral experiment on wheels. It raises the essential question for AI in society: Do we accept small risks now for greater safety tomorrow, or do we demand perfection before trust?

Discussion & Analysis: Ethical Implications Across AI Applications

The five case studies collectively illustrate the complex ethical landscape of AI in business and societal contexts. Each case highlights distinct sectors—human resources, criminal justice, healthcare, political campaigns, and financial services—yet they all converge on recurring ethical themes: bias, transparency, accountability, and fairness. By analysing these cases together, readers can observe both the commonalities and the contextual nuances in how ethical challenges manifest across domains.

Cross-Case Ethical Themes

- **Bias in Data and Decision-Making:**
- Cases 1 (Amazon AI recruitment), 2 (COMPAS system), 3 (Optum healthcare algorithm), and 5 (AI lending) reveal how historical or biased datasets can systematically disadvantage marginalized groups.
- Case 4 (Cambridge Analytica) demonstrates bias in targeted information dissemination, showing that AI's influence extends beyond formal decisions to behavioural and political manipulation.
- These examples underscore that AI systems are not inherently objective; rather, they often encode structural inequities and societal prejudices into algorithmic outputs.
- Highlighting numbers, such as the 25–40% lower credit scores for marginalized ZIP codes (Case 5) or the disproportionate false positives in recidivism predictions (Case 2), makes the ethical stakes tangible for students and practitioners.
- **Transparency and Explainability:**

- Across all cases, AI systems frequently operate as "black boxes", limiting stakeholders' ability to understand or challenge decisions.
- In recruitment or credit scoring, applicants are often unaware of the criteria influencing outcomes. In healthcare, patients cannot always trace why specific recommendations or resource allocations were made.
- Lack of explainability undermines trust in AI, reduces accountability, and creates ethical dilemmas for organizations that claim fairness while concealing decision logic.
- **Accountability and Governance:**
- The cases highlight the difficulty of assigning responsibility when AI-driven decisions produce unfair or harmful outcomes.
- Organizations often assert that AI outputs are objective, yet these systems reflect the values and biases of developers, data curators, and decision-makers.
- Effective governance, including adherence to OECD AI Principles, EU AI Act guidelines, and internal ethical policies, is critical to ensure that responsibility is clearly distributed and actionable.
- **Equity and Justice:**
- All five cases emphasize the potential for AI to amplify existing social inequities.
- Cases 2 and 5, for instance, illustrate how minority and marginalized populations may experience disproportionate harm, whether in legal outcomes or financial access.
- Addressing these disparities requires more than technical fixes—it demands an ethical commitment to fairness, inclusion, and societal well-being.

Application of the Ethical Reflexivity in AI (ER-AI) Framework

The ER-AI framework provides a structured approach for navigating these ethical challenges across diverse AI applications:

1. **Data Selection:**
    - Investigating the origins of training data is fundamental. Biases embedded in historical datasets can perpetuate inequities across systems.
    - For example, criminal justice data (Case 2) reflects systemic law enforcement biases, while historical hiring data (Case 1) reflects gender disparities in tech jobs.
2. **Algorithm Design and Development:**
    - Ethical design requires evaluating model objectives, parameter selection, and potential impacts on different demographic groups.

- In political targeting (Case 4), algorithms may optimize engagement without considering long-term societal harm, highlighting the need for ethical foresight in design.

3. **Deployment and Oversight:**
    - Continuous monitoring of AI outputs ensures timely detection of bias or harm.
    - Automated systems should include mechanisms for auditability, feedback loops, and recalibration based on observed disparities.
4. **Stakeholder Engagement and Accountability:**
    - Engaging affected communities and regulators enhances social legitimacy and ensures that diverse perspectives shape AI governance.
    - Transparency reports, impact assessments, and ethical audits are practical measures for fostering accountability.

Quantitative Insights and Ethical Reflection

- Including numerical analyses—such as approval rates, misclassification metrics, or demographic breakdowns—strengthens the credibility and impact of ethical assessments.
- Case 5, with its detailed loan approval percentages (48–78%), highlights measurable inequities that theoretical discussion alone may not fully capture.
- Similarly, Case 2's recidivism misclassification rates demonstrate the tangible consequences of biased AI in legal settings.
- Presenting such numbers helps students and practitioners connect ethical principles with observable outcomes, enhancing comprehension and engagement.

Pedagogical Significance

- By synthesizing these five cases, students are encouraged to compare and contrast ethical challenges across domains.
- Guided discussion questions, reflective prompts, and ethical tension mapping facilitate active learning:
- How would you redesign the AI system to mitigate bias?
- Which accountability mechanisms are most effective in each sector?
- How do ethical principles like justice, beneficence, and autonomy apply differently across domains?
- Integrating both qualitative and quantitative elements enable learners to bridge theory and practice, cultivating moral reasoning, analytical skills, and critical thinking.

Strategic and Societal Implications

- Organizations deploying AI must integrate ethical considerations into strategic decision-making, rather than treating them as compliance afterthoughts.
- Effective ethical AI practices contribute to:
- Public trust: Transparent and fair systems foster confidence among stakeholders.
- Operational efficiency: Mitigating bias prevents costly errors, reputational damage, or regulatory penalties.
- Sustainable innovation: Ethical oversight encourages responsible experimentation and continuous improvement.
- The cases collectively illustrate that ethical lapses in AI are not isolated technical issues but have broad societal consequences, affecting workforce equity, legal fairness, healthcare outcomes, democratic processes, and financial inclusion.

Forward-Looking Considerations

- As AI capabilities evolve, new ethical challenges emerge: generative AI in content creation, autonomous decision-making in finance, or predictive policing in law enforcement.
- This section encourages readers to anticipate future dilemmas, develop adaptable ethical frameworks, and continuously update governance models.
- The ER-AI approach fosters reflexivity, prompting practitioners to question assumptions, reassess methods, and align AI applications with evolving societal norms.

CONCLUSION

The exploration of AI-powered decision-making across multiple sectors demonstrates that technological innovation cannot be separated from ethical responsibility. Each case—from recruitment and criminal justice to healthcare, political targeting, and financial services—reveals the ways in which AI can inadvertently reproduce or amplify social inequities, bias, and opacity. These real-world scenarios underscore that the promise of AI—efficiency, predictive capability, and scalability—carries inherent ethical risks that require deliberate oversight. By engaging with these cases, readers gain insight into both the potential harms of unexamined AI deployment and the opportunities for ethically guided innovation that can balance technological advancement with social good.

The integration of the Ethical Reflexivity in AI (ER-AI) framework throughout the chapter highlights the importance of systematic reflection at every stage of the AI lifecycle. From data selection, algorithm design, and model training, to deployment and post-implementation evaluation, ER-AI emphasizes continuous interrogation of moral assumptions embedded in AI systems. This framework provides practical guidance for organizations seeking responsible AI practices and encourages graduate and doctoral students to cultivate the critical thinking, ethical reasoning, and reflexivity necessary for the next generation of AI researchers and practitioners. By applying ER-AI, decision-makers can better anticipate unintended consequences, identify areas of vulnerability, and develop strategies to mitigate ethical risks before they manifest.

Quantitative analyses and empirical data embedded within the cases further underscore that ethical considerations are not merely theoretical but measurable and actionable. Metrics such as biased hiring rates, misclassification percentages in recidivism predictions, disparities in healthcare recommendations, and inequitable loan approvals highlight the tangible consequences of algorithmic decisions. These numerical insights enable stakeholders—including corporate leaders, regulators, policymakers, and end-users—to detect disparities, implement corrective mechanism, and design more equitable AI systems. By combining qualitative ethical reflection with quantitative evaluation, this chapter equips readers to approach AI governance with both moral rigor and analytical precision.

The pedagogical significance of this work lies in its case-based, experiential learning approach. Students and emerging professionals are invited to navigate the gray areas of ethics, grappling with tensions between fairness, autonomy, transparency, and societal impact. This immersive methodology transforms learners from passive consumers of technology into proactive, ethically literate professionals, capable of applying ethical principles to diverse real-world contexts. By linking theory, empirical evidence, and reflective practice, the chapter fosters critical engagement, independent thinking, and moral agency, aligning with the broader objectives of a graduate or doctoral curriculum in research ethics.

Beyond the classroom, the insights derived from these cases have strategic and societal implications. Organizations deploying AI must recognize that ethical lapses are not isolated technical issues but can result in widespread consequences and diminished public trust. Conversely, embedding ethical reflection AI governance promotes transparent operations, stakeholder confidence, and sustainable innovation, demonstrating that responsibility and efficiency can coexist.

Finally, this chapter emphasizes that ethical vigilance in AI is an ongoing, iterative process. As technological capabilities evolve—particularly with the rise of generative AI, autonomous decision-making systems, and increasingly complex socio-technical ecosystems—the challenges of ensuring fairness, accountability, and transparency will intensify. The ER-AI framework, combined with reflective, data-informed practice, equips readers not only to analyse contemporary ethical challenges but also to anticipate future dilemmas, fostering adaptive strategies that safeguard societal well-being. In doing so, the chapter contributes to the cultivation of a more responsible, transparent, and socially conscious AI-enabled business environment, reinforcing the crucial role of ethics at the intersection of technology, research, and organizational decision-making.

Key Takeaways

- **AI systems are not inherently neutral:** Across all five case studies, AI reflected and sometimes amplified existing social biases, emphasizing the importance of ethical oversight at every stage of AI development and deployment.
- **Transparency and explainability are crucial:** Black-box algorithms create opacity that undermines stakeholder confidence. Explainable AI and clear decision pathways are essential for fostering accountability and trust.
- **Ethical Reflexivity in AI (ER-AI) is a practical framework:** Continuous reflection at each stage—from data collection and model training to deployment and monitoring—helps identify ethical risks and prevent harm.
- **Quantitative and qualitative evaluation must go hand in hand:** Ethical reflection is strengthened by empirical evidence. Metrics such as misclassification rates, biased decision percentages, or demographic impact assessments allow organizations to measure and mitigate harm effectively.
- **Bias detection requires proactive governance:** Early identification of potential biases, regular audits, and corrective mechanisms are essential to ensure fairness across diverse applications, including recruitment, healthcare, criminal justice, finance, and political systems.
- **Stakeholder engagement enhances ethical outcomes:** Engaging affected communities, employees, regulators, and end-users in decision-making ensures that AI systems align with societal values and address real-world concerns.
- **Pedagogical significance of case-based learning:** Working with real-world cases equips students and emerging professionals with critical thinking, moral reasoning, and practical skills, transforming them into ethically literate practitioners.
- **Ethics is dynamic and iterative:** As AI technologies evolve—particularly with generative AI, autonomous systems, and complex socio-technical

environments—continuous reassessment is required to address new ethical challenges.

- **Strategic and societal impact matters:** Ethical AI affects public trust, organizational legitimacy, and long-term sustainability. Businesses that prioritize ethics can gain competitive advantage and enhance stakeholder confidence.
- **Cross-domain applicability of ethical principles:** Lessons learned in one sector (e.g., criminal justice or healthcare) can inform ethical practices in others, reinforcing the universality of principles such as fairness, beneficence, autonomy, and accountability.
- **Integration of ethics into business strategy:** Ethical reflection should not be an afterthought; it must be embedded in organizational culture, AI governance, and decision-making processes to ensure sustainable innovation.
- **Forward-looking ethical foresight is essential:** Anticipating future dilemmas, understanding emerging AI capabilities, and preparing adaptive ethical frameworks ensures long-term alignment with societal norms and legal standards.
- **AI accountability requires multi-level responsibility:** Ethical oversight should involve organizational leadership, technical teams, and external auditors to ensure distributed accountability across all layers of decision-making.
- **Data ethics is foundational:** Proper handling of data, including privacy, security, consent, and representativeness, underpins responsible AI practices and prevents systemic harm.
- **Promoting inclusivity and fairness:** Organizations must intentionally design AI systems that are equitable and inclusive, reducing systemic bias and ensuring that marginalized populations are not disproportionately affected.

REFERENCES

Angwin, J., Larson, J., Mattu, S., & Kirchner, L. (2016). *Machine bias*. ProPublica. https://www.propublica.org/article/machine-bias-risk-assessments-in-criminal-sentencing

Beauchamp, T. L., & Childress, J. F. (2019). *Principles of biomedical ethics* (8th ed.). Oxford University Press.

Bryson, J. J., & Winfield, A. F. T. (2017). Standardizing ethical design for artificial intelligence and autonomous systems. *Computer*, *50*(5), 116–119. DOI: 10.1109/MC.2017.154

Dastin, J. (2018, October 10). Amazon scraps secret AI recruiting tool that showed bias against women. *Reuters*. https://www.reuters.com/article/us-amazon-com-jobs-automation-insight-idUSKCN1MK08G

Emanuel, E. J., Wendler, D., & Grady, C. (2000). What makes clinical research ethical? *Journal of the American Medical Association*, *283*(20), 2701–2711. DOI: 10.1001/jama.283.20.2701 PMID: 10819955

European Commission. (2021). *Proposal for a regulation laying down harmonised rules on artificial intelligence (Artificial Intelligence Act)*. https://eur-lex.europa.eu/legal-content/EN/TXT/?uri=CELEX%3A52021PC0206

Floridi, L., & Cowls, J. (2019). A unified framework of five principles for AI in society. *Harvard Data Science Review*, *1*(1). Advance online publication. DOI: 10.1162/99608f92.8cd550d1

Herreid, C. F. (1994). Case studies in science-A novel method of science education. *Journal of College Science Teaching*, *23*, 221–221.

IEEE Global Initiative on Ethics of Autonomous and Intelligent Systems. (2019). Ethically aligned design: A vision for prioritizing human well-being with autonomous and intelligent systems (Version 2). IEEE. https://standards.ieee.org/wp-content/uploads/import/documents/other/ead_v2.pdf

Isaak, J., & Hanna, M. J. (2018). User data privacy: Facebook, Cambridge Analytica, and privacy protection. *Computer*, *51*(8), 56–59. DOI: 10.1109/MC.2018.3191268

Jobin, A., Ienca, M., & Vayena, E. (2019). The global landscape of AI ethics guidelines. *Nature Machine Intelligence*, *1*(9), 389–399. DOI: 10.1038/s42256-019-0088-2

Mittelstadt, B. D., Allo, P., Taddeo, M., Wachter, S., & Floridi, L. (2016). The ethics of algorithms: Mapping the debate. *Big Data & Society*, *3*(2), 1–21. DOI: 10.1177/2053951716679679

National Institute of Standards and Technology. (2023). *Artificial Intelligence Risk Management Framework (AI RMF 1.0) (NIST AI 100-1)*. U.S. Department of Commerce., DOI: 10.6028/NIST.AI.100-1

O'Neil, C. (2016). *Weapons of math destruction: How big data increases inequality and threatens democracy*. Crown.

Obermeyer, Z., Powers, B., Vogeli, C., & Mullainathan, S. (2019). Dissecting racial bias in an algorithm used to manage the health of populations. *Science*, *366*(6464), 447–453. DOI: 10.1126/science.aax2342 PMID: 31649194

Organisation for Economic Co-operation and Development. (2019). *OECD principles on AI*. https://www.oecd.org/going-digital/ai/principles/

Resnik, D. B. (2015). *Ethics of research with human subjects: Protecting people, advancing science, promoting trust*. Springer.

Tesla, Inc. (2023). *Tesla vehicle safety report Q1–Q4 2023*. Tesla Investor Relations. https://www.tesla.com/presskit

Thorne LeClair, D., Ferrell, O. C., & Ferrell, L. (1997). Federal Sentencing Guidelines for Organizations: Legal, ethical, and public policy issues for international marketing. *Journal of Public Policy & Marketing*, *16*(1), 26–37. DOI: 10.1177/074391569701600104

Wachter, S., Mittelstadt, B., & Floridi, L. (2017). Why a right to explanation of automated decision-making does not exist in the General Data Protection Regulation. *International Data Privacy Law*, *7*(2), 76–99. DOI: 10.1093/idpl/ipx005

World Medical Association. (1964). *Declaration of Helsinki: Ethical principles for medical research involving human subjects*. https://www.wma.net/policies-post/wma-declaration-of-helsinki-ethical-principles-for-medical-research-involving-human-subjects/

Zwitter, A. (2014). Big data ethics. *Big Data & Society*, *1*(2), 1–6. DOI: 10.1177/2053951714559253

Chapter 13
Role of Artificial Intelligence in Public Administration Research Ethics

Daniel Uriel Scott
http://orcid.org/0009-0004-9925-7435
Niagara University, USA & NWCU Law School, USA

EXECUTIVE SUMMARY

The use of artificial intelligence (AI) in public administration research provides transformative opportunities and complex ethical challenges. In this chapter, the focus is on the role of AI in public administration research ethics. Despite the prevalent use of the term AI, there is no generally accepted definition of the concept. However, in this chapter, AI was used to mean technology that enables machines to imitate different complex human competencies related to public administration research. The discussion is focused on how AI impacts key ethical principles, specifically, informed consent, autonomy, privacy, confidentiality, equity, bias, transparency, accountability, scientific validity, and public interest. In this chapter a cases study on issues and dilemmas associated with the application of AI in public administration research was included to assists in explaining that ethics in research are indispensable because the pursuit of knowledge should not come at the expense of public trust and equitable governance outcomes.

DOI: 10.4018/979-8-3693-8302-5.ch013

INTRODUCTION

The focus in this chapter is on exploring the role of Artificial Intelligence (AI) in research ethics in public administration. The need to focus on the topic was supported by the concept that the core attribute of AI is that it enables the automation of some tasks that have traditionally required human input. The application of AI has resulted in a transformation that has provided new opportunities for individuals, the economy, and society. Notably, the adoption of AI has disrupted the manner things are done globally and in the United States (Collins et al., 2021; Xu et al., 2021; Zuiderwijk et al., 2021). One area where AI has caused a disruption is in research. As such, the focus in this chapter is the role of AI in public administration research ethics.

AI is not a novel discipline, but the increased focus and the adoption of innovation across different sectors, including public administration, have made it a modern phenomenon. There is a consensus among scholars on the definitional explanation of AI (Sheikh et al., 2023). For instance, Nilsson (2009) defined AI as a technology that functions appropriately and with foresight in its environment. Comparably, the High-Level Expert Group on Artificial Intelligence (2019) defined AI as systems that have intelligent behavior by assessing their environments and taking actions, with some degree of autonomy to fulfill certain goals. Hence, in this chapter, AI was defined as technologies such as machine learning, predictive analytics, and natural language processing that can be used in public administration research.

The increased adoption of AI in government has resulted in numerous opportunities for enhancing the quality of public services, fostering citizens' trust, and promoting effectiveness and efficacy (Zuiderwijk et al., 2021). For instance, AI can result in aggregate productivity growth through automation of routine tasks, process optimization, and data-informed decision-making (Lee, 2016). In public administration research, AI can be applied to optimize operations such as fraud detection in social security systems (Criado et al., 2025; Noordt & Tangi, 2023). In addition, AI's capacity to analyze large data sets can promote evidence-based policymaking, but there is a need to address ethical concerns to ensure scientific validity is achieved and prevent biases, ensuring compliance with research principles. The potential positive impact of AI needs to be effectively managed, ensuring that benefits are equitable.

The adoption of AI in public administration can result in decreased trust in government by citizens, because of a lack of transparency, privacy concerns, and vague responsibility and accountability (Kuziemski & Misuraca, 2020; Zuiderwijk et al., 2021). The use of AI can cause changes in the skills demanded by the job market, with increased demand for technical skills in data science and programming, causing a shift in public administration roles towards policy analysis (Lee, 2016; Organization for Economic Co-operation and Development [OECD], 2019). The changes could result in the potential for uneven distribution of effects across sectors and populations that underpin inequities in resource allocation toward marginalized populations. Accordingly, understanding the societal implications is indispensable because AI will be applied in searching for evidence to inform policies, making it essential to understand the manner the technology can be applied to promote equity and justice, consistent with the public's interest.

Ethical practices are globally acknowledged as a fundamental component of the research process (Islam et al., 2024). Research ethics have a core role in ensuring that studies are performed in adherence to the proper methods and approaches. Research ethics, such as informed consent, respect for persons, justice, beneficence, and confidentiality, are critical in all disciplines, including public administration. In public administration, research ethics are essential, particularly because studies in the field involve collecting and assessing sensitive data such as citizen feedback that directly informs policy and governance. Ethical practices ensure that researchers comply with the rigorous methodological standards, promote transparency, and decrease harm to participants, ensuring the legitimacy of findings and their application in public policies (Islam et al., 2024; Zuniga et al., 2024).

In public administration, ethical research practices are essential in promoting trust, accountability, and fairness (Zuniga et al., 2024). It is essential for any research that influences policy decisions, such as service improvements and resource allocation, to be conducted with fairness to mitigate exacerbating inequities or cause mistrust in public institutions. Research ethics are essential because they guide scholars in conducting valid and reliable research.

Institutional Review Boards (IRBs) safeguard the ethical dimension of research. The role of IRBs includes reviewing, approving, and monitoring studies involving human subjects (Capili & Anastasi, 2024). In public administration research, the role of IRBs is to ensure that the studies conducted comply with ethical standards, protecting participants' rights and well-being. The specific roles include verifying that informed consent is transparent, minimizing risks, and handling participants' data with confidentiality. Conversely, the adoption of AI in public administration research has resulted in new ethical considerations, which create a need for IRBs to update their processes. In essence, despite the advantages associated with the adoption of AI, there are ethical, authenticity, and privacy concerns associated with the deployment of the technology (Kieslich et al., 2022). Similarly, irrespective of the risks and vulnerabilities associated with AI advancements, AI has the potential to positively impact research (Zuniga et al., 2024).

In this chapter, the aim is to explain the role of AI in public administration research ethics. Particularly, the focus is on the ethical implications of applying AI core research activities, including informed consent, data collection, and data analysis. The chapter is categorized into seven major sections. The first section contains an essential background on AI, public administration research, and research ethics. Section two contains a discussion of the impact of AI on public administration research ethics, providing a discussion on transformative opportunities and complex ethical challenges. The focus in the section is research ethics principles, specifically informed consent and autonomy, privacy and confidentiality, equity and bias, transparency and accountability, and scientific validity and public interests, which are discussed using concrete examples of AI-driven methodological challenges and strategies for public administration research.

Section three of the chapter contains a case study on the application of AI in public administration research. In the case study, a description of how AI was applied and the ethical dilemmas that emerged in hypothetical public administration research is discussed. The lessons from the case study are integrated throughout section two to assist the reader in understanding the practical implications of ethical principles. In this section, questions are also included, which can be used by faculty to engage learners in class discussions. Section four contains a conclusion to the book chapter. Section five has a discussion on future research directions, which are based on the gaps in knowledge that were identified when writing this book chapter. In section six, the key terms used in this chapter are defined according to how they were applied in this study. Section seven contains three recommended readings on the case studies that support the importance of adhering to research ethics.

The content in this chapter is valuable to researchers, students, faculty, and practitioners in public administration. For researchers, the content in the chapter contains information about the increasing knowledge on AI ethics in the context of public administration, providing a foundation for additional scholarship. The future research recommendations are based on the identified gaps, which can be explored by conducting primary studies.

After reading this chapter, students will understand the historical and technical evolution of AI and its application in public administration research. Students will also be able to recognize the core ethical principles governing research, including informed consent, privacy, transparency, equity, and scientific validity, and their application to public administration. Reading the chapter will impart to students an understanding of the ethical challenges of using AI in research, such as bias in data analysis, transcription errors, and privacy risks associated with cloud-based systems. Students will also understand the ASPA code of ethics and the Balmont Report principles, with a focus on their impact on research.

At the end of the chapter, there are questions that faculty can use to engage students in discussions on the importance of balancing the application of AI with human-centric research practices as an approach for overcoming the ethical complexities of the research process. For public administration practitioners, the content in the chapter provides insights into the ethical principles that should be considered when designing an AI-driven research protocol, underpinning in provision of tailored recommendations. The chapter is both an educational resource and a call to action, urging researchers to apply AI to uphold public administration research, promoting trust and advancing governance.

The first strength of this chapter is that the information included in the discussion was from credible sources, which are peer-reviewed articles, government publications, and reports from research institutions. The use of credible evidence ensured that the discussion was robust. A second strength is that in the book, actionable insights and discussion questions are included to encourage readers' critical thinking. The third strength is that the chapter is categorized into clear sections, improving accessibility of the information. The sections provide readers with foundational knowledge to promote their understanding of all the concepts discussed.

The first weakness associated with the chapter is that much of the information used in developing the chapter was retrieved from secondary sources, which are limited. The reliance on evidence from secondary sources introduced biases, such as an overemphasis on the positive impacts of AI applications. Also, depending on secondary sources meant that the discussion on the role of AI in public administration research ethics was limited by the published literature on the topic. The second weakness is that the adoption of AI in public administration research is still a novel concept, meaning that there will be new insights on how technology is adopted in the field. The application of AI is rapidly evolving, impacting public administration in different ways. The third limitation is that the author does not claim to provide an exhaustive discussion on the role of AI in public administration research ethics. For instance, in this book, we focus on the application of AI in public administration research ethics only. Hence, we will not discuss how AI is applied in the macro, meso, and micro levels of public administration. There is a lack of a standardized definition of AI, which creates another reason why the chapter cannot be categorized as exhaustive. The fourth weakness is that the findings are based mainly on secondary research because of the limited availability of primary and recent applied research on the role of AI in public administration research ethics.

SECTION 1: BACKGROUND OF THE CORE CONCEPTS

In this section, a background of AI, public administration research, adoption of AI in public administration, and research ethics in public administration are provided. The purpose of the background section is to provide readers with a foundational understanding of the concepts. The background understanding is important because it provides the theoretical and practical context needed to comprehend the intersections between AI and public administration research ethics.

Section 1:1: AI

Artificial intelligence includes technologies such as machine learning, natural language processing, and automation that enable machines to perform tasks that traditionally need human intelligence. The premise of modern AI in research can be traced to John McCarthy, who is attributed to having coined the term AI at a conference in Dartmouth College in 1956 (McCarthy et al., 1956). Earlier AI was mainly focused on high-level cognition, as opposed to the ability to recognize concepts and perform complex motor skills (Collins et al., 2021). Significant breakthroughs occurred in 2006 when Hinton et al. proposed approaches for creating deeper neural networks as a method for mitigating gradient vanishing during training (Hinton et al., 2006). Deep learning is a subset of machine learning that is based on numerous layers of neural networks, which enables AI to match or exceed human capabilities in areas such as computer vision and natural language processing (Xu et al., 2021).

In 2020, OpenAI introduced Generative Pre-trained Transformer (GPT)-3, a state-of-the-art autoregressive language model that uses deep learning to create a variety of computer codes and language tasks exceptionally similar and often indistinguishable from those written by humans (Grzybowsk et al., 2024). OpenAI debuted ChatGPT, an AI chatbot, which resulted in a debate among the public on the widespread use of AI and its possible impact on society. Other AI chatbots such as Gemini, Grok, Claude, Copilot, DeepSeek, Poe, and Perplexity continue to advance and be adopted in different disciplines and contexts, including public administration research.

Section 1:2: Public Administration Research

The role of administration emerged upon the development of the first formal governments (McDonald et al., 2022). The first notable initiative to promote public management practices happened in China in the second century BC. However, the modern study of public administration began in the 20th century. In the United States, the origin of public administration study was Hull House in Chicago and the establishment of the New York Bureau of Municipal Research. (McDonald et al., 2022). It is often challenging to determine whether public administration is a science or an art because of its multidisciplinary influence and dominance by Anglo-American traditions.

Public administration research is the systematic evaluation of the manner public organizations, governance systems, and policies can be improved (Carboni et al., 2019; Christensen & Lægreid, 2025). The scope of public administration has become more dynamic and multidisciplinary because of policy initiatives, administrative reforms, governance, and the use of information and communication technology

(Islam et al., 2024). The focus in public administration research is on examining structures, processes, and behaviors within government agencies, nonprofits, and institutions that promote government interests. Public administration is informed by disciplines such as political science, economics, sociology, and management to assess topics such as policy implementation, public service delivery, bureaucratic efficiency, and decision-making (Islam et al., 2024).

There has been a significant increase in methodological rigor in public administration research, causing an emergence of advanced quantitative approaches, mixed methods research, and experiments. There has been an increased emphasis on methodological pluralism, which is related to conducting mixed-methods studies to promote a better understanding of complex issues and enhance data triangulation (Schwartz-Shea, 2021). Comparative studies are also common in public research because the research allows learning from different global contexts to adopt the identified solutions to local needs (McCandless et al., 2022).

The focus in public administration research is on informing better governance, promoting efficiency, and addressing societal issues such as inequality (Carboni et al., 2019; Christensen & Lægreid, 2025). Public administration research is focused on problem-driven and interdisciplinary frameworks, such as integrative governance, to address complex societal problems. Additionally, in public administration research, there has been recognition of the need for increased inclusion, particularly of indigenous perspectives and historically marginalized groups ensuring in addressing the past injustices committed by public institutions (McDonald et al., 2022). The advancement of technologies, including AI, digital tools, and big data, results in both ethical challenges and opportunities for public management and service delivery.

Section 1:3: Adoption of AI in Public Administration

The adoption of AI in public administration dates to the inception of expert systems in the 1980s (Vatamanu & Tofan, 2025). The AI was implemented to support tax auditing, detect fraud, and assist with basic decisions in sectors such as municipal services and social security (Mergel et al., 2019). The increased evolution of AI policy frameworks in the early 2000s resulted in the emergence of electronic government initiatives that underpinned data-driven public services (Criado & Gil-Garcia, 2019). The continued breakthroughs in natural language processing, machine learning, cloud computing, and big-data analytics have increased the integration of AI into governance, resulting in a transformation (Madan & Ashok, 2023).

In public administration, AI technologies are gaining increased attention because of the benefits associated with the technology on government operations (Criado et al., 2025; Noordt & Tangi, 2023). The adoption of AI has the potential to enhance governmental efficacy, optimizing the decision-making process and improving public service delivery (Vatamanu & Tofan, 2025). For example, the use of AI-driven analytics can assist in policy formulation by supporting the prediction of outcomes based on archival data and current trends. Conversely, the translation of technology into meaningful public value is limited (Vatamanu & Tofan, 2025). In addition, there is the potential of data privacy concerns and ethical considerations. The ethical considerations are related to the possibility of biases embedded within AI algorithms that can cause inequitable outcomes in public service delivery. The biases emerge from training data that is not representative of the diverse populations, reinforcing social disparities (Vatamanu & Tofan, 2025).

Understanding the impact of AI on public administration requires assessing the complex association between the innovation's capabilities and foundational paradigms that influence governance theories and practices (Aboud et al., 2025). The use of AI acts as a catalyst and could disrupt the established administrative models. The core functionalities relevant to public administration include machine learning, natural language processing, and automation. Machine learning allows systems to learn from data to conduct activities such as predicting, optimizing, recognizing patterns, performing classifications, and converting large datasets into actionable knowledge. The use of machine learning can promote the development of evidence-based policymaking and predictive governance (Aboud et al., 2025).

The use of natural language processing allows computers to understand, interpret, and generate human language (Aboud et al., 2025). In public administration, natural language processing is used to power chatbots that promote interactions among citizens, facilitate information extraction, enhance sentiment analysis of public feedback, and support the delivery of multilingual services. Artificial intelligence promotes the automation of routine administrative tasks and workflows such as managing permits, processing applications, and handling standardized inquiries, which enhances operational efficacy (Aboud et al., 2025).

In the United States, the Department of Government Efficiency has made claims that the federal government can and will be automated (Shorey, 2025). The use of AI in public administration is on the rise because, in an October 2023 survey, it was found that 2% of cities were using AI (Bloomberg Philanthropies, 2023). In 2024, it was reported that more than 50% of federal, state, and local government workers at senior levels were regularly using AI applications (Fagan, 2024). There are different AI use cases in public administration, but there is a limited understanding of how the technology is impacting public service and research.

Section 1:4: Research Ethics

Ethics are essential because they are the guiding principles that influence researchers' conduct (Miteu, 2024). In research, ethical considerations include the management of data, responsible use of resources, respect for human rights, the treatment of human and animal subjects, integrity, social responsibility, and dissemination of results. Ethics in research ensures that scientific research is not conducted at the expense of societal or individual wellness. Adhering to ethical standards creates an environment where research can occur responsibly.

In general, research ethics began to gain prominence in the mid-20th century because of the need to protect human subjects and promote integrity (Miteu, 2024; Nagai et al., 2022). Research ethics are influenced by past events, where a lack of ethical considerations had negative consequences. For instance, the Tuskegee Syphilis Study that was conducted between 1932 and 1972, where African American men in Alabama were not informed about their syphilis diagnosis and were denied effective treatment despite penicillin being available in the 1940s (see recommended reading; Centers for Disease Control and Prevention, 2024).

The occurrence of unethical practices, such as infecting hepatitis on institutionalized children with disability in the Willowbrook Study to research the viral infection despite the Nuremberg Code being implemented, creates concerns and the need for adhering to research ethics (see recommended reading; Khuroo & Sofi, 2020). The Nuremberg Code was developed to prevent unethical experimentation, emphasizing consent and participants' safety. The Nuremberg Code provided the foundation for The Declaration of Helsinki, which ensures human participants' protection in medical research, promoting a balance between scientific progress with respect for participants' rights, dignity, and safety. The National Commission for the Protection of Human Subjects of Biomedical and Behavioral Research [NCPHSBBR] (1978) was established, which produced the Belmont Report of 1979 that provides the principles of informed consent, respect for persons, beneficence, and justice that underpin ethical research practices. Institutional Review Boards were also established to oversee research involving human subjects (Miteu, 2024).

Institutional review boards have a critical role in promoting research ethics (Miteu, 2024). An IRB is made up of a committee that is responsible for reviewing, approving, and monitoring research involving human subjects (Capili & Anastasi, 2024). The main role of the IRB is to ensure that participants' rights and welfare are protected. In the instance where any ethical violations occur, the IRB is authorized to impose sanctions that can range from requiring the study to be modified or stopped in its entirety. Informed consent requires participants to be fully informed about the nature of the research, disclosing the objectives, procedures, potential risks, and benefits. It is essential for participants to be allowed to ask questions and voluntarily agree to participate without being coerced (NCPHSBBR, 1978).

Researchers should avoid harm (non-maleficence) and maximize possible benefits while decreasing potential harm (beneficence) (Miteu, 2024; NCPHSBBR, 1978). Researchers need to ensure that studies do not result in undue risks to participants.

The research ethic of justice involves the fair selection and treatment of research participants, ensuring that the burdens and benefits of research are equally distributed among various cohorts in society, to prevent the exploitation of vulnerable populations. Confidentiality is important in research involving human subjects. Participants' personal information should be protected from unauthorized access or disclosure. Researchers should take initiatives to preserve participants' anonymity and privacy to promote trust.

Adhering to research ethics ensures that a study is rigorous, transparent, and compliant with the established protocols, resulting in reliable and valid findings (Miteu, 2024). Compliance with research ethics decreases the probability of studies being impacted by biases or results being fabricated, which promotes credibility. The compliance with principles of justice and beneficence ensures that research contributes positively to society. Additionally, adhering to ethics can promote trust between the public and the scientific community. When research is guided by a respect for societal norms and values, it reinforces the perception that studies are conducted to serve the public good.

Section 1:5: Research Ethics in Public Administration

The importance of ethics supports the essence of adhering to the codes in public administration (Islam et al., 2024; Radhika, 2012). The introduction of the new public administration and reinventing government has underpinned the prioritization of ethical concepts in public administration for practitioners and academicians. The background of research ethics in public administration is founded on international frameworks such as The Declaration of Helsinki, The Belmont Report, and the International Association for Public Administration. In the United States, organizations such as the American Society for Public Administration (ASPA) and, American Political Science Association provide codes of ethics that emphasize integrity, accountability, and respect for persons in practice and research.

For instance, the ASPA (2025) has a code of ethics that was adopted in 1984 and last updated in 2013. The ASPA code contains eight ethical research principles that guide public administrators. The first principle is that public administration research should advance public interest, serving the public good over personal or organizational interests. As a result, researchers need to design studies that result in societal benefits, such as improving policy outcomes. The second principle is that public administration should be conducted in compliance with law and ethical guidelines from institutions such as IRBs. The third principle is that public administration research should promote democratic participation. Democratic participation involves the use of inclusive and transparent processes (ASPA, 2025).

The fourth principle is the need for public administration research to strengthen social equity (ASPA, 2025). Strengthening social equity involves promoting fairness and justice in policies and practices, ensuring vulnerable populations are protected from exploitation, and ensuring equitable recruitment of research participants. The fifth principle supports the need for public administration researchers to fully inform and advise participants by offering accurate and transparent information, which can be achieved through informed consent. The informed consent should clearly outline the purpose of the study, risks, and benefits (ASPA, 2025).

The sixth principle requires researchers to promote personal integrity by acting with fairness, honesty, and accountability (ASPA, 2025). The principle requires avoiding data falsification, plagiarism, or manipulation of results, and underpins research integrity. The seventh principle is that public administration should promote ethical cultures in public institutions. Researchers should ensure that studies are conducted in compliance with the ethical standards, such as non-coercion and confidentiality. The eighth principle requires researchers to advance professional excellence when conducting public administration research. Advancing professional excellence requires conducting research consistent with current ethical standards.

Public administration research often involves some unique ethical challenges because of its focus on governance and public trust (Trondal, 2025; Yadav, 2024). The unique ethical challenges include power imbalances, confidentiality and privacy, public trust, vulnerable populations, and policy impact. Power imbalance can occur, which makes it essential for researchers studying employees, citizens, or marginalized groups to avoid exploiting hierarchical relationships or coercing participants. Studies that focus on sensitive data, such as citizen feedback on policies and employee performance, make it essential to promote confidentiality and privacy to prevent misuse or breaches (Trondal, 2025; Yadav, 2024).

Promoting confidentiality and privacy is essential in public administration research because it influences public trust. Hence, in the instance that public administration research lacks transparency and accountability, it could erode public trust. Research in public administration often involves individuals from marginalized and disadvantaged groups, creating a need to safeguard and prevent exploitation or harm. Public administration research can impact public policy, which makes it essential for it to be accurate, free from bias, and considerate of societal consequences. In essence, ethics is foundational in public administration research because the concepts influence public value (Romanelli, 2023). Adhering to research ethics supports enhancing the legitimacy of public administration studies. Research ethics in public administration are important because they inform policy decisions that impact communities and populations, making ethical considerations essential. Ethical research practices can promote public trust in governmental institutions by supporting research outcomes credibility.

SECTION 2: IMPACT OF AI ON RESEARCH ETHICS IN PUBLIC ADMINISTRATION

Artificial intelligence has resulted in the revolutionization of research by promoting the management of complex ideas and extensive information (Khalifa & Albadawy, 2024). The importance of considering research ethics is to ensure participants' security, enhance researchers' capability to ensure quality, promote trust, foster integrity, and improve professionalism (Islam et al., 2024). The adoption of AI in the research process has resulted in the emergence of complex ethical challenges that require consideration to align with the principles established in the ASPS code of ethics, the Nuremberg Code, and the Belmont Report. When conducting scientific studies, public administration researchers need to deal with administrative and political environments to access information (Islam et al., 2024).

Accordingly, public administration researchers need to consider the impact of ethics, specifically at four major stages. The stages are research design, data collection, data analysis, and research presentation and sharing. In this section, the focus is on discussing the impact of AI on research ethics, specifically informed consent and autonomy, privacy and confidentiality, equity and bias, transparency and accountability, and scientific validity and public interests (see Table 1).

Section 2:1 Informed Consent and Autonomy

Informed consent is essential in ethical research because it is a requirement for participants to understand and voluntarily agree with the study's purpose, risks, and benefits (Islam et al., 2024). Autonomy is closely related to informed consent and involves an individual's capacity for self-determination and right to independent decision-making relating to participation in research. The use of AI can improve the informed consent processes in public administration research through personalization. The application of generative AI to create informed consent that is consistent with participants' comprehension levels and cultural backgrounds signifies a core advancement.

The use of AI in public administration research can complicate the informed consent process when data are sourced from digital platforms or administrative records. In studies where AI is used, big data from government databases, social media, or Internet of Things devices is used, where individuals may not explicitly consent to their data being applied, could result in ethical challenges. For example, when using AI in public administration to analyze citizen feedback from administrative databases, participants may not recognize that their municipal complaints about housing conditions could be aggregated with their demographic data, resulting in the identification of neighborhood-level patterns. Researchers need to explicitly explain that AI will link these data sources and outline the insights that might be derived.

The application of AI in public administration can make it challenging for researchers to explain to participants how their data can be applied. Researchers often do not fully understand the complex algorithms of some of the generative AI applied in data analysis, which makes it challenging for professionals to provide clear explanations to participants and enable them to provide informed consent. The problem is acute in public administration, where decisions informed by AI output could directly affect citizens' access to benefits and services.

Comparably, in qualitative research, AI tools are increasingly being applied to transcribe data from interviews and focus groups (Samuel & Wassenaar, 2025). There is a debate on whether the use of AI transcription software, such as Otter. AI, requires informed consent from participants (Samuel & Wassenaar, 2025). Traditionally, qualitative data was transcribed by researchers, or, with participants' express permission, the activity was outsourced to a third party, who usually signs a confidentiality agreement. The incorporation of AI in the transcription process raises ethical issues about whether the practices should be disclosed to participants. When AI is used in research activities such as transcription, researchers need to include an explicit clause in the information sheets and consent forms. Researchers should also provide participants with the right to decline or withdraw consent for the use of AI in transcribing data. The application of AI in transcribing qualitative data and analyzing big data challenges ASPS's principle of fully informing stakeholders that their data is being used.

As illustrated in the case study in section 3, Bio Info's failure to explicitly disclose Policy-GPT's application in the informed consent forms violated participants' autonomy. The IRB correctly identified phrases such as advanced computational methods hider as opposed to clarifying AI use. The instance supports the essence of specific, transparent language in consent documents.

Adhering to the principle of informed consent is fundamental to ethical research. For this reason, it will be fundamental for researchers to ensure that participants are fully aware of the use of AI in processing their responses. An example is if a qualitative researcher intends to use AI in transcribing the collected data, the information about the software should be included in the informed consent, increasing participants' awareness of how their data will be handled. An example of the language that can be included in the informed consent is "Your interview responses will be transcribed using AI software, which will process your audio recordings through cloud-based servers. You may decline AI transcription and request the manual approach instead." In essence, ensuring informed consent and autonomy when AI is applied in public administration research should be the responsibility of all involved in the research process.

Section 2:2 Privacy and Confidentiality

In public administration, researchers often must deal with some sensitive and critical issues that could impact politicians and government executives (Islam et al., 2024). The breach of privacy or confidentiality in public administration research can result in significant consequences beyond individual harm, which can undermine public trust. Ensuring confidentiality and privacy is essential in ensuring participants feel comfortable sharing internal issues. The use of AI in public administration research can increase the risk because of the technology's capacity to aggregate and assess data on a scale (Martin & Zimmermann, 2024).

The emergence of AI is fundamentally transforming the manner privacy is protected in public administration research. The use of generative AI in academic research can result in the collection and processing of personal data, which is a concern for privacy and security (Khalifa & Albadawy, 2024). For instance, AI-powered anonymization techniques represent a major advancement over traditional approaches. Machine learning algorithms can be used to identify, remove, or alter information with much precision and awareness (Yang et al., 2024). The use of natural language processing can be applied to identify or redact information in unstructured text data, such as interviews or focus group interviews and policy documents, without altering semantic meaning (Yechuri, 2024).

Comparably, advanced AI techniques pose a risk to privacy and confidentiality because they can be used to identify information in datasets that have been anonymized (El Mestari et al., 2024). Machine learning can be applied to identify patterns and correlations that result in linking anonymized data with external datasets that could potentially expose participants' identities. For instance, in public administration, research assessing multiple employee satisfaction surveys, even if personally identifiable data such as names are removed, AI can potentially link responses to individuals by matching patterns in demographic information with publicly available data, which increases the risk of re-identification, especially in studies conducted in smaller municipalities or specialized departments.

The application of AI in transcribing qualitative data can result in data privacy and confidentiality issues (Samuel & Wassenaar, 2025). The use of AI to perform transcription often involves uploading a recording to cloud-based services. It is essential to inform participants about the data transfer and the potential risk that could be associated with a third-party having access to the information. Researchers in the consent forms should specify (a) the name of the AI service provider, (b) where data will be stored geographically, (c) the length that the data will be retained, (d) the security measures implemented, and (e) whether data will be used for any other purpose, such as improving their algorithms. Providing participants with adequate information on the application of AI and how the researcher intends to address any concerns related to data security, potential data breaches, and the handling of sensitive information is essential. It is important for IRBs to ensure that the application of AI in public administration research does not adversely impact privacy or confidentiality. It is essential for IRBs to require researchers to document adherence with privacy laws, where applicable, in public administration.

Section 2:3 Equity and Bias

According to the ASPA, promoting social equity in research promotes the need to mitigate discrimination and foster fairness in research (ASPA, 2025). AI can be integrated to promote idea development and design research. In general, the application of ChatGPT and other chatbots in research is associated with advantages such as improved efficiency, enhanced accuracy, better objectivity, and greater consistency (Khalifa & Albadawy, 2024). Generative AI such as ChatGPT and similar technologies are trained on large datasets, which may contain biases and inaccuracies that could be identified in the output model (Khalifa & Albadawy, 2024). The output of generative AI could be biased or perpetuate existing stereotypes. Artificial Intelligence systems can have racial, gender, and cultural biases, which makes it essential to consider such ethical considerations when applying AI in the research process (Huriye, 2023).

Generative AI, such as ChatGPT, can help researchers in performing data analysis (Khalifa & Albadawy, 2024). The use of AI in data analysis can help enhance the handling and interpretation of complex datasets (Khalifa & Albadawy, 2024). The use of generative AI improves efficiency and accuracy, but consistent monitoring is needed to assist in maintaining data integrity and ethical standards. The quality of the data analysis output is significantly reliant on the effectiveness of generative AI in producing non-biased, incomplete, or flawed output (Khalifa & Albadawy, 2024).

The limitation that AI can perpetuate or exacerbate the biases that are present during training, leading to inequitable outcomes, could result in a violation of research ethics. For example, in studies where researchers analyze public hiring practices, the use of AI could inadvertently reinforce historical biases in the training data containing discriminatory patterns. The outcome violates the Belmont Report's principle of justice, which demands equitable treatment of research participants. In the View Land case described in section 3, the risk of equity and bias is evident. Policy-GPT relied on historical data that contained past discriminatory practices in service delivery, which led to biased recommendations that would have resulted in inequities.

Comparably, generative AI improves the efficiency of the research process by automating data extraction and analysis, resulting in the generation of research findings summaries. The improved efficiency can enhance the rate at which research is conducted. The use of generative AI can enhance the accuracy of research by detecting and correcting errors in the data or analysis (Khalifa & Albadawy, 2024). Notably, the use of generative AI can also result in errors and inconsistencies and accuracy in the output. The enhanced objectivity associated with the use of generative AI is achieved because technology provides researchers with objective insights by eliminating the potential influence of bias or subjective judgment during analysis. The advantage of greater consistency is achieved because generative AI adheres to standardized procedures and protocols (Khalifa & Albadawy, 2024).

Section 2:4 Transparency and Accountability

Generative AI relies on complex algorithms and processes that lack transparency and might not be comprehensively understood (Khalifa & Albadawy, 2024). The lack of transparency and accountability makes it challenging to hold technologies accountable for errors or biases produced. The complexity associated with AI makes it difficult for researchers to explain how the technology arrived at the conclusion. The lack of interpretability can erode public trust in research findings, particularly when informing policy decisions. The use of AI in the research process can result in erroneous output, which has resulted in a debate as to who should be accountable for any unintended consequences that might emerge.

The OECD (2019) AI principles support the need for transparency, accountability, and human-centered values that are consistent with research ethics requirements. The principles require AI systems to be transparent to underpin the understanding of the outcomes. The European Union, in the AI Act, classifies public sector applications as high-risk, which requires analysis, oversight, and documentation of algorithmic decision-making (European Commission, 2025). In the United States, organizations such as the National Institute of Standards and Technology (NIST) support the essence of fairness, accountability, and transparency, which are frameworks that public administration researchers can adapt (NIST, 2025).

As a result, researchers need to disclose that AI has been applied in the research process to promote the overall transparency of the research methodology. Transparency promotes trust between researchers and participants (Khalifa & Albadawy, 2024). Also, when participants understand the tools that are being used in the research process, it increases the likelihood of their being comfortable to share their experiences and insights. Transparency and accountability are essential in ensuring accurate and meaningful data are collected and analyzed. Researchers should take accountability for their findings and have clear protocols for validating AI outputs and addressing them as an approach for mitigating any unintended consequences.

Section 2:5 Scientific Validity and Public Interests

The use of AI in research can enhance scientific validity. For instance, during idea generation, the use of AI promotes brainstorming processes by offering insights derived from current trends, cross-disciplinary studies, and archival or historical data (Alshater, 2022; Khalifa & Albadawy, 2024). Generative AI, such as ChatGPT, can assess large datasets and create quality content, which makes it indispensable to prevent the production of fraudulent material. When generative AI is applied in idea generation, technology could assist researchers in identifying gaps in research (Khalifa & Albadawy, 2024). Natural language processing can allow for scrutinizing numerous documents to identify overlooked or under-researched areas. Generative AI can be used by researchers to generate hypotheses, which can be tested to determine causal relationships or correlations. Also, generative AI can be used to suggest the appropriate sample sizes and statistical approaches that can be applied to derive accurate results (Khalifa & Albadawy, 2024).

Despite the possible beneficial impact of AI on scientific validity and public interests, the depth and accuracy of the information generated by AI are an issue of concern (Khalifa & Albadawy, 2024). Notably, the output from generative AI could mislead researchers towards unintended areas that deviate from the target objective. Another example is that the application of AI in the transcription of qualitative data has been associated with increased accuracy (Samuel & Wassenaar, 2025). A limitation is that AI can still make errors during transcription, especially if participants have accents, technical terms are used, or the quality of the audio is poor. In the View Land case study in section 3, Policy-GPT's transcription errors with regional accents compromised data accuracy, demonstrating that AI efficacy gains must not hinder scientific rigor.

Limited understanding of the study context could hinder generative AI from being applied in analyzing or interpreting findings in public administration research (Khalifa & Albadawy, 2024). Despite the suitability of applying generative AI in idea generation, it is important to consider the essence of human intelligence in research. Human intelligence can help mitigate overreliance on AI, a risk that can result in prioritizing technical efficiency over human-centric outcomes, possibly undermining the public good. It is the researchers' responsibility to ensure that AI-driven studies are methodologically accurate and consistent with ethical principles.

Table 1: Key Ethical Challenges of AI in Public Administration Research

Ethical Principle	AI-Related Challenge	Implications for Public Administration	Mitigation Strategy
Informed Consent	Complex AI processes that are difficult to explain	Participants' limited ability to make informed decisions	Use of plain language in informed consent to explain AI tools, data processing, and risks
Privacy and confidentiality	Cloud-based storage	Sensitive citizen data can be linked to participants, which can compromise participants' privacy.	Data can be stored in private clouds.
Equity and Bias	Use of data with historical bias in AI training data	Biased data reinforces historical discrimination against marginalized populations	Bias detection protocols should be adopted.
Transparency and accountability	Lack of clear responsibility for errors	Undermines research credibility and public trust	AI use should be documented and applied in compliance with the NIST framework and the EU AI Act.
Scientific validity and public interest	Transcription errors with accents.	Inaccurate findings can result in misguided policies	Human verification protocols should be used.

SECTION 3: CASE STUDY: EMERGING ETHICAL DILEMMAS

Policymakers at View Land, a mid-sized city with a population of approximately 20,000 people, partnered with a research consultant called Big Info. Big Info was mandated by the city to conduct a study to understand citizens' feedback on the municipal services delivered, specifically waste management, housing, public transportation, and social welfare programs. As an innovative company, Big Info decided to apply its generative AI called Policy-GPT to assist with the large volume of data, such as social media comments, city surveys, and administrative records, that will be collected in the study. Policy-GPT uses natural language processing and machine learning to process and analyze data. Technology also has a feature to first transcribe any audio data from interviews, focus groups, or videos posted on social media, before the analysis of the content.

Big Info aims to use Policy-GPT to automate data collection, transcription, cleaning, and analysis to enhance the efficiency of the process. The need to use Policy-GPT was supported by the perception that the technology would result in identifying unique insights that would have been missed if the research process were conducted manually. The data to be analyzed contains sensitive information, such as citizens' personal opinions on public services, demographic details, and responses to different government initiatives for promoting equity. Big Info submitted the study proposal to the city's ethical research department, and they have received a verdict from the IRB. The IRB response contains some concerns related to the application of Policy-GPT in the research process.

The major concepts of concern are related to informed consent, data privacy, and potential biases. The IRB also raised concerns about the fact that Policy-GPT is hosted is a cloud-based platform that is managed by a third-party vendor. The IRB required Big Info to address its concerns about data security. An additional factor identified by the IRB is the limited explanation in the study proposal about how Policy-GPTs will process and analyze the data to provide insights that can be used to inform policies related to the public services delivered. The case description below contains a description of Big Info's IRB submission, which they need to review and adjust, ensuring they address the identified concerns.

Section 3:1: Case Description

Big Info proposes to begin the study by collecting data from three data sources. The decision to have three data sources was supported by the need to achieve data triangulation. The data sources will be interviews, survey responses, and customer feedback from the respective government service providers' social media. The researchers plan to apply a mixed-methods methodology as an approach for achieving methods triangulation. The mixed-methods methodology will involve collecting and analyzing both quantitative and qualitative data.

In their IRB application, Big Info indicated that during the informed consent process, participants would be informed that the data they provide would be assessed using advanced computational methods. Conversely, the researchers from Big Info do not explicitly outline how Policy-GPT will be used. Also, in the IRB application, there was a lack of information about cloud-based storage. In their response, the IRB acknowledges that the study will involve collecting data from marginalized populations, but Big Info did not outline how this would be handled. The IRB requires Big Info to explain how they will involve marginalized communities in the research process and handle any expressions of participants having discomfort with sharing personal feedback. Big Info provides a clause assuring participants that their data will be anonymized, by Policy-GPT can link anonymized data with external databases, which creates a risk of re-identification.

Section 3:2: Peer-review Committee Feedback

After addressing the IRB's concerns, the study received approval, and Big Info began the data collection. After data collection, Big Info's Policy-GPT was used in preparing the output for analysis. Preparing the qualitative data required performing data transcription. Policy-GPT was applied to perform the transcription, and a review of the transcripts by the peer-review committee resulted in identifying some inaccuracies. According to the peer-review committee, Policy-GPT had challenges accurately transcribing interviews by participants with strong regional accents, which led to errors in the qualitative data assessed. The errors in the transcription process could have skewed the qualitative study results and affected the policy recommendations.

In the reported findings, Big Info found that certain neighborhoods in View-Land had predominantly low-income and minority communities, who reported dissatisfaction with different services delivered, including waste management, transportation, and housing. Policy-GPT's recommendation was to prioritize resource allocation in the minoritized communities. Conversely, during the peer-review process of the study output, it was identified that the outcomes were not consistent with current evidence. In the peer-review report, it was noted that AI hallucination is the cause of the significant variance in the research findings from the current situation. The AI hallucination can be attributed to the fact that Policy-GPT used outdated historical data that contained past discriminatory practices, such as underfunded services. If the peer-review committee had not reviewed the report by Big Info's Policy-GPT and the recommendations were adopted, it would have resulted in perpetuating bias, as opposed to addressing inequities.

The researchers identified numerous emerging issues in the manner the study was conducted, which resulted in them recommending another IRB review. The need for the IRB review was rationalized by the peer review committee as essential because of the subsequent ethical violations that occurred. The IRB review of how Big Info applied Policy-GPT resulted in identifying different issues and ideas, which are discussed in the subsequent section.

Section 3:3: Issues and Ideas Identified

The first issue was related to informed consent and autonomy. In the public administration research, Policy-GPT complicated the informed process because participants did not fully understand how their data would be analyzed. Data security was also an issue because of the potential risks associated with cloud-based storage. In addition, the issues adversely impact the principles of autonomy as participants did not have sufficient information to make an informed decision about their involvement in the study. The peer-review committee's idea for addressing the issue is for Big Info to revise the consent forms to explicitly disclose AI use, including the potential risks and the participants' right to opt out of the Policy-GPT processed data analysis.

The second issue is related to privacy and confidentiality. Policy-GPT is cloud-based, which creates a risk of the collected data being breached and bad actors having unauthorized access. The AI could also identify all anonymized data, which creates a risk to participants' confidentiality. The peer-review committee proposed the need for advanced anonymized techniques and the implementation of additional security measures to improve data protection, ensuring adherence to the standards.

The third issue is focused on equity and bias. Policy-GPT relied on historical data, which the review committee identified to be embedded with biases and risks related to inequities in policy recommendations. If the findings were used to inform policies, it would result in inequities. As such, the review committee recommended that the Big Info team add a bias identification feature and involve community stakeholders in validating the AI output to ensure that the findings promote equity and fairness.

The fourth issue is related to transparency and accountability. Policy-GPT is proprietary, which limits Big Info in explaining how the data is processed and the conclusions are achieved. The lack of transparency undermined public trust and research legitimacy. An idea for addressing the issue is for Big Info to provide as much information as possible and be accountable for the findings.

The fifth issue is related to scientific validity and public interest. Errors in the transcription process occurred, which affected the study's validity. The inaccurate findings could result in misguided policy decisions, harming public interest. The idea for addressing the problem is for Big Info to have data verification approaches. The incorporation of human oversight can help ensure that methodological rigor and alignment with the public good are achieved.

Section 3:4: Discussion Questions

1. How can researchers ensure informed consent and participant autonomy when using AI tools in the research process?
2. What steps can be taken to protect participants' privacy and confidentiality associated with the integration of AI in the research process?
3. How can a research team identify and mitigate biases related to the output of AI data, particularly when historical data reflects past discriminatory practices?
4. What strategies can researchers use to promote transparency and accountability in AI-driven public administration research?
5. How can researchers ensure the accuracy of the data transcribed using AI and promote scientific validity?
6. What specific guidelines should IRBs develop to ensure the use of AI in public administration research is consistent with ethical considerations?

SECTION 4: CONCLUSION

Artificial intelligence is increasingly becoming an essential part of human lives, with technology having applications in different fields, including public administration. The increased application of AI has been underpinned by its potential to promote productivity and revolutionize the manner research is conducted. Applying AI in public administration research has significant potential to improve efficiency, enhance effectiveness, and promote in identification of unique insights into complex societal challenges. Conversely, the application of AI in public administration results in ethical complexities that should be overcome to uphold the ethical principles of informed consent, privacy, equity, transparency, and scientific rigor. In the presented case study of View-Land, the use of AI in public administration can improve the efficiency of the process; however, research ethical challenges can occur. The application of Policy-GPT underpinned efficiency, but it introduced transcription errors, perpetuated historical biases, and hindered accountability, which can undermine policy recommendations. As such, the application of AI in public administration research should be accompanied by ethical safeguards and human oversight. The occurrence of ethical challenges can undermine trust in public administration research findings. Hence, ethical considerations should be prioritized, especially with the increasing adoption of AI in research processes. It is the researchers' responsibility to ensure participants are fully informed about the role of AI in protecting sensitive data, mitigating bias, and ensuring transparency, which are consistent with ASPA's codes and foundational frameworks such as the Belmont Report. The IRBs also have a core role in ensuring that the application of AI in public administration research aligns with ethical standards. It is also important for researchers to acknowledge that the use of AI cannot replace human authors' intellectual contributions or accountability for their publications. Hence, achieving a balance between technological innovation with oversight is indispensable to ensure that public administration research is consistent with ethical standards. The chapter provides researchers with knowledge on the importance of practicing ethical strategies to ensure that the use of AI in public administration research is principled.

SECTION 5: FUTURE RESEARCH DIRECTIONS

First, future researchers should come up with a standard definition of AI. The lack of a standard definition of AI limits in explaining how technology can be applied in research. Second, future researchers should focus on establishing ethical frameworks for AI use in public administration research. The development of AI-specific ethical guidelines is essential to address the unique challenges associated with the application of technology in public administration research. The guidelines can provide a framework for addressing the unique ethical challenges associated with the application of AI in public administration. The third recommendation for future practice is for future researchers to conduct studies exploring how AI-based public research can be free from bias, have better interpretability, and promote privacy and confidentiality. Fourth, future researchers should conduct more primary and applied research on the role of AI in public administration research ethics to demonstrate real-world applications.

REFERENCES

Aboud, B. O., Zacarias, T. H., & Oliver, C. E. E. (2025). Advancing smart public administration: Challenges and benefits of artificial intelligence. *Urban Governance, 1*(1), 1–10. DOI: 10.1016/j.ugj.2025.06.003

Alshater, M. (2022). Exploring the role of artificial intelligence in enhancing academic performance: A case study of ChatGPT. *Social Science Research Network, 1*(1), 1–10. https://papers.ssrn.com/sol3/papers.cfm?abstract_id=4312358

American Society for Public Administration. (2025). *Code of ethics*. ASPA. https://www.aspanet.org/ASPA/ASPA/Code-of-Ethics/Code-of-Ethics.aspx

Bloomberg Philanthropies. (2023). *State of cities: Generative AI in local governments*. https://cityaiconnect.jhu.edu/pdfs/Final-Gen-AI-In-Cities-Report_10.18.2023.pdf

Capili, B., & Anastasi, J. K. (2024). Ethical research and the institutional review board: An introduction. *The American Journal of Nursing, 124*(3), 50–54. DOI: 10.1097/01.NAJ.0001008420.28033.e8 PMID: 38386835

Carboni, J. L., Dickey, T., Moulton, S., O'keefe, S., O'leary, R., Piotrowski, S. J., & Sandfort, J. (2019). Start with the problem: Establishing research relevance with integrative public administration. *Perspectives on Public Management and Governance*, 2(4), 267–274. DOI: 10.1093/ppmgov/gvz021

Centers for Disease Control and Prevention. (2024). About The Untreated Syphilis Study at Tuskegee. CDC. https://www.cdc.gov/tuskegee/about/index.html

Christensen, T., & Lægreid, P. (2025). Future directions of public administration research—Addressing fundamental issues and questions. *International Journal of Public Administration, 48*(5-6), 299–305. DOI: 10.1080/01900692.2025.2470006

Collins, C., Dennehy, D., Conboy, K., & Mikalef, P. (2021). Artificial intelligence in information systems research: A systematic literature review and research agenda. *International Journal of Information Management, 60*, 1–13. DOI: 10.1016/j.ijinfomgt.2021.102383

Criado, J. I., & Gil-Garcia, J. R. (2019). Creating public value through smart technologies and strategies: From digital services to artificial intelligence and beyond. *International Journal of Public Sector Management, 32*(5), 438–450. DOI: 10.1108/IJPSM-07-2019-0178

Criado, J. I., Sandoval-Almazán, R., & Gil-Garcia, J. R. (2025). Artificial intelligence and public administration: Understanding actors, governance, and policy from micro, meso, and macro perspectives. *Public Policy and Administration*, *40*(2), 173–184. DOI: 10.1177/09520767241272921

El Mestari, S. Z., Lenzini, G., & Demirci, H. (2024). Preserving data privacy in machine learning systems. *Computers & Security*, *137*, 1–10. DOI: 10.1016/j.cose.2023.103605

European Commission. (2025). *AI Act*. EC. https://digital-strategy.ec.europa.eu/en/policies/regulatory-framework-ai

Fagan, M. (2024). *AI for the people: Use cases for government*. Harvard Kennedy School. https://www.hks.harvard.edu/sites/default/files/centers/mrcbg/working.papers/M-RCBG%20Working%20Paper%202024-02_AI%20for%20the%20People.pdf

Grzybowsk, A., Pawlikowska, K., & Lambert, C. (2024). A history of artificial intelligence. *Clinics in Dermatology*, *1*(1), 1–28. DOI: 10.1016/j.clindermatol.2023.12.016

High-Level Expert Group on Artificial Intelligence. (2019). *A definition of AI: Main capabilities and scientific disciplines*. European Commission. https://ec.europa.eu/newsroom/dae/document.cfm?doc_id=56341

Hinton, G. E., Osindero, S., & Teh, Y. W. (2006). A fast learning algorithm for deep belief nets. *Neural Computation*, *18*(7), 1527–1554. DOI: 10.1162/neco.2006.18.7.1527 PMID: 16764513

Huriye, A. Z. (2023). The ethics of artificial intelligence: Examining the ethical considerations surrounding the development and use of AI. *American Journal of Technology*, *2*(1), 37–44. DOI: 10.58425/ajt.v2i1.142

Islam, R., Shahan, A. M., & Rahman, A. (2024). Research ethics in public administration. In O. F. V. Feigenblatt & M, R. Islam. (Eds.), *Ethics in social science research* (pp. 155–166). Springer. DOI: 10.1007/978-981-97-9881-0_12

Khalifa, M., & Albadawy, M. (2024). Using artificial intelligence in academic writing and research: An essential productivity tool. *Computer Methods and Programs in Biomedicine Update*, *5*, 11–19. DOI: 10.1016/j.cmpbup.2024.100145

Khuroo, M. S., & Sofi, A. A. (2020). The discovery of hepatitis viruses: Agents and disease. *Journal of Clinical and Experimental Hepatology*, *10*(4), 391–401. DOI: 10.1016/j.jceh.2020.04.006 PMID: 32655240

Kieslich, K., Keller, B., & Starke, C. (2022). Artificial intelligence ethics by design. Evaluating public perception on the importance of ethical design principles of artificial intelligence. *Big Data & Society*, *9*(1), 1–10. DOI: 10.1177/20539517221092956

Kuziemski, M., & Misuraca, G. (2020). AI governance in the public sector: Three tales from the frontiers of automated decision-making in democratic settings. *Telecommunications Policy*, *44*(6), 1–10. DOI: 10.1016/j.telpol.2020.101976 PMID: 32313360

Lee, K. (2016). *Artificial intelligence, automation, and the economy*. The White House. https://obamawhitehouse.archives.gov/blog/2016/12/20/artificial-intelligence-automation-and-economy

Madan, R., & Ashok, M. (2023). AI adoption and diffusion in public administration: A systematic literature review and future research agenda. *Government Information Quarterly*, *40*(1), 1–13. DOI: 10.1016/j.giq.2022.101774

Martin, K. D., & Zimmermann, J. (2024). Artificial intelligence and its implications for data privacy. *Current Opinion in Psychology*, *58*, 1–10. DOI: 10.1016/j.copsyc.2024.101829 PMID: 38954851

McCandless, S., Bishu, S. G., Gomez Hernandez, M., Paredes Eraso, E., Sabharwal, M., Santis, E. L., & Yates, S. (2022). A long road: Patterns and prospects for social equity, diversity, and inclusion in public administration. *Public Administration*, *100*(1), 129–148. DOI: 10.1111/padm.12830

McCarthy, J., Minsky, M. L., Rochester, N., & Shannon, C. E. (1956). *A proposal for the Dartmouth summer research project and artificial intelligence*. https://home.dartmouth.edu/about/artificial-intelligence-ai-coined-dartmouth

McDonald, B. D.III, Hall, J. L., O'Flynn, J., & Van Thiel, S. (2022). The future of public administration research: An editor's perspective. *Public Administration*, *100*(1), 59–71. DOI: 10.1111/padm.12829

Mergel, I., Edelmann, N., & Haug, N. (2019). Defining digital transformation: Results from expert interviews. *Government Information Quarterly*, *36*(4), 1–10. DOI: 10.1016/j.giq.2019.06.002

Miteu, G. D. (2024). Ethics in scientific research: A lens into its importance, history, and future. *Annals of Medicine and Surgery (London)*, *86*(5), 2395–2398. DOI: 10.1097/MS9.0000000000001959 PMID: 38694399

Nagai, H., Nakazawa, E., & Akabayashi, A. (2022). The creation of the Belmont Report and its effect on ethical principles: A historical study. *Monash Bioethics Review*, *40*(2), 157–170. DOI: 10.1007/s40592-022-00165-5 PMID: 36357708

National Institute of Standards and Technology. (2025). *Artificial intelligence.* NIST. https://www.nist.gov/artificial-intelligence

Nilsson, N. J. (2009). *The quest for artificial intelligence.* Cambridge University Press. DOI: 10.1017/CBO9780511819346

Noordt, C., & Tangi, L. (2023). The dynamics of AI capability and its influence on the public value of AI within public administration. *Government Information Quarterly*, *40*(4), 1–10. DOI: 10.1016/j.giq.2023.101860

Organization for Economic Co-operation and Development. (2019). *AI principles.* OECD. https://www.oecd.org/en/topics/sub-issues/ai-principles.html

Radhika, D. (2012). Ethics in public administration. *Journal of Public Administration and Policy Research*, *4*(2), 23–31. DOI: 10.5897/JPAPR11.049

Romanelli, M. (2023). Advancing ethics within public administration. *Lumen Proceedings, 19*, 01-11. DOI: 10.18662/lumproc/gekos2022/01

Samuel, G., & Wassenaar, D. (2025). Joint editorial: Informed consent and AI transcription of qualitative data. *Journal of Empirical Research on Human Research Ethics; JERHRE*, *20*(1-2), 3–5. DOI: 10.1177/15562646241296712 PMID: 39600037

Schwartz-Shea, P. (2021). Under threat? Methodological pluralism in public administration. *Public Performance & Management Review*, *44*(5), 975–1005. DOI: 10.1080/15309576.2019.1694547

Sheikh, H., Prins, C., & Schrijvers, E. (2023). *Mission AI: The new system technology*. Springer. https://link.springer.com/book/10.1007/978-3-031-21448-6

Shorey, S. (2025). *AI and government workers: Use cases in public administration. Roosevelt Institute.* https://rooseveltinstitute.org/publications/ai-and government -workers/

Trondal, J. (2025). The multilevel administrative state and the future of public administration research. *International Journal of Public Administration*, *48*(5-6), 321–333. DOI: 10.1080/01900692.2024.2412273

United States. National Commission for the Protection of Human Subjects of Biomedical, & Behavioral Research. (1978). *The Belmont report: ethical principles and guidelines for the protection of human subjects of research (Vol. 2).* The Commission. https://www.google.com/books/edition/The_Belmont_Report/cAqtFohwYhIC?hl=en&gbpv=0

Vatamanu, A. F., & Tofan, M. (2025). Integrating artificial intelligence into public administration: Challenges and vulnerabilities. *Administrative Sciences*, *15*(4), 149–169. DOI: 10.3390/admsci15040149

Xu, Y., Liu, X., Cao, X., Huang, C., Liu, E., Qian, S., Liu, X., Wu, Y., Dong, F., Qiu, C. W., Qiu, J., Hua, K., Su, W., Wu, J., Xu, H., Han, Y., Fu, C., Yin, Z., Liu, M., & Zhang, J. (2021). Artificial intelligence: A powerful paradigm for scientific research. *Innovation (Cambridge (Mass.))*, *2*(4), 1–10. DOI: 10.1016/j.xinn.2021.100179 PMID: 34877560

Yadav, B. (2024). Ethics in public administration: Why integrity matters? *Research Review International Journal of Multidisciplinary*, *9*(1), 282–286. DOI: 10.31305/rrijm.2024.v09.n01.035

Yang, L., Tian, M., Xin, D., Cheng, Q., & Zheng, J. (2024). AI-driven anonymization: Protecting personal data privacy while leveraging machine learning. *Applied and Computational Engineering*, *71*(1), 7–13. DOI: 10.54254/2755-2721/71/2024MA0053

Yechuri, S. (2024). Enhanced utility-driven data anonymization: Leveraging AI and machine learning for sensitive data privacy. *Journal of Artificial Intelligence General Science*, *1*(1), 229–232. DOI: 10.60087/jaigs.v1i1.229

Zuiderwijk, A., Chen, Y. C., & Salem, F. (2021). Implications of the use of artificial intelligence in public governance: A systematic literature review and a research agenda. *Government Information Quarterly*, *38*(3), 1–12. DOI: 10.1016/j.giq.2021.101577

Zuniga, D. H., Goyanes, M., & Durotoye, T. (2024). A scholarly definition of artificial intelligence (AI): Advancing AI as a conceptual framework in communication research. *Political Communication*, *41*(2), 317–334. DOI: 10.1080/10584609.2023.2290497

ADDITIONAL READING

Centers for Disease Control and Prevention. (2024). *About the untreated syphilis study at Tuskegee*. CDC. https://www.cdc.gov/tuskegee/about/index.html

Grzybowsk, A., Pawlikowska, K., & Lambert, C. (2024). A history of artificial intelligence. *Clinics in Dermatology*, *1*(1), 1–28. DOI: 10.1016/j.clindermatol.2023.12.016

Khuroo, M. S., & Sofi, A. A. (2020). The discovery of hepatitis viruses: Agents and disease. *Journal of Clinical and Experimental Hepatology*, *10*(4), 391–401. DOI: 10.1016/j.jceh.2020.04.006 PMID: 32655240

KEY TERMS AND DEFINITIONS

Accountability: The process of researchers taking responsibility for AI outputs and ensuring they are accurate and ethical.

Artificial intelligence: Technologies such as machine learning, natural language processing, and automation that are used in public administration to analyze data, transcribe interviews, and interpret findings, and exemplified by Policy-GPT in the case study.

Autonomy: A person's capability for self-determination and the right to make an independent decision relating to participating in research.

Bias: Prejudice or inconsistencies in AI output often emerge when technologies are trained using unrepresentative data.

Equity: The process of promoting fairness and justice in research, ensuring each participant is treated equitably, results in avoiding exploitation, according to the ASPA codes of ethics.

Generative AI: An AI system that is used in public administration to perform idea generation, hypothesis creation, transcription, and data analysis.

Informed consent: A document used to inform participants about a study's purpose, procedures, risks, benefits, and voluntary nature of the process, ensuring they understand and agree to be part of the research without coercion.

Institutional review board: The committee is responsible for reviewing, approving, and monitoring research involving human subjects to ensure ethical compliance.

Machine learning: A subset of AI that allows systems to learn from data and improve, which can be used in public administration research for tasks related to predicting outcomes, pattern recognition, and assessing large datasets.

Natural language processing: A branch of AI that enables computers to comprehend, interpret, and generate human language. In public administration research, natural language processing can be applied in transcribing qualitative data and analyzing data.

Privacy and confidentiality: The protection of research participants' personal information from unauthorized access or disclosure, promoting trust in the process.

Public administration research: A multidisciplinary field informed by political science, sociology, economics, and management, focusing on areas such as policy analysis, public service delivery, and organizational behavior, with emphasis on addressing societal inequities.

Public interest: The overarching goal of research is to serve societal benefits such as promoting governance and fulfilling public needs.

Research ethics: Principles such as informed consent, beneficence, justice, respect for persons, and confidentiality, which are founded on the Belmont report, were essential in promoting trust and integrity in public administration research.

Scientific validity: The methodological rigor and accuracy of the research findings support the reliability and credibility of the results.

Social equity: A principle that relates to promoting fairness and justice in research practices, specifically by protecting vulnerable populations and ensuring equitable distribution of research benefits and burdens among participants.

Transparency: Transparency involves disclosing research-related activities, including AI use and required research activities, to promote trust and understanding among participants and stakeholders.

Chapter 14

Safeguarding Ethics and Academic Integrity in the Age of Generative Artificial Intelligence:

A Case Study on Plagiarism Prevention

Jaswinder Pal Singh
http://orcid.org/0000-0001-5515-6051
Chitkara University, India

Baljinder P. Kaur
http://orcid.org/0000-0002-4909-8948
Punjabi University, India

EXECUTIVE SUMMARY

This chapter examines the evolving challenges to academic integrity posed by the integration of generative artificial intelligence (AI) in higher education. Focusing on a case study of plagiarism prevention at a university, the chapter explores how AI's dual role as both a tool for enhancing research and a potential source of academic dishonesty necessitates a robust ethical framework. It discusses the importance of institutional support, advanced detection software, and continuous education on ethical research practices. By highlighting the collaboration between ethics committees, libraries, and research networks, the chapter underscores the need for universities to rigorously enforce academic integrity policies while adapting to the

DOI: 10.4018/979-8-3693-8302-5.ch014

rapidly changing digital landscape. Ultimately, this chapter provides insights into the balance required between leveraging AI's capabilities and preserving the core principles of academic honesty and originality in scholarly work.

INTRODUCTION

In the rapidly evolving landscape of higher education, the emergence of generative artificial intelligence (AI) presents both opportunities and challenges to maintaining academic integrity. As digital technologies continue to permeate academia, the traditional boundaries of ethical research and writing are being tested. This chapter delves into a case study of plagiarism prevention at a university, exploring the ethical implications of AI in academic practices. By examining the intersection of AI tools and academic integrity, the chapter aims to offer insights into the evolving strategies needed to uphold the values of honesty and originality in academic work amidst the growing influence of AI technologies.

It is highly imperative for the credibility and reliability of academic practice. Ethical research corroborates the findings of a given study and also substantiates the larger frame of academic integrity (Aita & Richer, 2005). All this academic ethos honesty, transparency, and productivity stand against academic deceit in any form. AI seems very transformative in bringing opportunities and challenges for academic integrity (Sullivan et al., 2023). This chapter attempts to deconstruct AI into research and writing to be carried out ethically. Plagiarism remains one of the major problems faced by students, professionals, educational institutions, and the publishing sector. Questions abound regarding how prevalent plagiarism is and how it can be prevented. People have varying views and definitions of plagiarism, especially in different parts of the world, including Africa, Asia, and Europe. While some consider it a criminal offense, others regard it as a professional mistake (AL Harrasi, 2023; Molnar & Jozsef, 2022). The rise in accessibility to information on the internet has only fueled the issue further, thus requiring that undergraduates, graduates, and faculty be given early-on orientation about what plagiarism are, its implications, and strategies of prevention. We should distinguish textual matching from broader concept theft and articulate the crime. The prevention tools available, and the consequences for offenders, must be defined.

OVERVIEW OF ACADEMIC INTEGRITY

Academic integrity encompasses the ethical principles and practices fundamental to teaching, research, and service within academic institutions. As defined by Macfarlane and colleagues (2017), it often highlights issues such as plagiarism and cheating, particularly in student conduct (Macfarlane, 2017). The Centre for Academic Integrity outlines core values of honesty, trust, fairness, respect, accountability, and courage, which are essential for a thriving academic community (Miron et al., 2022). These principles are crucial not only for maintaining the credibility of teaching, learning, and research but also for ensuring the effective functioning of universities. According to Gao (2023), academic integrity is a foundational element for most institutions, necessitating support from all members of the academic community (C. A. Gao et al., 2023). Fostering academic integrity requires active involvement and coordinated communication across the institution, highlighting that while measures may vary among Higher Education Institutions (HEI), common strategies include engaging students in discussions about academic integrity, providing clear definitions and examples of dishonesty, consistently communicating unacceptable behaviours and consequences, and implementing transparent guidelines for recording violations and enforcing appropriate responses (Bayaa Martin Saana et al., 2016; Groves & Mundt, 2021).

In an age of collaboration, sharing, and social media, educators, librarians, and administrators struggle to promote academic integrity. New technologies, millennial' different traits and perspectives, and a changing society and environment are among these issues. Each of these issues allows educators to adapt their teaching style (O'Connell, 2016). Many HEIs have reduced cheating and plagiarism. Their success seems to depend on student cooperation in defining academic dishonesty norms and achieving wide student acceptance. Many HEIs use academic honour codes to achieve these goals.

THE PLAGIARISM CHALLENGE

The term plagiarism, which elicits significant apprehension among educators and researchers, denotes a grave violation of academic and intellectual integrity (Apoorv et al., 2020; Jereb et al., 2018). Instances of plagiarism are prevalent in the academic realm, manifesting in many ways, but all characterized by the unauthorised dissemination of someone else's ideas, words, or work as one's own, without appropriate acknowledgment. The aforementioned activities not only damage the fundamental values of trust and honesty within the academic community, but also weaken the ideals of originality and the quest of knowledge (Wu et al., 2021). Direct copying, the most conspicuous manifestation of plagiarism is the act of exact reproduction of text, documents, or other material from external sources without appropriate quotation or citation. Although the most clearly identifiable and often denounced variety, other subtle variations frequently manage to avoid being detected. The acts of plagiarism encompass paraphrase without appropriate citation, when the original concepts are restated without enough modification in structure or language, and self-plagiarism, which is duplicating one's own prior work without appropriate recognition or disclosure (Tindall et al., 2021). Additional forms of academic dishonesty include mosaic plagiarism, which is the assemblage of material from different sources without appropriate reference, and fraudulent citation, which entails distorting sources. Another type of plagiarism that poses quite significant difficulties is collusion. It is the situation when several students perform the same written task for each of them to present it as individual work. Although the case described may also be relevant to this section, it is also a good illustration of the above-mentioned criterion, and such papers are not accepted (Wittenberg, 2024). Nevertheless, it would be interesting from the point of view of discussing the possibilities that higher education institutions may have to maintain and advance the level of academic integrity. In fact, plagiarism and a culture of cheating have a much broader impact on the system beyond individual dishonesty (Adzima, 2020; Bretag et al., 2019). Thus, first, it is primarily an educational institution faithful to the truth and originality that is subject to such disturbances. Higher education is undermined by plagiarism that is reinforced by an education system that gives the wrong signals. Indeed, students who cheat do not add the necessary skill set to plagiarism, which hinders the development of independent, critical thinking, responsible for this when writing academic papers, and the ability to convey their thoughts effectively (Apoorv et al., 2020). Moreover, the extinction of plagiarism also leads to changes in the evaluation criteria, which automatically distorts the assessment of students' knowledge, competencies, and skills. Together with the impact on the training quality for the specialist, they are directly responsible for the academic institutions that establish students to be ethical. Apart from that, the proportion of plagiarism in students' lives can have very serious repercussions for students' future employment or professional opportunities. So, the reputation may be severely damaged. All sciences need to work with integrity to preserve the integrity of research (Richards, 2016).

CAUSES AND MOTIVATIONS FOR PLAGIARISM

The problem of plagiarism that is widely-spread in various academic and professional spheres cannot be addressed unless the intricate reason and fundamental factors causing it are understood in a comprehensive manner (Pradhan & Kumar, 2023). In fact, instances of plagiarism are never born in a vacuum but rather emerge as a result of a number of interrelated academic, personal, and environmental factors. First and foremost, the academic load and consequently academic stress contributes to plagiarism, especially in terms of students (Salauddin, 2023). It is not a secret that the modern educational system is characterized by increasingly high demands and expectations regarding students who often experience an overwhelming pressure of strict timetables and dozens of substantial academic tasks. Under such conditions, some students may face the necessity to find fastest ways of meeting deadlines and academic requirements (Torres & Fons, 2018). In the light of the high amount of assignments and exams to handle, some students may resort to plagiarism as a way to achieve better grades with less effort on the grounds of a high fear of failure or poor performance (Mansoor et al., 2024). Time constraints are, indeed, linked to academic stress experienced by an increasing number of learners. Taking their multiple responsibilities and aspirations into account, students are often hard-pressed for a choice of dedicating much needed time to profound research and personal independent work, which makes them utilize the easiest way to get their coursework done in the shortest period of time (Barrientos et al., 2021).

Lack of knowledge of citation guidelines can be referred to one of the key reasons for plagiarism. Not only new students and researcher to the academic environment may get into trouble with citation but also ordinary students who have no experience with proper citation rules and methods. It can be too demanding to adhere to certain styles of citation like APA or MLA since the information is hard to find or understand. As a result, students may easily be misled and misattribute works or do not attribute them at all, which may be regarded as an arbitrary plagiarism form. Another reason for plagiarism development is digital technologies that provide people with a wealth of information (Alawad, 2021). Now, it is easy to find any piece of data on the internet and copy and paste this information to a personal paper. However, people still may not understand that the number of different works online requires proper citing as well. It can be too easy to access the works, and, as a result, people tend to overlook that the information they see online should also be correctly cited, which can be regarded as accidental plagiarism (Caminiti & De Fiore, 2023).

It is also important to note that incentives for plagiarizing are, to a certain extent, conditioned by cultural and age differences. Adhering to a specific position or understanding of any phenomenon may differ, and principles-related issues are cultural specifics issues (Roig, 2010). Often students from other cultures are not aware of the concepts of "source," "drawer" and are perplexed by the desire to find "someone else's information" on the Internet (Masic, 2014). Plagiarism, in general, is a phenomenon synthesized in an academic environment; it is not usually accepted to talk about the author and his/her rights in other cultures. While in Western academic culture, emphasis is placed on the individual authorship and uniqueness of its work. Another point here is the attitude of different generations towards technology and the exchange of information in general and young people, about the limits of plagiarism. To prevent plagiarism in an educational or professional environment, it is necessary to act on many fronts, and many factors should be considered when developing strategies (Patwardhan et al., 2020).

IMPACTS AND CONSEQUENCES OF PLAGIARISM

Students have a range of distinct and severe consequences. In the first place, students may have immediate academic consequences such as receiving failing scores or being placed on academic probation (Bernal Arellano et al., 2024). It is the most direct consequence for students, as it affects their academic achievement and progress. However, the consequences of cheating go far beyond these short-term disruptions in academic achievements. First, it impairs their education by preventing them from developing critical thinking skills and necessary expertise (Maryon et al., 2022). Cheating is defined as a serious problem that prevents them from excelling on the intended learning path. Second, plagiarism gradually damages personal ethics over time (Zhang et al., 2023). Attempting to cheat and copying work from others lead to deception as a rule, which is a problem standing in the way of their ethics (Ivanov, 2023; Zhang et al., 2023). The ethical dilemma is exacerbated by the fundamental discrepancy between personal ethics and the context of academic requirements. The most problematic is the long-term impact on one's career prospects. History will show that they have been involved in academic dishonesty, and potential employers will use this as a criterion to limit their future career opportunities and throw them into a potential educational path (Mahabeer & Pirtheepal, 2019).

In addition to the implementation issues concerning learners, another severe consequence of plagiarism is the myriad of possible consequences for instructors. The gradual erosion of trust between students and teachers is a severe concern since students who constantly engage in academic dishonesty are likely to conclude that their teachers are incompetent or apathetic, undermining the trust that is present as the founding bedrock of education (Dempere et al., 2023). Moreover, the increased labor from teachers adopting the rigorous responsibilities of uncovering, handling, and adjudicating these cases can focus their efforts away from their primary teaching endeavors. These include the restrictions in the amount of help that instructors can offer their learners and uphold of instructional professionalism. Otherwise, plagiarism may affect the career prospects of these instructors and the relationships with their colleagues and positions in the institution (McAlpine et al., 2023).

Large-scale plagiarism in education is fraught with devastating consequences, mainly due to the loss of reputation that poses the danger of both the falling in college rankings and the problem of attracting financial funds(Morocco-Clarke et al., 2024). Institutions that do not recognize the problem of large-scale plagiarism do not take responsibility for proper measures for its prevention and compromise the status of their organization in the academic world; moreover, they compromise the functioning of partnerships in the field of research (Désiron & Petko, 2023). Integrity of such organizations is lost, since the re-use of someone else's text already seriously affects the legal component of publication. As for the ethical component, such an act is considered deceitful and dishonest, infringing on the trust and reputation of the authors of the original text, to whom no mention is made during re-creation. In the field of academia, plagiarism itself is an obstacle to the learning process, because it prevents students from authentic interaction with the content of the course; as for the formation of research skills and critical thinking, they were severely undermined. However, this violation is not a one-time manipulation: unlike other errors, which can sometimes enhance the learning process, plagiarism and their students learning as it's perfect does not contribute to this in any way. Moreover, overall harming the quality of education, Poor preparation of specialists significantly impedes the development of industry. In addition, frequent plagiarism can ruin a student's future in all its sectors (Phyo et al., 2023). The value of reputation in business and academia is highly appreciated, and a history of plagiarism can significantly hinder a person's ability to receive scholarships, enroll, or get a job in the future.

PREVENTION AND DETECTION OF PLAGIARISM

The prevention and detection of plagiarism are crucial for educational institutions to uphold academic integrity. Given its scale, a comprehensive approach utilizing a range of resources and tactics is required (Saeed et al., 2011). Turnitin, Grammarly, and Copyscape are extensively utilized for their role in upholding academic integrity. These software applications use sophisticated algorithms to compare submitted works with an extensive collection of scientific materials, periodicals, and online content. Proficiency in identifying similarities and duplicate content is essential for detecting plagiarism. It is fascinating that these software systems possess the ability to detect and deter unwanted behaviours (Youmans, 2011). The assurance of comprehensive evaluation of their work by plagiarism detection software serves as a strong incentive for students and researchers to uphold academic integrity. By explicitly stating that academic misconduct is immoral and will be penalized, it fosters an ethical educational environment.

For academic institutions and educators, it is imperative to have a plagiarism detection system to immediately and effectively identify and respond to any incident of plagiarism (Patel et al., 2011). In this way, businesses facilitate the analysis of possible scenarios and preventions of such situations by imposing actions appropriate to their policy. It is safe to say that implementing this policy allows to make sure that academic misconduct is addressed on a constant basis and any misbehaviour is not accepted. When put in place and mandated, rigorous policies to encourage and support academic honesty provide institutions with the opportunity to prevent any inappropriate occurrences (Park, 2017). The codes of conduct explicitly define the nature of any action and the consequences for it if related to the issue of plagiarism. Moreover, the rules clearly set such procedures in place as reporting and responding to the issue of plagiarism in such a way as to provide a sense of uniformity to the students and their mentors (Siddhpura & Siddhpura, 2020). Academic institutions are bound to provide their students, and the professors with the relevant education on the issue of plagiarism and the importance of citation and referencing. Institutions apply a variety of approaches to find the most efficient way to reach the goal and arrange seminars, orientations, and distribute relevant literature (Tripathi & Patel, 2021). Such a scope of activities allows eliminating the ambiguity and empowering students to perform effective research and citation, and teachers with the means to prevent and spot the issue of plagiarism in their environment.

CONNECTION BETWEEN ACADEMIC INTEGRITY, PLAGIARISM DETECTION, AND ARTIFICIAL INTELLIGENCE IN ACADEMIC WRITING

Artificial intelligence based writing tools revolutionize academic writing thanks to their ability to enhance accuracy and productivity (Garg et al., 2023). Natural language processing technology allows authors to organize their work and improve it by generating and interpreting textual data that resembles human language (Med et al., 2023). Editors and reviewers benefit from automated peer-review and plagiarism detection systems, which help them, evaluate numerous submissions quickly and efficiently. AI enables improved writing efficiency, as the number of writing-related tasks with which authors have to deal reduces. They concentrate on creating better, evidence-based content, rather than on technical details. Being proficient in information analysis, AI tools can detect and correct mistakes in writing. AI enhances the efficiency of literature search and can accelerate other routine tasks including language translation or text summarization (Watanabe & Wiseman, 2023). The technology can also develop an outline for various types of academic or research documents, including a research protocol or a grant proposal. It can also suggest a better sample size, tests, or several other measures to improve the paper's quality. Artificial intelligence does not access the latest research and cannot replace human judgment fully. It can, however, create innovative ideas by spotting gaps in prior studies (Tortora, 2024). Using big data, AI quickly finds the data it needs to process. Unbiased by preconceived notions, it never misses occasional findings. It recognizes patterns and suggests rigorous methods of testing the newly identified links at a much faster rate. Finally, AI makes document editing much quicker, as it can address both grammatical and structural problems. The overall text quality could also be significantly improved when AI generates output, as in the case of ChatGPT which may identify grammar issues in users' documents and enhance their texts (Lo, 2023).

The scope of using artificial intelligence is actively expanding, especially regarding large language models. It can be compared to the internet in the 1990s in terms of widespread use. While AI enhances our efficiency in processing and managing data, it also brings similar problems related to early internet adoption, meaning risks and cons. Particularly, the main drawback of trusting AI to write for us is that it might suppress human creativity and critical thinking skills (Dempere et al., 2023). The works produced by AI are biased and flawed, as they are generated from biased and flawed data, based on their overall output. The two distinct methods can be identified: the "passive" method and the Socratic approach.

The passive method implies inputting the topic of a paper or a research problem into a large language model (LLM) and then including the generated text to one's own work with a little to no alteration (B. Li et al., 2024). Following the prompt for the position paper, I used this method with GP4 as to analyze the efficiency of LLM use in modern education. It noticeably simplified the writing process and improved the efficiency of my time-consuming. However, although the text generated by AI is satisfying at first sight, it is too cliché, inaccurate, and dry, as it represents the mean of the responses it got excited to when describing the input data. Indeed, as I employed GP4 to argue the effectiveness of LLM use, the latter claimed, "Ask any expert, and they'll tell you that large language models are a brilliant tool used to save time and aid productivity" (Goyal et al., 2024). In summary, passive use can be effective in limited contexts when high productivity is required, although it cannot be enough to improve one's writing. On the other hand, the Socratic approach is also partially productive, as researchers can generally critically develop the perspective. It can be explained through its similarity to the ancient Greek dialogue, which is noted as a reliable way of developing one's perspective. In our context, I utilized the Socratic approach after the passive one showed no tangible results in the context of our paper. I employed some of the key AI-generated concepts to fill the gaps in the research background, induced critical inquiries on how it interacts with readers, and whether it can be relatable, which helped develop the perspectives (Kuzlu et al., 2023). I then inputted these inquiries into GP4 and extracted notable details with additional questions based on them. In this way, Socratic dialogue helped understand AI writing's foundation through this detailed discussion. I received the ideas that influenced future research; conversely, it was more challenging to implement the conclusions, but it was more beneficial than the passive method due to its dialogical character following the example of the Socratic one.

APPLICATIONS OF GENERATIVE ARTIFICIAL INTELLIGENCE IN ACADEMIC WRITING

The term "technology benefits" (Berg, 2023) encompasses the advantages gained from utilizing specific technologies, including AI, in academic contexts (Berg, 2023). For students, these benefits involve leveraging AI to enhance their academic tasks such as term papers, research papers, and assignments (Kitamura, 2023). However, "actual technology use" refers to how individuals incorporate technology into their daily routines, with "AI usage" specifically addressing the applications of generative AI in academic work. Before adopting any technology, it is advisable for users to assess its benefits based on research. Technologies that are perceived as advantageous in facilitating tasks and boosting productivity are more likely to be embraced by users (Berg, 2023). Alkire (2024) suggest that students will be motivated to use new technologies in educational settings if they believe these tools simplify or enhance their academic work (Alkire et al., 2024). AI technologies offer numerous features that support various stages of research and writing, from sentence analysis to large-scale text processing. They handle vast amounts of textual data efficiently, execute complex computations, and provide valuable insights. Predictive modeling allows AI to generate coherent and contextually relevant language, while reasonable language creation involves producing text that closely resembles human comprehension. Generative AI combines predictions with analyzed data to produce polished, grammatically correct, and coherent writing that aligns with its intended meaning.

Generative AI writing tools contribute greatly to the writing and drafting aspects of a research paper by converting initial thoughts into well-structured arguments, and improving grammar, tone, and style (Crompton & Burke, 2024). AI systems have a comparative advantage in research organization and summarization. Through the analysis of vast databases quickly, AI is able to extract key findings from numerous articles, forming a holistic summarization without the need to check each source. Moreover, generative AI can recommend processes in the research by analyzing research already done and suggesting techniques to adjust your course of action and obtain better results. In terms of data, the AI is great for analyzing big amounts of data and can find connections and patterns hardly detectable by a human. Another useful capacity of the AI systems is their capability to help keep the immense amount of PDF documents in order. If you upload pdf files into the system, it can highlight all the texts and images, then categorize them into two columns including the marked-up texts and images. Furthermore, it can annotate objects, tagging titles, or adding footnotes. Generative AI internally covers a variety of writing tools, exercise organizers, and PDF management tools. The value generated for a researcher comes in the form of automating repetitive processes, more precise data treatment, improved writing through a better understanding and relevance checking of one's work with the research already done, and adjustments to internal processes. On the other hand, generative AI allows one to generate more insightful data analysis through comprehensive data review. The overall effect of these tools is more time on important work items, such as building hypotheses and experiments, and less time on auxiliary processing of the large amount of data and writing instruments.

GENERATIVE ARTIFICIAL INTELLIGENCE CHALLENGES

Technology use issues are the limitations, obstacles, or inefficiencies that hinder individuals from using technology. In this study, “artificial intelligence issues” refers to the various barriers that hinder graduate students from engaging with AI technologies (Razack et al., 2021). The adoption of technology has been acknowledged as a substantial obstacle, particularly for nations that are considered emerging or developing (Alkire et al., 2024). Technological infringement refers to the situation when prospective customers see a certain technology as difficult to use because of limitations in terms of availability, support, skill prerequisites, accessibility, and other associated aspects. Based on the findings of X. Gao who perceive technology as difficult often exhibit reduced usage or completely avoid its use (X. Gao et al., 2023). If postgraduate students consider AI to be a challenging endeavor, they will avoid using it in the educational setting. Nevertheless, if people are better predisposed to easily use artificial intelligence, they will engage it in their academic pursuits.

VIEWS OF GENERATIVE ARTIFICIAL INTELLIGENCE IN ACADEMIC WRITING

The capacity of generative AI models to generate content that closely emulates information created by humans has led to a significant increase in public interest in these models. In 2023, OpenAI unveiled two well acclaimed models, ChatGPT and DALL-E. The GPT architecture serves as the foundation for these models, which have obtained considerable appeal (Sallam et al., 2023). ChatGPT has shown remarkable proficiency in several application domains, including the generation of thoroughly organized essays and material, the provision of chatbot replies, the decoding of languages, the answering of queries, and the facilitation of coding. Previous research findings further demonstrate the creators' commitment to enhancing LLMs for certain tasks and investigating the potential of transfer learning in many domains. ChatGPT offers several advantages for academic and pedagogical purposes, benefiting both instructors and learners. Utilizing ChatGPT, students gain knowledge in a wide variety of disciplines, write essays, conquer difficult obstacles, and rapidly resolve issues in their academic endeavors. ChatGPT can improve students' programming abilities by assisting them in overcoming complex programming problems. Hung (2023) and Kotovich (2023) enumerates the advantages of ChatGPT for students (Hung & Chen, 2023; Kotovich & Oriol, 2023). The benefits encompass boosts in learners' intrinsic motivation to participate in study, deepening their comprehension of the subject matter, and fostering their development. Achieving proficiency and automation is crucial for improving instructor efficiency and streamlining the creation of lesson plans. Furthermore, educational institutions are embracing more efficient diagnostic methods such as oral assessments, written evaluations, and cooperative academic assignments. This shift in the assessment approach places more emphasis on the advancement of knowledge through the use of indicators and the derivation of conclusions. That demonstrates the correlation between critical and generative AI cognition.

CONCERNS ABOUT CHATGPT USE IN ACADEMIC WRITING

Although ChatGPT has some potential benefits, its usage in an educational setting is not without its drawbacks and limitations. There is increasing apprehension among many individuals that the study and education on artificial intelligence systems would play a role in the development and furtherance of prejudice and discrimination against certain populations (Guleria et al., 2023). Not only might alterations or consequences cause present damage, but they can also result in biased or faulty estimate of artificial intelligence systems (Sidorkin, 2024). The literature evaluation undertaken by Qasam (2023) was to investigate the possible uses of ChatGPT in several fields, including education, and to analyze the concerns expressed by experts during the initial three months of its deployment. The findings indicate that ChatGPT demonstrated varying degrees of performance across different subjects. Furthermore, apart from his profound knowledge in mathematics, he has also exhibited remarkable proficiency in the scientific domains of economics and programming. There are several important problems that algorithms can help solve (Talyshinskii et al., 2023). For example, ChatGPT can direct people, answer their questions, help develop learning materials, conduct online classes for students, and support teachers, inspiring them. However, there are several limitations associated with the implementation of the algorithm. These are the inability to generate and change data, as well as the ability to evade plagiarism detection.

ETHICAL ISSUES IN GENERATIVE ARTIFICIAL INTELLIGENCE

The use of generative AI in different spheres implies the consideration of a range of ethical and practical issues. According to Zhu et al., use of ChatGPT in the educational curriculum may contribute to the improvement of such outcomes as the retention of information, provision of personal learning, and critical thinking development. However, the use of generative AI for educational purposes poses the risks for independent thought and creativity development. In addition, the probability of misinformation, bias, and plagiary is high. Perkins (2024) warns that the implementation of AI in education is the serious threat to fundamental educational principles (Perkins & Roe, 2024). For this reason, the relevant ethical themes should be discussed continuously in the context of evolving use of generative AI in education.

Artificial intelligence generated content, especially deep fakes, has problematic concepts of authenticity and veracity by using a combination of real and falsified elements (Favela & Amon, 2023). Widespread implementation of AI technologies complicates the process of ensuring content's reliability, especially with the use of deep fakes that can create a distorted and false reality. Depending on AI to generate texts automatically implies the possibility of committing plagiarism since some of the content may unintentionally repeat the information from examined sources without proper citation. There is a need for more sophisticated methods of authenticating data and distinguishing between human and AI authors. The authenticity of the content is inherently defective since AI lacks genuine life experiences that would constitute the basis for accurate information on the subject. The same ability is also reflected in data bias and fairness as essential elements of AI ethics. Generative models can create content based on biased themes present in the training data. Biases in algorithms can perpetuate and even exacerbate unequal social relations. Both users from underprivileged backgrounds have equal rights to unbiased treatment. Remedies for discrimination and discrimination data provide for the transparency and fairness in AI's functioning to promote public trust and improve inclusivity.

Explainability and transparency are essential when artificially intelligent decisions lead to significant impacts on lives. On the one hand, the users have to understand the principles behind the decision they are the recipient of, and the technology developers need to disclose information about how exactly AI models are built, trained, and operated (M. Li et al., 2024). On the other hand, the outcomes generated by AI must remain an explicit justification prompting a tangible response, and the emerging question of accountability and responsibility determines whose failure will be associated with harm or underperformance caused by automate decision-making. To provide solutions to many problems, clear standards and responsibilities must be fixed for the design, development, and operation of technology. Moreover, attributing AI-created content is also associated with intellectual property ethics because the possible advancement of technologies leading to the creation of AI-generated art, music, or literature feeds into the increasing concern that those cannot be classed as creations and undermines ownership and all the effects on intellectual property law (Ara & Ara, 2024). AI also has economic impacts as it can lead to the creation of new jobs in tech-related fields and generate unemployment of outdated job positions, caused mostly by the automation of the routine. Arguments can also be made for AI influence on creativity and efficiency in domains like completing repetitive tasks data processing as well as design, revolutionizing how data is stored and businesses and education sectors as a result (Zlateva et al., 2024). The ethics of security and privacy come into play, when generated content may be manipulated or used to deceive users. Lastly, the failure to cultivate human skills in the light of constantly developing AI means a civilization crisis as the level at which humanity could function is lowered.

ADDRESSING PLAGIARISM IN UNIVERSITIES

Addressing the issue of plagiarism requires a thorough approach that includes both preventive and corrective strategies. At the core of these measures is the creation of a strong culture of academic integrity inside the learning institutions (Moya, 2024). To encourage such culture, the institutions should support and promote ethical principles, making them part of their nature. The most important preventative measure is the active dissemination of knowledge and resources. Learning institutions, accredited by a recognized body, should organize thorough orientations and on-site workshops and seminars for both teachers and students. The student orientation should focus on the drawbacks of plagiarism, such as the legal implications and reputation's lower. Moreover, the students should be taught the necessary skills for proper research, essay writing, and referencing. Such resources have a two-fold benefit as they will help the students with their learning and the teacher with building an understanding of ethical requirements of research and writing, thus overwhelming the need for adopting unethical practices. Additionally, the colleges should also take a proactive stance on the issue of plagiarism by implementing plagiarism detection software (Moya & Eaton, 2024). The software has a twofold effect in this situation: it proves to be a deterrent of plagiarism and those who might fall in it, as they know they have a high chance of getting caught, and it greatly helps in the rapid detection of possible cases, thus allowing the institutions to act quickly according to their internal regulation. Moreover, learning institutions should establish a set of formulated conduct rules and vigorously enforce them. The rules should specify what is expected of students in terms of academic honesty and what happen to those who do not follow the guidelines, establishing a standard for reporting and investigating plagiarism incidents (Mphahlele & McKenna, 2019).

One of the main peculiarities of creating a culture of academic integrity is an idea of the mutual responsibility of all parts of this culture such as students, professors, and administration. Therefore, universities and colleges rely on their ability to develop honesty and integrity in education. Once an educational institution adopts academic integrity as the principal for maintaining and promoting, it creates a climate in which intellectual theft is seen as contradictory. The mission of colleges and universities is to create such an environment where plagiarism cannot exist as it contradicts the promotion of such principles as creativity and honesty in research. Professors are in the key position to safeguard and uphold the standards of academic integrity. Furthermore, they act not only as educators but also, they are those who help their students to become creative writers and develop ethics. This framework obliges college professors to present guidelines on principles of fair depiction, reading, and research. In addition, they have to teach the principles of ethical research and the relevance of accurate citation. It is possible to suggest that educators provide learners with tools such as impartial and accurate materials which facilitate the performance of adequate and adequate research.

The dedication with which faculty members approach their work sets the standard for ethical behaviour in academia. Their enthusiasm for research and personal commitment to academic integrity and innovation serve as role models for students and make it clearer that these values are essential for academic work. This is why faculty members should focus on preventive measures against plagiarism, namely, creating tasks that facilitate critical thinking, innovation, and original content. When students realize that their genuine engagement with the content rather than the grade is the most important for an instructor, they feel encouraged to produce unique results. Also, when faculty members use plagiarism detection software, they can find cases of wrongdoing and demonstrate their commitment to academic integrity while making it clear that any kind of unethical behaviour will be exposed.

CASE STUDY: PLAGIARISM PREVENTION AT UNIVERSITY

Chitkara University is steadfast in upholding its core values and principles, which guide the academe recognizes the value of integrity in academic work. An organization that prides itself on the practice and upholding of academic integrity is the Chitkara University. Chitkara University's practice of promoting integrity is evident in all the activities of employees, students, and their academic and administrative units. The schools have integrated it into its essential values, including leadership, responsibility, honesty, transparency, accountability, fairness, respect, teamwork, creativity, and quality. As this organization is geared towards fostering an environment of integrity, this goal mirrors the standards because academic integrity is based on a fundamental ethical principle and is upheld by existing laws and regulations. To support this, the Chitkara University had set specific rules and regulations that all of its students observe. However, faculty members play a significant role in this university, as they are the prime movers that persuade students to adhere to these regulations and be accountable for promoting, monitoring, and committing to live as a class community. The Chitkara University polices these regulations in a structured approach: conducting workshops informing students and the faculty of existing regulations against academic dishonesty and the penalties due delinquent students; sponsoring educational activities that promote academic integrity; and, addressing questions from individual students.

- Ensuring Educational Quality: Maintaining high standards in the educational process and supporting activities to foster a rigorous academic environment.
- Encouraging Positive Student Behavior: Guiding students to become proactive and responsible members of society through adherence to academic integrity principles.
- Addressing Violations: Implementing appropriate sanctions for violations to address issues educationally and academically, ensuring that consequences are proportional to the severity of the infraction.

The given measures highlight Chitkara University's attempt to secure its dedication to high academic standards and to continue promoting a culture of academic honesty in particular. The Chitkara University Research and Innovation Network can be regarded as a vibrant center of innovative ideas and intellectual explorations at Chitkara University. Driven by a desire to foster academic and technological progress, CURIN aims to cultivate innovation and facilitate research. CURIN aspires to explore innovative opportunities, render support to research pursuits and, subsequently, form a basis for the development of partnership networks and rapport which target crossing conventional disciplinary borders and which aspire to deliver a wide spectrum of outcomes. For this purpose, both academics, as well as industry professionals, engage in research efforts, integrating a plethora of fields which would lead to technological breakthroughs and innovative solutions, novel ambitions, and initiatives. Practically speaking, the Chitkara Innovation Centre creates environments across five different thematic tracks which foster collaboration and knowledge-sharing. As a result, interaction across scientific communities is facilitated, thereby, allowing scientific sharing with the other parts of academia and society. In addition, CURIN both emphasizes the importance of mentorship and institutional context to enhance the experience of undergraduate and postgraduate students and integrates a wide range of research opportunities open to both categories of students, which are likely to contribute to the quality and results of research.

The Chitkara University Research and Innovation Network (CURIN) exemplify how imaginative and groundbreaking research can catalyze revolutionary change. The primary objective of CURIN is to cultivate research that has a significant and influential effect on both academic and commercial domains. The Chitkara University Research Ethics Committee (REC) is essential in this undertaking, as its primary objective is to maintain strict ethical measures in all academic research conducted at the university. In accordance with the July 23, 2018, guidelines established by the University Grants Commission (UGC) on academic integrity and plagiarism prevention, the REC guarantees that all research outputs, including research articles, book chapters, and complete books, adhere to these standards (COMMISSION, 2018). Furthermore, the REC is dedicated to guaranteeing that faculty members strictly comply with the utmost levels of academic integrity. Upon identification of deficiencies, the REC offers guidance and suggests appropriate measures to resolve and fix the problems, therefore strengthening a climate of academic honesty and preserving the reputation of the academic community. The REC's endeavors not only maintain academic standards but also foster a shared dedication to ethical practice in scholarship.

Chitkara University's Central Library stands as a beacon of academic excellence, playing a pivotal role in the university's research initiatives and providing the foundational support crucial for their success. It stimulates both intellect and creativity through its impressive collection of print and digital resources, state-of-the-art facilities, and a team of dedicated professionals. The library offers an extensive range of academic materials, including journals, textbooks, and specialized databases, which are essential for conducting original and impactful research. Moreover, the library's modern infrastructure and skilled staff facilitate seamless collaboration and knowledge sharing among researchers. It fosters an environment where individuals from diverse fields can engage and communicate effectively. As a cornerstone of the university's research mission, the Central Library underpins the creation of a culture of academic excellence by delivering an exceptional array of resources, services, and programs designed to support and enhance the success of its academic community.

CURIN has formed a partnership with the Research Ethics Committee (REC) and Central Library at Chitkara University to prevent any compromise to academic and research integrity. A comprehensive approach has been adopted by Chitkara University to effectively tackle and deter unethical behavior in academic writing. The objectives of academic integrity at our institution are manifold. This policy establishes explicit expectations and criteria for academic work for instructors and scholars. The document outlines the potential for improper conduct, elucidates the procedures for reporting and resolving violations of academic integrity and dependability, and delineates the several penalties that can be enforced. Prior to publishing any documents such as papers, book chapters, books, or patents, professors must obtain a specific number known as the Unique Identification Number before communication (UINC). This number is also issued after publication as the Unique Identification Number after publication (UINP) by the Dean Research's office. This procedure is specifically employed to expedite the publishing process. This clear identification facilitates the monitoring of updates to the findings, as well as the detection of duplicate and any unauthorized dissemination of the data. This approach successfully attained both the transparency and accountability of the research dissemination. Evidence of plagiarism is only found at Chitkara University Library. The Main Library conducts plagiarism checks on manuscripts based on the following criteria: papers with multiple authors but similar wording, papers including concealed quotation marks or image text, manuscripts with white text or concealed letters, documents developed by Letter Substitution or using characters resembling letters, a collection of documents submitted to several journals, papers with different titles but similar content and a high similarity index, and papers generated by artificial intelligence.

Enhancing the reliability and credibility of academic research is a primary advantage of this proactive strategy. Reports of unethical conduct, such as research misconduct or plagiarism, are promptly brought to the attention of the Research Ethics Committee (REC). This is to ensure that the REC possesses all the essential information regarding the subject under consideration. This involves delivering a report. The REC will inaugurate a comprehensive inquiry by collecting evidence and obtaining testimonials from individuals with knowledge about the case. Upon receiving evidence and testimonies, if the committee determines that the case is actually factual, it will proceed with the requisite measures to penalize the individuals implicated. The whole procedure is characterized by a strong commitment to transparency and justice, which enables accused academics or faculty members to confront their accusations and present a defense. The strict measures used by the REC ensure their responsibility and adherence to academic integrity requirements. In conclusion, the collaboration among the Central Library, REC, and CURIN exemplifies Chitkara University's commitment to maintaining the highest standards of academic and research integrity. The goal is to construct a residence conducive to ethical and innovative behavior. All the affiliated groups have endeavored to establish rigorous criteria, efficient mechanisms to detect cases of plagiarism, and somewhat expedited reactions when wrongdoing is first found. Significantly, all of these groups have created forums for the purpose of discussing and exchanging information, therefore providing a shared platform from which they may together engage in the noble pursuit of research and maintaining ethical standards in the workplace. The collective efforts of these organizations have made it a priority to ensure the excellence of the content they publish within their networks. This commitment to high-quality content has led them to implement rigorous verification and peer-review processes. As a result, their journals and books consistently meet the stringent criteria required to achieve recognition in the top tiers of academic publishing. Specifically, these works often qualify for Q1, Q2, and Q3 rankings, which are indicators of the highest standards in scholarly publishing. These rankings signify that the publications are not only well-respected in their respective fields but also widely cited and influential in advancing knowledge and research. Through these measures, the organizations have solidified their reputation as leading contributors to academic and professional literature, ensuring that the works they publish are of exceptional quality and credibility.

CONCLUSION

As AI continues to advance and integrate into academic practices, the challenges surrounding academic integrity are set to become more complex. This chapter's examination of plagiarism prevention at a university highlights the critical role of ethical guidelines, institutional support, and AI-driven tools in safeguarding academic honesty. While AI can significantly enhance the detection of plagiarism and streamline academic processes, it also necessitates a renewed commitment to educating students and faculty about the ethical use of technology. Ultimately, the balance between leveraging AI's capabilities and preserving the core principles of academic integrity will be key to fostering a culture of ethical scholarship in the digital age.

The integration of generative artificial intelligence (AI) into academic settings presents a dual-edged sword, offering both powerful tools for enhancing research and writing while simultaneously posing significant challenges to maintaining academic integrity. As explored in this case study, the rapid adoption of AI technologies necessitates a robust framework for preventing and detecting plagiarism, underpinned by a strong culture of academic ethics. Institutions must proactively engage with these emerging technologies, not only by implementing advanced detection software but also by fostering a deep understanding among students and faculty of the ethical implications of AI use. The collaboration between educational institutions, such as Chitkara University, and their dedicated ethics committees, libraries, and research networks exemplifies a comprehensive approach to upholding academic standards. By rigorously enforcing academic integrity policies, providing ongoing education on ethical research practices, and leveraging AI responsibly, universities can safeguard the credibility of academic work in this new digital era. The ongoing dialogue and adaptation to these technologies will be crucial in ensuring that the core values of honesty, originality, and intellectual integrity remain central to academic endeavors, thereby preserving the trust and reliability that are the bedrock of scholarly communication.

REFERENCES

Adzima, K. (2020). Examining online cheating in higher education using traditional classroom cheating as a guide. *Electronic Journal of e-Learning*, *18*(6), 476–493. DOI: 10.34190/JEL.18.6.002

Aita, M., & Richer, M.-C. (2005). Essentials of research ethics for healthcare professionals. *Nursing & Health Sciences*, *7*(2), 119–125. DOI: 10.1111/j.1442-2018.2005.00216.x PMID: 15877688

Alawad, A. A. (2021). Students' attitudes toward plagiarism in the interior design field. *Civil Engineering and Architecture*, *9*(4), 1179–1188. DOI: 10.13189/cea.2021.090419

Alkire, L., Bilgihan, A., Bui, M., Buoye, A. J., Dogan, S., & Kim, S. (2024). RAISE: Leveraging responsible AI for service excellence. *Journal of Service Management*, *35*(4), 490–511. Advance online publication. DOI: 10.1108/JOSM-11-2023-0448

Apoorv, R., Dahiya, A., Sreeram, U., Patil, B. R., Irish, I., Graziano, R., & Starner, T. (2020). Examinator: A Plagiarism Detection Tool for Take-Home Exams. *L@S 2020 - Proceedings of the 7th ACM Conference on Learning @ Scale*, 261–264. DOI: 10.1145/3386527.3406723

Ara, A., & Ara, A. (2024). Exploring the ethical implications of generative AI. In *Exploring the Ethical Implications of Generative AI*. IGI Global., DOI: 10.4018/979-8-3693-1565-1

Barrientos, A., Cuadros, M., Alba, J., & Cruz, A. S. (2021). Implementation of a remote system for the supervision of online exams through the use of cameras with artificial intelligence. *Proceedings of the 2021 IEEE Engineering International Research Conference, EIRCON 2021*. DOI: 10.1109/EIRCON52903.2021.9613352

Bayaa Martin Saana, S. B., Ablordeppey, E., Mensah, N. J., & Karikari, T. K. (2016). Academic dishonesty in higher education: Students' perceptions and involvement in an African institution. *BMC Research Notes*, *9*(1), 234. Advance online publication. DOI: 10.1186/s13104-016-2044-0 PMID: 27112550

Berg, C. (2023). The Case for Generative AI in Scholarly Practice. SSRN *Electronic Journal, April*, 1–15. DOI: 10.2139/ssrn.4407587

Bernal Arellano, W. M., Galarza Parra, J. N., Villavicencio Reinoso, J. M., & Quito Ochoa, J. F. (2024). A study on academic dishonesty among English as a foreign language students. *Heliyon*, *10*(13), e33876. DOI: 10.1016/j.heliyon.2024.e33876 PMID: 39050424

Bretag, T., Harper, R., Burton, M., Ellis, C., Newton, P., Rozenberg, P., Saddiqui, S., & van Haeringen, K. (2019). Contract cheating: A survey of Australian university students. *Studies in Higher Education, 44*(11), 1837–1856. DOI: 10.1080/03075079.2018.1462788

Caminiti, C., & De Fiore, L. (2023). Fake paper factories: Root cause solutions are needed. *Recenti Progressi in Medicina, 114*(3), 148–153. DOI: 10.1701/3981.39638 PMID: 36815417

COMMISSION, U. G. (2018). *UNIVERSITY GRANTS COMMISSION (PROMOTION OF ACADEMIC INTEGRITY AND PREVENTION OF PLAGIARISM IN HIGHER EDUCATIONAL INSTITUTIONS) REGULATIONS, 2018. 2018*(D).

Crompton, H., & Burke, D. (2024). The Educational Affordances and Challenges of ChatGPT: State of the Field. *TechTrends, 68*(2), 380–392. DOI: 10.1007/s11528-024-00939-0

Dempere, J., Modugu, K., Hesham, A., & Ramasamy, L. K. (2023). The impact of ChatGPT on higher education. *Frontiers in Education, 8*, 1206936. Advance online publication. DOI: 10.3389/feduc.2023.1206936

Désiron, J. C., & Petko, D. (2023). Academic dishonesty when doing homework: How digital technologies are put to bad use in secondary schools. *Education and Information Technologies, 28*(2), 1251–1271. DOI: 10.1007/s10639-022-11225-y PMID: 35912037

Favela, L. H., & Amon, M. J. (2023). The ethics of human digital twins: Counterfeit people, personhood, and the right to privacy. *2023 IEEE 3rd International Conference on Digital Twins and Parallel Intelligence, DTPI 2023*. DOI: 10.1109/DTPI59677.2023.10365409

Gao, C. A., Howard, F. M., Markov, N. S., Dyer, E. C., Ramesh, S., Luo, Y., & Pearson, A. T. (2023). Comparing scientific abstracts generated by ChatGPT to real abstracts with detectors and blinded human reviewers. *NPJ Digital Medicine, 6*(1), 75. Advance online publication. DOI: 10.1038/s41746-023-00819-6 PMID: 37100871

Gao, X., Shen, H., & Duan, J. H. (2023). A Study on ChatGPT to Alleviate the Difficulties in Writing Scientific Technology Papers. *2023 International Conference on Cognitive Computing and Complex Data, ICCD 2023*, 118–122. DOI: 10.1109/ICCD59681.2023.10420820

Garg, R. K., Urs, V. L., Agarwal, A. A., Chaudhary, S. K., Paliwal, V., & Kar, S. K. (2023). Exploring the role of ChatGPT in patient care (diagnosis and treatment) and medical research: A systematic review. *Health Promotion Perspectives, 13*(3), 183–191. DOI: 10.34172/hpp.2023.22 PMID: 37808939

Goyal, N., Chang, M., & Terry, M. (2024). Designing for Human-Agent Alignment: Understanding what humans want from their agents. *Conference on Human Factors in Computing Systems - Proceedings*. DOI: 10.1145/3613905.3650948

Groves, M., & Mundt, K. (2021). A ghostwriter in the machine? Attitudes of academic staff towards machine translation use in internationalised Higher Education. *Journal of English for Academic Purposes, 50*, 100957. Advance online publication. DOI: 10.1016/j.jeap.2021.100957

Guleria, A., Krishan, K., Sharma, V., & Kanchan, T. (2023). ChatGPT: Ethical concerns and challenges in academics and research. *Journal of Infection in Developing Countries, 17*(9), 1292–1299. DOI: 10.3855/jidc.18738 PMID: 37824352

Harrasi, A. L. (2023). Developing a needs-based plagiarism management in second-language writing in a higher education institute: Practice-oriented research. *Instructional Science, 51*(6), 1079–1115. DOI: 10.1007/s11251-023-09628-6

Hung, J., & Chen, J. (2023). The Benefits, Risks and Regulation of Using ChatGPT in Chinese Academia: A Content Analysis. *Social Sciences (Basel, Switzerland), 12*(7), 380. Advance online publication. DOI: 10.3390/socsci12070380

Ivanov, S. (2023). The dark side of artificial intelligence in higher education. *Service Industries Journal, 43*(15–16), 1055–1082. DOI: 10.1080/02642069.2023.2258799

Jereb, E., Urh, M., Jerebic, J., & Šprajc, P. (2018). Gender differences and the awareness of plagiarism in higher education. *Social Psychology of Education, 21*(2), 409–426. DOI: 10.1007/s11218-017-9421-y

Kitamura, F. C. (2023). ChatGPT Is Shaping the Future of Medical Writing But Still Requires Human Judgment. *Radiology, 307*(2), e230171. Advance online publication. DOI: 10.1148/radiol.230171 PMID: 36728749

Kotovich, J., & Oriol, M. (2023). Is ChatGPT 3 Safe for Students? In C. A., E. S., B. J., & M. B. (Eds.), *Lecture Notes in Computer Science (including subseries Lecture Notes in Artificial Intelligence and Lecture Notes in Bioinformatics)* (Vol. 14387, pp. 100–107). Springer Science and Business Media Deutschland GmbH. DOI: 10.1007/978-3-031-48639-5_8

Kuzlu, M., Xiao, Z., Sarp, S., Catak, F. O., Gurler, N., & Guler, O. (2023). The Rise of Generative Artificial Intelligence in Healthcare. *12th Mediterranean Conference on Embedded Computing, MECO 2023*. DOI: 10.1109/MECO58584.2023.10155107

Li, B., Chen, Q., Lin, J., Li, S., & Yen, J. (2024). Assessing GPT-4 Generated Abstracts: Text Relevance and Detectors Based on Faithfulness, Expressiveness, and Elegance Principle. In T. Y. & S. Y. (Eds.), *Communications in Computer and Information Science: Vol. 2017 CCIS* (pp. 165–180). Springer Science and Business Media Deutschland GmbH. DOI: 10.1007/978-981-97-0837-6_12

Li, M., Enkhtur, A., Cheng, F., & Yamamoto, B. A. (2024). Ethical Implications of ChatGPT in Higher Education: A Scoping Review. *Journal of Interdisciplinary Studies in Education*, *13*(1), 55–68. DOI: 10.32674/jise.v13i1.6072

Lo, C. K. (2023). What Is the Impact of ChatGPT on Education? A Rapid Review of the Literature. *Education Sciences*, *13*(4), 410. Advance online publication. DOI: 10.3390/educsci13040410

Macfarlane, B. (2017). The ethics of multiple authorship: Power, performativity and the gift economy. *Studies in Higher Education*, *42*(7), 1194–1210. DOI: 10.1080/03075079.2015.1085009

Mahabeer, P., & Pirtheepal, T. (2019). Assessment, plagiarism and its effect on academic integrity: Experiences of academics at a university in South Africa. *South African Journal of Science*, *115*(11–12). Advance online publication. DOI: 10.17159/sajs.2019/6323

Mansoor, F., Ameen, K., & Arshad, A. (2024). An exploratory study of university librarians' perceptions on causes and deterrents of plagiarism: A Pakistani perspective. *Global Knowledge, Memory and Communication*, *73*(4–5), 634–649. DOI: 10.1108/GKMC-04-2022-0074

Maryon, T., Dubre, V., Elliott, K., Escareno, J., Fagan, M. H., Standridge, E., & Lieneck, C. (2022). COVID-19 Academic Integrity Violations and Trends: A Rapid Review. *Education Sciences*, *12*(12), 901. Advance online publication. DOI: 10.3390/educsci12120901

Masic, I. (2014). Plagiarism in Scientific Research and Publications and How to Prevent It. *Materia Socio-Medica*, *26*(2), 141. DOI: 10.5455/msm.2014.26.141-146 PMID: 24944543

McAlpine, J. B., Ferreira, D., Pauli, N. E., Gafner, S., & Pauli, G. F. (2023). The Ethics of Publishing Biomedical and Natural Products Research. *Journal of Natural Products*, *86*(9), 2228–2237. DOI: 10.1021/acs.jnatprod.3c00165 PMID: 37638654

Med, D. G., Nohl, S., Wolf, C., Rupert, Y., Rimkus, L., Ehlers, J., Breuckmann, F., & Sellmann, T. (2023). A SWOT (Strengths, Weaknesses, Opportunities, and Threats) Analysis of ChatGPT in the Medical Literature: Concise Review. *Journal of Medical Internet Research*, *25*, e49368. Advance online publication. DOI: 10.2196/49368 PMID: 37865883

Miron, J., Wilson, R., Freeman, J., & Sears, K. (2022). Academic integrity in upper year nursing students' work-integrated settings. *International Journal for Educational Integrity*, *18*(1), 16. Advance online publication. DOI: 10.1007/s40979-022-00107-y

Molnar, G., & Jozsef, C. (2022). AI Based Plagiarism Checking : Ease of use and appli7cable system for teachers to find similarities in students' assessments. *CANDO-EPE 2022 - Proceedings: IEEE 5th International Conference and Workshop in Obuda on Electrical and Power Engineering*, 187–192. DOI: 10.1109/CANDO-EPE57516.2022.10046379

Morocco-Clarke, A., Sodangi, F. A., & Momodu, F. (2024). The implications and effects of ChatGPT on academic scholarship and authorship: A death knell for original academic publications? *Information & Communications Technology Law*, *33*(1), 21–41. DOI: 10.1080/13600834.2023.2239623

Moya, B. A. (2024). Academic Integrity Policies in Hispanic South American Higher Education: Status and Recommendations. In *Springer International Handbooks of Education: Vol. Part F2304* (pp. 911–933). Springer Nature. DOI: 10.1007/978-3-031-54144-5_122

Moya, B. A., & Eaton, S. E. (2024). Academic Integrity Policy Analysis of Chilean Universities. *Journal of Academic Ethics*, *22*(4), 639–663. Advance online publication. DOI: 10.1007/s10805-024-09515-w

Mphahlele, A., & McKenna, S. (2019). The use of turnitin in the higher education sector: Decoding the myth. *Assessment & Evaluation in Higher Education*, *44*(7), 1079–1089. DOI: 10.1080/02602938.2019.1573971

O'Connell, J. (2016). Networked participatory online learning design and challenges for academic integrity in higher education. *International Journal for Educational Integrity*, *12*(1), 4. Advance online publication. DOI: 10.1007/s40979-016-0009-7

Park, C. (2017). In other (people's) words: Plagiarism by university students-literature and lessons. In *Academic Ethics* (pp. 525–542). Taylor and Francis., DOI: 10.4324/9781315263465-42

Patel, A., Bakhtiyari, K., & Taghavi, M. (2011). Evaluation of cheating detection methods in academic writings. *Library Hi Tech, 29*(4), 623–640. DOI: 10.1108/07378831111189732

Patwardhan, B., Desai, A., Chourasia, A., Nag, S., & Bhatnagar, R. (2020). Good Academic Research Practices. *University Grants Commission, September*, 56. https://www.ugc.ac.in/e-book/UGC_GARP_2020_Good Academic Research Practices.pdf

Perkins, M., & Roe, J. (2024). Academic publisher guidelines on AI usage: A ChatGPT supported thematic analysis. *F1000Research, 12*. DOI: 10.12688/f1000research.142411.2

Phyo, E. M., Lwin, T., Tun, H. P., Oo, Z. Z., Mya, K. S., & Silverman, H. (2023). Knowledge, attitudes, and practices regarding plagiarism of postgraduate students in Myanmar. *Accountability in Research, 30*(8), 672–691. DOI: 10.1080/08989621.2022.2077643 PMID: 35686819

Pradhan, T. K., & Kumar, A. (2023). Plagiarism in the Research Reports of Indian Doctoral Students: Causes and a Remedial Action Plan. *Center for Educational Policy Studies Journal, 13*(3), 185–212. DOI: 10.26529/cepsj.1494

Razack, H. I. A., Mathew, S. T., Saad, F. F. A., & Alqahtani, S. A. (2021). Artificial intelligence-assisted tools for redefining the communication landscape of the scholarly world. *Science Editing, 8*(2), 134–144. DOI: 10.6087/kcse.244

Richards, S. (2016). Academic integrity, plagiarism, and mercenary authorship in online learning environments. In N. J. & J. A. (Eds.), *Proceedings of the European Conference on e-Learning, ECEL* (Vols. 2016-Janua, pp. 575–578). Academic Conferences Limited. https://www.scopus.com/inward/record.uri?eid=2-s2.0-85000786211&partnerID=40&md5=caaa2f7156a580c7be5bd728a6ed469b

Roig, M. (2010). Plagiarism and self-plagiarism: What every author should know. *Biochemia Medica, 20*(3), 295–300. DOI: 10.11613/BM.2010.037

Saeed, S., Aamir, R., & Ramzan, M. (2011). Plagiarism and its implications on higher education in developing countries. *International Journal of Teaching and Case Studies, 3*(2/3/4), 123. DOI: 10.1504/IJTCS.2011.039552

Salauddin, N. (2023). Factors Influencing Plagiarism Among Faculty Members and Research Scholars in the Higher Education System. *DESIDOC Journal of Library and Information Technology, 43*(5), 364–374. DOI: 10.14429/djlit.43.05.19113

Sallam, M., Salim, N. A., Barakat, M., & Al-Tammemi, A. B. (2023). ChatGPT applications in medical, dental, pharmacy, and public health education: A descriptive study highlighting the advantages and limitations. *Narra J, 3*(1), e103. Advance online publication. DOI: 10.52225/narra.v3i1.103 PMID: 38450035

Siddhpura, A., & Siddhpura, M. (2020). Plagiarism, contract cheating and other academic misconducts in online engineering education: Analysis, detection and prevention strategies. In M. H., G. Y., O. Y., R. Ma.M.T., S. J., V. N., W. G., Y. M., & C.-U. L. L. (Eds.), *Proceedings of 2020 IEEE International Conference on Teaching, Assessment, and Learning for Engineering, TALE 2020* (pp. 112–119). Institute of Electrical and Electronics Engineers Inc. DOI: 10.1109/TALE48869.2020.9368311

Sidorkin, A. M. (2024). Embracing Chatbots in Higher Education: The Use of Artificial Intelligence in Teaching, Administration, and Scholarship. In *Embracing Chatbots in Higher Education: The Use of Artificial Intelligence in Teaching, Administration, and Scholarship*. Taylor and Francis., DOI: 10.4324/9781032686028

Sullivan, M., Kelly, A., & McLaughlan, P. (2023). ChatGPT in higher education: Considerations for academic integrity and student learning. *Journal of Applied Learning and Teaching*, *6*(1), 31–40. DOI: 10.37074/jalt.2023.6.1.17

Talyshinskii, A., Naik, N., Hameed, B. M. Z., Zhanbyrbekuly, U., Khairli, G., Guliev, B., Juilebø-Jones, P., Tzelves, L., & Somani, B. K. (2023). Expanding horizons and navigating challenges for enhanced clinical workflows: ChatGPT in urology. *Frontiers in Surgery*, *10*, 1257191. Advance online publication. DOI: 10.3389/fsurg.2023.1257191 PMID: 37744723

Tindall, I. K., Fu, K. W., Tremayne, K., & Curtis, G. J. (2021). Can negative emotions increase students' plagiarism and cheating? *International Journal for Educational Integrity*, *17*(1), 25. Advance online publication. DOI: 10.1007/s40979-021-00093-7

Torres, C. G., & Fons, I. T. (2018). Cyber-plagiarism in Spanish higher education: Exploring the causes in order to reach specific solutions. *Synergies Espagne, 11*, 195–212. https://www.scopus.com/inward/record.uri?eid=2-s2.0-85059908937&partnerID=40&md5=ba93a4a01eeaaf564061a3ba18f47dbb

Tortora, L. (2024). Beyond Discrimination: Generative AI Applications and Ethical Challenges in Forensic Psychiatry. *Frontiers in Psychiatry*, *15*, 1346059. Advance online publication. DOI: 10.3389/fpsyt.2024.1346059 PMID: 38525252

Tripathi, S., & Patel, D. (2021). Academic integrity: Looking beyond plagiarism. *Annals of Library and Information Studies*, *68*(2), 190–197. https://www.scopus.com/inward/record.uri?eid=2-s2.0-85108826989&partnerID=40&md5=53e0df4e6cebc6e399c5b3eab8289a03

Watanabe, A., & Wiseman, S. M. (2023). A New Era in Surgical Research: The Evolving Role of Artificial Intelligence. *American Journal of Surgery, 226*(6), 923–925. DOI: 10.1016/j.amjsurg.2023.06.040 PMID: 37419728

Wittenberg, H. (2024). Turn-It-Off: Reflections on Technology, the Knowledge Commons and the Academic Plagiarism Industry. *Scrutiny2*. DOI: 10.1080/18125441.2024.2320892

Wu, J.-S., Chien, T.-H., Chien, L.-R., & Yang, C.-Y. (2021). Using artificial intelligence to predict class loyalty and plagiarism in students in an online blended programming course during the covid-19 pandemic. *Electronics (Basel), 10*(18), 2203. Advance online publication. DOI: 10.3390/electronics10182203

Youmans, R. J. (2011). Does the adoption of plagiarism-detection software in higher education reduce plagiarism? *Studies in Higher Education, 36*(7), 749–761. DOI: 10.1080/03075079.2010.523457

Zhang, X., Li, D., Wang, C., Jiang, Z., Ngao, A. I., Liu, D., Peters, M. A., & Tian, H. (2023). From ChatGPT to China' Sci-Tech: Implications for Chinese Higher Education. *Beijing International Review of Education, 5*(3), 296–314. DOI: 10.1163/25902539-05030007

Zlateva, P., Steshina, L., Petukhov, I., & Velev, D. (2024). A Conceptual Framework for Solving Ethical Issues in Generative Artificial Intelligence. In T.-B. A.J., C.-A. E., & L.-G. D.A. (Eds.), *Frontiers in Artificial Intelligence and Applications* (Vol. 381, pp. 110–119). IOS Press BV. DOI: 10.3233/FAIA231182

Chapter 15
Future Trends in Research Ethics:
Navigating New Ethical Frontiers

Debajit Karmakar
http://orcid.org/0000-0001-9272-0627
Lakshmibai National Institute of Physical Education, India

Anindita Das
http://orcid.org/0000-0002-6502-9619
Lakshmibai National Institute of Physical Education, India

Tashi Ongmu Bhutia
http://orcid.org/0009-0000-0127-5709
Government B.Ed. College, Soreng, India

Karuppasamy Govindasamy
http://orcid.org/0000-0002-3019-5545
Symbiosis International University, India

Atisha Mandal
http://orcid.org/0009-0003-2731-5985
Lakshmibai National Institute of Physical Education, India

EXECUTIVE SUMMARY

The chapter explores the transformative evolution of research ethics in the 21st century, shaped by rapid technological innovation, global collaboration, and shifting societal values. It examines how traditional ethical principles such as respect, beneficence, justice, and nonmaleficence must adapt to address emerging challenges in artificial intelligence, automation, immersive technologies, and big data. Key concerns including algorithmic accountability, informed consent in digital

DOI: 10.4018/979-8-3693-8302-5.ch015

environments, participant vulnerability, data governance, and intellectual property are critically analyzed. The discussion highlights tensions between global harmonization and cultural relativism in ethical governance, while emphasizing the need for anticipatory and adaptive frameworks. Special attention is given to ecological responsibility, participant care, and the growing role of ethics education in ensuring human dignity, scientific integrity, and social responsibility.

INTRODUCTION

The landscape of research ethics stands at a critical inflection point, driven by revolutionary technological developments, unprecedented global connectivity, and rapidly evolving societal demands. As we advance further into the digital age, established ethical frameworks that have governed research practice for generations must undergo significant adaptation to confront the intricate moral complexities arising from digital innovation, artificial intelligence integration, worldwide collaborative networks, environmental stewardship, and transforming models of human engagement in research activities (Sachs & Siegler, 1993).

The core principles of research ethics, which include respect for people, doing good, avoiding harm, and fairness, continue to be essential for providing ethical guidance. Nevertheless, these core tenets require substantial reinterpretation and broadening to address contemporary ethical challenges that span national borders, cross disciplinary boundaries, and challenge conventional notions of research risk and benefit. The proliferation of big data systems, machine learning technologies, cloud-based infrastructures, and international research partnerships has generated new ethical domains requiring innovative frameworks, comprehensive policies, and robust governance mechanisms. Central to addressing these challenges is the implementation of systematic algorithm auditing processes that can evaluate the fairness, transparency, and accountability of automated decision-making systems used in research contexts (Benatar & Singer, 2000).

Modern research ethics must confront complex issues surrounding data sovereignty, algorithmic accountability, environmental stewardship, cross-cultural competency, and the dynamic nature of informed consent within digital ecosystems. The conventional research ethics paradigm, traditionally centered on safeguarding individual participants in controlled experimental conditions, must broaden its scope to encompass collective risks, societal ramifications, ecological consequences, and global equity considerations (Sachs & Siegler, 1993). This transformation demands both conceptual innovation and practical application through modernized institutional review processes, international policy coordination, and comprehensive ethics training for researchers navigating increasingly sophisticated environments. Algorithm auditing emerges as a critical tool in this evolution, providing systematic methods to assess whether AI-powered research tools operate fairly across diverse populations and maintain transparency in their decision-making processes (Fischer, 2006)

The convergence of technological innovation and ethical obligation presents both remarkable opportunities and significant challenges for the research community. While digital platforms facilitate extraordinary collaboration, seamless data exchange, and powerful analytical capabilities, they simultaneously introduce risks concerning privacy protection, cybersecurity, algorithmic bias, and digital exclusion. The widespread availability of research tools through artificial intelligence and automation holds promise for accelerating scientific breakthroughs while generating fundamental questions about intellectual ownership, professional accountability, and humanity's role in knowledge creation. Rigorous algorithm auditing becomes essential in this context, serving as a safeguard to ensure that automated research systems do not perpetuate existing biases or create new forms of discrimination in scientific inquiry (Benatar & Singer, 2000).

As we look toward future developments, research ethics must proactively identify and prepare for emerging challenges while preserving fundamental commitments to human dignity, scientific rigor, and social justice. This necessitates creating forward-looking governance systems capable of rapid response to technological advances, promoting international collaboration in ethical standard development, and establishing adaptive infrastructure within research institutions and regulatory frameworks. Algorithm auditing will play an increasingly vital role in this future landscape, providing ongoing oversight mechanisms to ensure that as research becomes more automated and data-driven, it remains aligned with ethical principles and serves the broader interests of humanity.

ARTIFICIAL INTELLIGENCE, AUTOMATION, AND RESEARCH ETHICS

The integration of artificial intelligence and automation technologies into research processes represents one of the most transformative developments in contemporary research ethics. These technologies promise to accelerate scientific discovery, improve research efficiency, and enable new forms of inquiry while simultaneously challenging fundamental assumptions about human agency, accountability, and the nature of knowledge generation. The ethical implications of AI in research extend beyond simple tool use to encompass questions of authorship, bias, fairness, and the evolving relationship between human researchers and intelligent systems.

- **Autonomous Decision-Making in Research**

The increasing sophistication of AI systems enables autonomous decision-making in various aspects of research, from study design and participant recruitment to data analysis and hypothesis generation. While this autonomy can improve research efficiency and reduce human bias, it also creates new ethical challenges related to accountability, transparency, and human oversight of research processes. Algorithmic study design raises questions about the appropriate level of human involvement in fundamental research decisions. AI systems can optimize research protocols, select methodologies, and determine sample sizes based on statistical criteria and resource constraints. However, these decisions embody values and priorities that may not be explicitly programmed or easily auditable. The delegation of study design decisions to AI systems may inadvertently embed biases or priorities that conflict with research ethics principles or participant interests (Benatar & Singer, 2000).

Automated participant recruitment and screening systems can improve efficiency and reduce human bias in participant selection, but they may also introduce new forms of discrimination or exclusion. Machine learning algorithms used for participant recruitment may develop selection criteria that systematically exclude certain demographic groups or inadvertently perpetuate existing health disparities. The complexity of these systems makes it difficult for researchers to understand or control the selection processes fully (Fischer, 2006).

Dynamic consent and adaptive protocols enabled by AI systems allow research studies to evolve in response to emerging data patterns or participant preferences. While this adaptability can improve research relevance and participant experience, it also complicates informed consent processes and ethical oversight. Participants may not fully understand how their consent applies to dynamically changing research protocols or how AI systems will modify their participation over time. Real-time data analysis and intervention systems can identify patterns and trigger responses more quickly than human researchers, potentially improving participant safety or research outcomes. However, these systems may also make decisions about participant care or study modifications without adequate human review or understanding. The speed of AI decision-making may outpace human oversight capabilities, creating risks of inappropriate interventions or missed opportunities for human judgment (Benatar & Singer, 2000).

Quality control and error detection through AI systems can improve research reliability by identifying data anomalies, protocol violations, or analytical errors that human researchers might miss. However, AI quality control systems may also introduce their own errors or biases, particularly when they encounter data patterns that differ from their training experiences. Over-reliance on AI quality control may reduce human vigilance and critical thinking about research processes.

- **AI as a Co-Author: Ethical and Legal Implications**

The emergence of AI systems capable of generating research hypotheses, conducting analyses, and even writing research papers has created unprecedented questions about authorship, intellectual property, and credit attribution in scientific research. Traditional concepts of authorship assume human agency, creativity, and responsibility that may not apply to AI-generated contributions to research. Authorship criteria in academic publishing typically require substantial intellectual contributions, responsibility for research content, and accountability for research integrity. AI systems may make significant contributions to research through data analysis, pattern recognition, or hypothesis generation, but they cannot take responsibility for research accuracy or respond to questions about research conduct. This creates ambiguity about whether AI systems can or should be credited as authors on research publications (Lee, 2021).

Intellectual contribution assessment becomes complex when AI systems generate insights, discover patterns, or produce creative solutions that human researchers might not have developed independently. The originality and creativity traditionally associated with research authorship may be demonstrated by AI systems, but the absence of consciousness or intentionality in AI decision-making raises questions about the nature and significance of these contributions (Levine, 1979).

Legal liability and responsibility for research content remain with human researchers even when AI systems contribute significantly to research processes or outputs. Human researchers must take responsibility for verifying AI-generated results, ensuring research integrity, and addressing any errors or ethical violations. This responsibility may be difficult to fulfill when AI systems operate through complex processes that are not fully transparent or understandable to human users. Disclosure requirements for AI involvement in research are evolving as publishers and institutions grapple with appropriate transparency standards. Some journals now require explicit disclosure of AI use in research or writing processes, while others prohibit AI-generated content entirely. Researchers must navigate varying requirements while maintaining transparency about the role of AI systems in their work (Lee, 2021).

Intellectual property implications of AI-generated research contributions remain legally and ethically uncertain. Patent law and copyright frameworks were designed for human creators and may not adequately address AI-generated innovations or creative works. The ownership of AI-generated research outputs may depend on complex relationships between researchers, institutions, and AI system developers. Plagiarism and originality concerns arise when multiple researchers use similar AI systems that may generate similar results or insights. The possibility that AI systems could produce identical or nearly identical outputs for different users raises questions about research originality and potential inadvertent plagiarism. Academic integrity frameworks must evolve to address these new forms of similarity in research outputs (Levine, 1979).

- **Bias and Fairness in AI-Driven Research**

AI systems used in research can perpetuate or amplify existing biases while also introducing new forms of unfairness or discrimination. Understanding and addressing these biases is crucial for maintaining research integrity and ensuring that AI-driven research serves the interests of all populations rather than reinforcing existing inequalities. Training data bias represents a fundamental challenge in AI-driven research. Machine learning systems learn patterns from historical data that may reflect past discrimination, underrepresentation of certain groups, or systematic measurement errors. When these biased datasets are used to train AI systems for research applications, the resulting systems may perpetuate or amplify existing inequalities in ways that are difficult to detect or correct (Fischer, 2006).

Algorithmic fairness in research applications requires careful consideration of how AI systems treat different demographic groups or populations. Fairness can be defined in multiple ways, including equal treatment, equal outcomes, or equal opportunity, and these different fairness criteria may conflict with each other. Researchers must make explicit choices about fairness objectives and evaluate whether their AI systems achieve these goals across different populations (Benatar & Singer, 2000).

Representation and inclusion challenges arise when AI systems are trained on datasets that inadequately represent diverse populations or when AI-driven research methods systematically exclude certain groups. Underrepresented populations may receive lower-quality results from AI systems or may be excluded from research benefits due to biased algorithmic decision-making. Intersectionality considerations become particularly important in AI-driven research because machine learning systems may exhibit different biases for individuals who belong to multiple marginalized groups. Traditional bias detection methods that examine single demographic characteristics may miss complex forms of discrimination that affect individuals with intersectional identities (Lee, 2021).

Bias detection and mitigation in AI systems require ongoing monitoring and adjustment rather than one-time fixes. AI systems may develop new biases over time as they encounter new data or as the populations they serve change. Researchers must implement continuous bias monitoring and be prepared to adjust their systems to maintain fairness and accuracy. Stakeholder engagement in AI fairness assessment requires involving affected communities in defining fairness criteria, evaluating system performance, and identifying potential harms. Technical definitions of fairness may not align with community values or lived experiences of discrimination, making community input essential for meaningful fairness assessment (Fischer, 2006).

- **Intellectual Property and Machine-Generated Knowledge**

The generation of knowledge through AI systems challenges traditional intellectual property frameworks and raises questions about ownership, attribution, and commercialization of machine-generated discoveries. These challenges are particularly acute in research contexts where knowledge generation is a primary objective and intellectual property rights may affect future research directions. Patent eligibility for AI-generated inventions remains legally uncertain in most jurisdictions. Patent systems typically require human inventors, but AI systems may independently generate patentable innovations through processes that exceed human creative capacity. The absence of clear legal frameworks for AI-generated inventions creates uncertainty about intellectual property protection and commercialization opportunities (Levine, 1979).

Copyright protection for AI-generated written content, including research papers, reports, and documentation, varies across jurisdictions and continues to evolve. Some legal systems require human authorship for copyright protection, while others may recognize AI-generated works under certain circumstances. Researchers must understand the intellectual property implications of using AI systems for content generation. Trade secret protection may apply to AI algorithms, training data, or system configurations used in research, but this protection may conflict with research transparency and reproducibility requirements. Researchers may need to balance intellectual property protection with scientific openness and peer review requirements.

Licensing and technology transfer of AI-generated research outputs require careful consideration of ownership rights, attribution requirements, and ongoing responsibilities. Universities and research institutions must develop policies for managing intellectual property rights in AI-generated discoveries while ensuring appropriate benefit-sharing with human researchers and research participants (Soehartono et al., 2022). Open science and AI-generated knowledge create tensions between intellectual property protection and research accessibility. AI systems may generate valuable knowledge that could benefit society broadly, but commercial interests may limit access to AI-generated discoveries or the systems that produce them. Balancing innovation incentives with public access remains a significant challenge. Collaborative ownership models may be necessary when AI systems generate knowledge through interactions with multiple human researchers, datasets, or institutional resources. Traditional intellectual property frameworks may not adequately address the complex relationships and contributions involved in AI-assisted research collaborations (Benatar & Singer, 2000).

- **Ethical Governance of AI Research Tools**

The governance of AI systems used in research requires new frameworks for oversight, accountability, and quality assurance that address the unique characteristics and capabilities of AI technologies. Traditional research governance mechanisms may be inadequate for AI systems that can make autonomous decisions, learn from experience, and operate in ways that are not fully predictable or controllable by human researchers. Institutional oversight of AI research tools requires research institutions to develop policies and procedures for evaluating, approving, and monitoring AI systems used in research (Davies, 2020). These oversight mechanisms must address technical capabilities, ethical implications, and ongoing performance monitoring while avoiding unnecessarily restrictive approaches that could inhibit beneficial research applications. Algorithm auditing and validation processes for research AI systems must go beyond traditional software testing to include evaluation of bias, fairness, transparency, and alignment with research ethics principles. Regular auditing may be necessary to detect emerging problems or changes in system performance over time (Schuppli & Fraser, 2007).

Transparency and explainability requirements for AI research tools must balance the need for research transparency with the technical complexity of AI systems. Researchers should be able to understand and explain how AI systems contribute to their research, even when complete technical understanding may be impractical or impossible. Risk assessment frameworks for AI research applications should consider both technical risks related to system performance and ethical risks related to bias, discrimination, or inappropriate decision-making. Risk assessment should be ongoing rather than one-time evaluation, as AI systems may develop new capabilities or encounter new situations over time (Faden et al., 2002). Ethics review processes for research involving AI systems may require specialized expertise in AI technology, algorithmic fairness, and computational ethics. Traditional institutional review boards may need additional training or expert consultation to evaluate AI-related research proposals adequately. International coordination of AI research governance can help ensure consistency in ethical standards while avoiding regulatory fragmentation that could impede beneficial research collaborations. International organizations and professional societies play important roles in developing guidelines and best practices for AI research governance (Schuppli & Fraser, 2007).

ETHICAL DIMENSIONS OF HUMAN PARTICIPATION IN FUTURE RESEARCH

The evolving landscape of human participation in research reflects broader technological and social changes that are transforming how researchers engage with participants, collect data, and manage ongoing relationships throughout the research process. Traditional models of human subjects research, developed primarily for clinical trials and laboratory studies, must adapt to accommodate new forms of participation enabled by digital technologies, remote monitoring, immersive experiences, and long-term data collection. These developments create both opportunities for more inclusive and participant-centered research and new risks that require careful ethical consideration.

- **Informed Consent in Complex Studies**

The increasing complexity of contemporary research studies challenges traditional informed consent processes in multiple ways. Studies may involve multiple phases, adaptive designs, secondary data use, and integration with other research projects in ways that make it difficult to provide comprehensive information about research participation at the time of initial consent. Multi-phase and adaptive study designs allow research protocols to evolve based on interim results or changing circumstances, but this flexibility complicates informed consent because participants cannot be fully informed about future study phases that have not yet been designed (Beaulieu & Estalella, 2012). Dynamic consent models that allow participants to provide or withdraw consent for specific study components as they are developed may address some of these challenges while creating new administrative and ethical complexities. Broad consent frameworks attempt to address the challenge of complex studies by obtaining participant permission for general categories of research use rather than specific protocols. While broad consent can facilitate valuable secondary research and long-term studies, it may not provide participants with meaningful information about how their data will actually be used or what risks they may face from future research activities (Thakur & Lahiry, 2019).

Layered consent approaches provide different levels of detail for participants with varying interests in research information. Brief summaries for general participants can be supplemented with detailed technical information for those who desire more complete understanding. However, layered consent must ensure that all participants receive adequate information for meaningful decision-making regardless of their chosen level of detail. Ongoing consent processes for longitudinal studies must maintain participant engagement and informed decision-making over extended time periods. Participant circumstances, preferences, and risk tolerance may change over time, requiring researchers to provide updated information and renewal opportunities for consent (Lange et al., 2013). The administrative burden of ongoing consent must be balanced against participant autonomy and protection. Cultural and linguistic adaptation of consent processes becomes increasingly important as research becomes more global and inclusive. Informed consent must be meaningful across different cultural contexts, educational levels, and language preferences while maintaining scientific and ethical standards. Simple translation may be inadequate when consent concepts do not translate directly across cultural frameworks. Capacity assessment for complex research participation requires careful evaluation of participants' ability to understand and appreciate the implications of research participation. Complex studies involving multiple data types, long-term participation, and evolving protocols may exceed some participants' capacity for informed decision-making even when they have capacity for simpler research participation (Thakur & Lahiry, 2019).

- **Protecting Vulnerable Populations**

Vulnerable populations in research face evolving challenges as new technologies and research methods create different types of risks and benefits. Traditional approaches to vulnerability protection must adapt to address digital divides, online risks, and new forms of exploitation while avoiding paternalistic approaches that unnecessarily limit research participation opportunities. Digital vulnerability emerges when participants lack the technical skills, resources, or infrastructure necessary to participate safely in digital research. This digital divide can create new forms of exclusion from research benefits while potentially exposing participants to risks they cannot fully understand or manage. Researchers must consider how digital research methods may differentially affect participants with varying levels of technological access and literacy (Komić et al., 2015).

Economic vulnerability in research participation may be exacerbated by digital research methods that require participants to have access to smartphones, reliable internet connections, or other technological resources. Research that assumes universal access to digital technologies may inadvertently exclude economically disadvantaged populations or create financial burdens for participants who must acquire technology for research participation. Social vulnerability can be amplified through digital research methods that may expose participants to cyberbullying, harassment, or social exclusion based on their research participation or the data they provide. Social media research, online surveys, and digital health monitoring may create risks for social harm that traditional research protection measures do not adequately address (Beaulieu & Estalella, 2012).

Cognitive and emotional vulnerability requires special consideration in research involving immersive technologies, artificial intelligence, or intensive data collection that may affect participants' psychological well-being or decision-making capacity. New research methods may create forms of psychological pressure or manipulation that traditional vulnerability assessments do not anticipate. Institutional vulnerability affects participants in contexts such as prisons, schools, or healthcare facilities where digital monitoring or research participation may be required or strongly encouraged.

of VR/AR research participation because scientific understanding of these effects is still developing (Behr et al., 2005).

- **Remote and Digital Health Research Ethics**

The expansion of health research into remote and digital contexts has accelerated dramatically, particularly following the COVID-19 pandemic, creating new opportunities for inclusive participation while introducing novel ethical challenges related to data security, participant safety, and quality of care.

Remote monitoring and data collection enable continuous health data gathering through wearable devices, smartphone applications, and home-based sensors. While this technology can provide rich longitudinal data and reduce participant burden, it also creates privacy risks and may generate data that reveals sensitive health information or behavioural patterns that participants did not intend to share. Telemedicine integration with research blurs the boundaries between clinical care and research participation, potentially creating conflicts of interest or confusion about whether participants are receiving clinical care or contributing to research. Participants may not understand when their healthcare interactions involve research components or how research participation may affect their clinical care (Nebeker et al., 2019).

Data quality and validity concerns arise in remote research when researchers cannot directly observe or control data collection conditions. Participants may misunderstand instructions, use devices incorrectly, or provide data in contexts that affect validity. Remote research must develop new methods for ensuring data quality while respecting participant autonomy and privacy. Emergency response protocols for remote research must address situations where digital monitoring or research interactions reveal urgent health or safety concerns. Researchers may have limited ability to respond to emergencies or provide immediate assistance to remote participants, creating ethical obligations to develop appropriate referral and response systems (Shen et al., 2022).

Digital divide considerations become particularly important in health research because health disparities may be correlated with limited access to digital technologies. Research that relies heavily on digital participation methods may systematically exclude populations that are most in need of health research benefits. Cybersecurity risks in digital health research include potential exposure of sensitive health information through data breaches, device hacking, or inadequate security measures. Participants may face risks of identity theft, discrimination, or harassment if their health research data is compromised (Nebeker et al., 2019).

- **Post-Trial Responsibilities and Long-Term Participant Care**

The evolution of research methods toward longer-term engagement, continuous monitoring, and complex interventions has expanded researchers' responsibilities for participant care beyond traditional study endpoints. These extended relationships create new ethical obligations while challenging traditional boundaries between research and clinical care. Continued access to beneficial interventions after study completion raises questions about researchers' obligations to provide ongoing access to treatments, technologies, or services that participants received during research. When research demonstrates that interventions are beneficial, participants may develop expectations for continued access that researchers may not be prepared to meet (Perakslis, 2018).

Long-term follow-up and monitoring obligations may extend for years or decades after active research participation ends, particularly for studies involving genetic information, chronic conditions, or interventions with delayed effects. Researchers must balance the scientific value of long-term follow-up with participant burden and privacy concerns. Incidental findings from research activities, particularly genetic research or comprehensive health monitoring, create obligations for researchers to determine what information should be shared with participants and how to provide appropriate counseling or referral services. The complexity of research-generated information may exceed participants' understanding or ability to use effectively (Brall et al., 2019).

Data stewardship responsibilities continue long after active research participation ends when researchers maintain participant data for future analysis or sharing. Researchers must ensure ongoing protection of participant privacy and interests even as institutional affiliations change, funding ends, or research priorities evolve (Perakslis, 2018). Community relationships and obligations may persist when research involves entire communities or has community-wide implications. Researchers may have ongoing responsibilities to share results, provide community benefits, or maintain relationships with community partners beyond individual participant interactions (Brall et al., 2019).

Institutional transitions and researcher mobility can disrupt long-term participant relationships when researchers change institutions or retire. Research institutions must develop policies for maintaining participant relationships and honouring long-term commitments even when individual researchers are no longer available.

DATA ETHICS AND GOVERNANCE

The proliferation of data collection, storage, and analysis capabilities in research has created complex ethical challenges that extend far beyond traditional concerns about privacy and confidentiality. Contemporary data ethics must address questions of ownership, control, transparency, and justice in data systems while navigating technical complexity, regulatory diversity, and evolving social expectations about data use. The governance of research data requires frameworks that can balance competing interests while promoting both scientific progress and respect for individual and collective rights.

- **Data Ownership and Control**

The question of who owns and controls research data has become increasingly complex as data collection methods diversify, data sharing becomes more common, and multiple stakeholders claim interests in research data. Traditional models that assigned data ownership to researchers or institutions are being challenged by participants' growing expectations for data control and the recognition of collective interests in certain types of data. Participant data ownership claims reflect growing awareness of data value and concern about commercialization of personal information. Some participants argue that they should retain ownership rights in data derived from their bodies, behaviours, or experiences, particularly when research data is used for commercial purposes or leads to profitable discoveries. However, implementing participant data ownership faces practical challenges related to data aggregation, analysis, and the collective nature of many research insights (Hilgartner & Brandt-Rauf, 1994).

Institutional data policies typically assign data ownership to research institutions or funding organizations based on their investment in research infrastructure and personnel. These policies aim to ensure data preservation, responsible use, and institutional accountability for data management. However, institutional ownership may conflict with participant expectations or community interests in data control. Community data sovereignty movements, particularly among indigenous populations, assert collective rights to control data about community members and traditional knowledge. These movements challenge individualistic models of data ownership and consent while demanding recognition of collective interests in data governance. Research institutions must develop approaches that respect community sovereignty while maintaining research capabilities (Mirchev et al., 2020).

Commercial data interests arise when research data has potential commercial value or when private companies participate in research collaborations. The involvement of commercial entities in research data systems raises questions about profit-sharing, data use restrictions, and conflicts between commercial interests and research ethics principles (Hicks, 2023). International data governance creates additional complexity when research data crosses national boundaries or involves participants from multiple countries. Different national approaches to data ownership, privacy protection, and access rights may conflict when applied to international research collaborations.

Temporal dimensions of data ownership become important as the value and use of research data may change over time. Data that initially appears to have limited value may become highly valuable as new analytical methods develop or as data is combined with other sources. Long-term data governance must anticipate changing data value and use patterns (Wiltshire & Alvanides, 2022).

- **Transparency in Data Collection and Usage**

Transparency in research data systems requires clear communication about data collection practices, usage purposes, sharing arrangements, and governance structures. However, achieving meaningful transparency faces challenges related to technical complexity, competitive concerns, and the difficulty of predicting future data uses. Data collection transparency requires researchers to provide clear information about what data is being collected, how it is being gathered, and what technologies or methods are being used. However, comprehensive transparency may be technically complex or may compromise research validity when participants modify their behaviour based on knowledge of data collection methods (Mikk et al., 2017).

Purpose specification and limitation principles require researchers to clearly communicate the intended uses of collected data and to limit actual use to specified purposes. However, the value of research data often emerges from unexpected uses or combinations with other data sources, creating tensions between purpose limitation and research innovation (Hicks, 2023).

Data sharing transparency involves communicating with participants and stakeholders about how research data will be shared, with whom, and under what conditions. Transparency about data sharing must balance participant interests in privacy protection with scientific interests in collaboration and reproducibility. Algorithmic transparency becomes important when automated systems are used for data collection, analysis, or decision-making. Participants and stakeholders may have interests in understanding how algorithmic systems process their data, but providing meaningful algorithmic transparency faces technical and competitive challenges (Mikk et al., 2017).

Financial transparency about data monetization, commercialization, or economic benefits may be important for maintaining trust and ensuring fair benefit-sharing. However, commercial sensitivity and competitive concerns may limit willingness to provide complete financial transparency about data use. Governance transparency requires clear communication about who makes decisions about data use, what processes are followed for decision-making, and how stakeholders can participate in or influence data governance. Transparent governance helps build trust and accountability while enabling stakeholder participation in data management decisions (Hicks, 2023).

- **Ethical Use of Open-Access Data**

Open-access data policies aim to maximize the scientific and social benefits of research data by making it freely available for secondary analysis and reuse. However, ethical use of open-access data requires attention to privacy protection, appropriate use conditions, and recognition of original data contributors' interests and rights.

Privacy protection in open-access data requires careful attention to de-identification methods, re-identification risks, and appropriate access controls. Data that appears anonymous may be re-identifiable through combination with other available datasets or through sophisticated analytical techniques. Open-access policies must balance data accessibility with adequate privacy protection. Attribution and citation practices for open-access data help ensure that original data contributors receive appropriate recognition for their work while enabling users to understand data provenance and quality. However, traditional citation models may not adequately address the complex relationships involved in data reuse, particularly when data is modified, combined with other sources, or used for purposes very different from original research (Mauthner & Parry, 2013).

Usage conditions and restrictions for open-access data may be necessary to prevent inappropriate use while maintaining broad accessibility. Common restrictions might include prohibitions on commercial use, re-identification attempts, or research that could harm the populations from which data was derived. However, usage restrictions must be carefully designed to avoid unnecessarily limiting beneficial research uses. Quality assurance for open-access data requires mechanisms for documenting data quality, limitations, and appropriate use conditions. Data users need adequate information to assess whether data is suitable for their intended purposes and to interpret results appropriately. However, providing comprehensive data documentation requires significant resources and ongoing maintenance (Wiltshire & Alvanides, 2022).

Benefit-sharing considerations may be important when open-access data generates significant commercial or social benefits. Original data contributors, particularly research participants and communities, may have legitimate claims to share in benefits derived from their data contributions. However, implementing benefit-sharing for open-access data faces practical challenges related to tracking data use and distributing benefits (Mauthner & Parry, 2013).

Ethical review for secondary use of open-access data may be necessary when planned uses differ significantly from original research purposes or when secondary use might create new risks for original participants. However, requiring ethical review for all open-access data use could create barriers to beneficial research while potentially being unnecessary when appropriate protections are already in place.

- **Consent for Secondary Use of Data**

The reuse of research data for purposes beyond those originally intended creates challenges for traditional consent models while offering opportunities for maximizing the value of research investments. Developing appropriate consent frameworks for secondary data use requires balancing participant autonomy with research benefits and practical feasibility. Broad consent models attempt to address secondary use challenges by obtaining participant permission for general categories of future research rather than specific protocols. Broad consent can facilitate valuable secondary research while reducing administrative burden and participant contact requirements. However, broad consent may not provide participants with meaningful information about actual future uses or risks (Mikk et al., 2017).

Specific consent for each secondary use provides participants with detailed information about proposed research and clear opportunities to accept or decline participation. This approach maximizes participant control and information but may be impractical for large-scale data sharing or may create barriers to beneficial research when participants cannot be contacted for new consent (Stommel & Rijk, 2021).

Opt-out consent models assume participant agreement to secondary use unless they explicitly decline participation. This approach can facilitate research while providing participants with control over their data use. However, opt-out consent may not ensure active participant engagement with consent decisions and may not meet ethical standards for research involving sensitive data or vulnerable populations. Tiered consent approaches allow participants to provide different levels of permission for different types of secondary use. Participants might consent to some types of research while declining others, providing nuanced control over data use. However, tiered consent can create complex administrative challenges and may limit the utility of data for certain types of research (Wiltshire & Alvanides, 2022).

Dynamic consent systems enable participants to modify their consent preferences over time, providing or withdrawing permission for different types of secondary use as their preferences change. Dynamic consent can maintain participant control over extended time periods but requires ongoing participant engagement and sophisticated technical infrastructure. Community consent models may be appropriate when research data affects entire communities or when individual consent is not culturally appropriate. Community consent involves community representatives in decision-making about data use while potentially overriding individual preferences. This approach must carefully balance collective and individual interests while respecting cultural values and governance structures (Mauthner & Parry, 2013).

POLICY, REGULATION, AND FUTURE ETHICAL GOVERNANCE

The rapidly evolving landscape of research ethics requires adaptive governance mechanisms that can respond to technological developments, changing social values, and emerging ethical challenges while maintaining stable foundations for research conduct. Traditional regulatory frameworks, developed primarily for clinical research, must evolve to address digital research, artificial intelligence, global collaborations, and other contemporary research contexts. Future ethical governance must balance innovation facilitation with protection of human subjects and research integrity.

- **Evolution of Institutional Review Boards (IRBs)**

Institutional Review Boards have served as the primary mechanism for ethical oversight of human subjects research for several decades, but they face significant challenges in adapting to contemporary research contexts. The traditional IRB model, designed primarily for clinical trials and laboratory-based research, must evolve to address digital research methods, artificial intelligence applications, global collaborations, and other emerging research paradigms. Expertise requirements for IRB members have expanded dramatically as research becomes more technologically sophisticated and interdisciplinary. Traditional IRB composition, focused on medical and scientific expertise, may be inadequate for reviewing research involving artificial intelligence, big data analytics, social media research, or other specialized methods. IRBs must develop mechanisms for accessing specialized expertise while maintaining broad representation and community perspectives (Milne, 2005).

Review procedures and timelines designed for traditional research protocols may be inadequate for iterative research methods, adaptive study designs, or rapidly changing digital research environments. Fixed review schedules and lengthy deliberation processes may impede time-sensitive research while potentially failing to provide adequate oversight for dynamically changing research protocols. Risk assessment frameworks used by IRBs must expand beyond traditional concerns about physical and psychological harm to address digital risks, privacy violations, algorithmic bias, and collective harms. IRB members need training and tools for evaluating new types of risks while avoiding overly restrictive approaches that could impede beneficial research (Torous & Nebeker, 2017).

Continuing review and monitoring requirements may need modification for research involving continuous data collection, artificial intelligence systems, or long-term participant relationships. Traditional models of periodic review may be inadequate for research that evolves continuously or that involves automated decision-making systems. Community engagement and stakeholder participation in IRB processes may become more important as research increasingly affects entire

Digital research methods may create new opportunities for coercion or inappropriate surveillance in institutional settings (Benatar & Singer, 2000).

Legal vulnerability emerges when research participation creates risks for legal consequences, immigration status, or other legal protections. Digital research methods may create data trails that could be accessed by law enforcement or other authorities in ways that participants do not anticipate or understand.

- **Psychological Risks of Immersive Technologies (VR/AR)**

Virtual reality and augmented reality technologies enable new forms of research that can create realistic simulations, immersive experiences, and interactive environments for studying human behaviour and responses. However, these technologies also create novel psychological risks that traditional research ethics frameworks have not addressed comprehensively. Presence and immersion in virtual environments can create psychological responses that are as intense as those experienced in real-world situations. Participants may experience genuine fear, anxiety, or trauma from virtual experiences, particularly when research involves simulated dangerous or distressing situations. The psychological impact of virtual experiences may persist beyond the research session and require ongoing monitoring or support. Reality confusion and dissociation may occur when participants have difficulty distinguishing between virtual and real experiences, particularly after extended exposure to immersive technologies. This confusion may be more likely in participants with certain psychological conditions or cognitive vulnerabilities and could create risks for decision-making capacity or psychological well-being (Behr et al., 2005).

Motion sickness and physical discomfort from VR/AR technologies can create barriers to research participation and may disproportionately affect certain demographic groups. These physical effects may also influence research results in ways that create bias or reduce data quality. Privacy concerns in immersive research extend beyond traditional data collection to include biometric monitoring, behavioural tracking, and environmental scanning that may capture information about participants' homes, families, or personal spaces. Immersive technologies may collect data about participants' physical movements, eye tracking, emotional responses, and spatial environments without explicit awareness or consent (Denisov & Ильич, 2019).

Psychological manipulation through immersive technologies raises ethical concerns about the potential for researchers to influence participant behaviour, emotions, or decisions in ways that exceed traditional research influence. The persuasive power of immersive experiences may create risks for coercion or inappropriate influence over research participants. Long-term effects of immersive technology exposure remain largely unknown, creating challenges for risk assessment and informed consent. Participants cannot be fully informed about potential long-term consequences communities or involves participatory research methods. Traditional IRB composition may not adequately represent community interests or provide mechanisms for meaningful community input into research oversight (Milne, 2005).

International coordination between IRBs becomes critical as research becomes increasingly global and collaborative. IRBs must develop mechanisms for coordinating review processes, sharing expertise, and ensuring consistent ethical standards across different institutional and cultural contexts.

- **National vs. International Research Policies**

The globalization of research has created tensions between national sovereignty in research regulation and the need for international coordination and harmonization of ethical standards. Different countries have developed research ethics policies that reflect their cultural values, legal frameworks, and scientific priorities, but these differences can create barriers to international collaboration and may lead to ethical arbitrage (Grigis et al., 2025).

Regulatory harmonization efforts attempt to align research ethics requirements across different countries and jurisdictions to facilitate international collaboration while maintaining appropriate protections for research participants. However, harmonization must balance efficiency and consistency with respect for cultural diversity and national sovereignty in research governance. Jurisdictional conflicts arise when international research projects must comply with conflicting requirements from different national regulatory systems. Researchers may face situations where compliance with one country's requirements violates another country's laws or ethical standards, creating impossible compliance situations (Madani et al., 2020).

Extraterritorial application of national research ethics policies may occur when countries attempt to regulate research conducted by their institutions or researchers in other countries. These extraterritorial policies can create conflicts with local regulations and may represent inappropriate extensions of national authority over international research activities. Cultural relativism versus universal standards debates reflect fundamental disagreements about whether research ethics principles should be universally applied or adapted to local cultural contexts. While some argue for universal human rights and research protections, others contend that meaningful ethical standards must be culturally sensitive and locally appropriate (Grigis et al., 2025).

Capacity building and technical assistance programs can help developing countries strengthen their research ethics infrastructure while avoiding inappropriate imposition of external standards. However, capacity building must be designed to support local autonomy and cultural adaptation rather than simply reproducing Western research ethics models. Trade agreements and economic policies increasingly include provisions related to data flows, intellectual property, and research collaboration that affect research ethics governance. Economic interests in research and data may conflict with ethical principles and may create pressure for lowest-common-denominator ethical standards (Madani et al., 2020).

- **Regulatory Gaps in Emerging Research Fields**

New research fields and methods often emerge faster than regulatory frameworks can adapt, creating periods of regulatory uncertainty and potential gaps in ethical oversight. Emerging research areas such as artificial intelligence, synthetic biology, digital health, and environmental geoengineering may not fit well within existing regulatory categories and may require new governance approaches. Artificial intelligence research faces regulatory uncertainty because existing frameworks were not designed to address algorithmic decision-making, machine learning bias, or AI system autonomy. Regulators must develop new approaches for overseeing AI research while avoiding stifling innovation through overly restrictive or inappropriate requirements. Digital health research operates at the intersection of healthcare regulation, research oversight, and technology policy, creating complex regulatory landscapes with potential gaps or overlaps. The convergence of clinical care and research in digital health applications challenges traditional boundaries between regulated and unregulated activities (Schuppli & Fraser, 2007).

Environmental research and geoengineering may have global implications that exceed the authority of national regulators while potentially requiring international coordination and oversight. The global nature of environmental systems means that research activities in one country can affect populations worldwide (Zimmer, 2018).

Synthetic biology and genetic engineering technologies are developing rapidly and may not be adequately addressed by existing biotechnology regulations. New capabilities for genetic modification, organism design, and biological system engineering may require new regulatory approaches that address novel risks and ethical considerations. Space research and astrobiology face regulatory uncertainty due to the international nature of space activities and the novel environments and conditions involved in space-based research. Traditional research ethics frameworks may not be adequate for research conducted in space or involving potential extra-terrestrial life (Strosberg et al., 2014).

Anticipatory governance approaches attempt to identify and address regulatory gaps before they create problems for research conduct or participant protection. These approaches require ongoing monitoring of research developments and proactive policy development to address emerging ethical challenges.

ETHICS EDUCATION AND TRAINING FOR FUTURE RESEARCHERS

The increasing complexity and ethical sophistication of contemporary research requires comprehensive ethics education and training programs that prepare researchers to navigate complex ethical landscapes while maintaining research integrity and participant protection. Traditional research ethics training, focused primarily on regulatory compliance, must evolve to address critical thinking, ethical reasoning, and practical decision-making skills. Curriculum development for research ethics education must address both foundational ethical principles and contemporary challenges such as digital research methods, artificial intelligence applications, global collaborations, and environmental considerations. Ethics curricula should be interdisciplinary and adaptable to different research contexts while maintaining core competencies in ethical reasoning and decision-making (Lange et al., 2013).

Experiential learning approaches can help researchers develop practical skills for addressing ethical challenges through case studies, role-playing exercises, and real-world problem-solving experiences. Traditional lecture-based ethics training may be inadequate for developing the complex thinking skills necessary for ethical research conduct in challenging situations. Continuing education requirements reflect the need for ongoing ethics development as research methods evolve and ethical understanding advances. Researchers may need periodic updating on new ethical challenges, regulatory changes, and best practices throughout their careers rather than relying solely on initial training (Schuppli & Fraser, 2007).

Cultural competency training becomes increasingly important as research becomes more global and involves diverse populations with different cultural values and expectations. Researchers need skills for working respectfully and effectively across cultural boundaries while maintaining ethical standards. Leadership development in research ethics focuses on preparing researchers to serve as ethics leaders within their institutions and disciplines. Research ethics leaders need skills for policy development, institutional change, and mentoring other researchers in ethical conduct. Assessment and evaluation of ethics education programs requires measures of learning outcomes, behaviour change, and practical application of ethical knowledge and skills. Traditional compliance-based assessments may be inadequate for evaluating complex ethical reasoning and decision-making capabilities (Strosberg et al., 2014).

ANTICIPATORY GOVERNANCE OF SCIENCE AND TECHNOLOGY

The rapid pace of scientific and technological development requires governance mechanisms that can anticipate and prepare for future ethical challenges rather than simply responding to problems after they emerge. Anticipatory governance involves ongoing monitoring of research developments, stakeholder engagement, and adaptive policy-making that can respond to emerging challenges while supporting beneficial innovation. Technology assessment and forecasting methods can help identify potential ethical challenges associated with emerging research technologies before they become widespread. Early identification of ethical issues enables proactive policy development and stakeholder engagement while there is still time to influence technology development trajectories (Lange et al., 2013).

Stakeholder engagement processes for anticipatory governance must involve diverse perspectives including researchers, policymakers, community representatives, and potential users of research technologies. Meaningful stakeholder engagement requires ongoing dialogue and collaboration rather than one-time consultation exercises. Adaptive regulation approaches enable policy frameworks to evolve in response to changing circumstances and new information while maintaining core protections and principles. Adaptive regulation may involve sunset clauses, regular review requirements, or flexible implementation mechanisms that can be adjusted as experience accumulates (Blažun Vošner et al., 2017).

International coordination in anticipatory governance becomes critical for emerging technologies that transcend national boundaries or that require global coordination for effective oversight. International organizations and professional societies play important roles in facilitating coordination and developing shared approaches to emerging ethical challenges. Public engagement and communication about emerging research technologies can help build understanding and support for appropriate governance while identifying public concerns and values that should inform policy development. However, public engagement must be designed to provide meaningful input opportunities rather than simply educating the public about predetermined policies. Innovation sandboxes and pilot programs can provide controlled environments for testing new research approaches while gathering information about ethical implications and governance needs. These programs can enable beneficial innovation while providing learning opportunities for future policy development (Schuppli & Fraser, 2007).

CONCLUSION

The future of research ethics is at a decisive point where traditional frameworks must evolve to meet new challenges while upholding the core principles of human dignity, scientific integrity, and social responsibility. Rapid digital transformation has changed the ethical landscape by raising issues of privacy, consent, algorithmic accountability, and data governance. While big data and artificial intelligence offer vast opportunities, they also introduce risks that require new ethical approaches. Similarly, globalization has expanded research across borders, demanding sensitivity to cultural values, fair benefit sharing, and equitable collaboration. Ethical frameworks must move beyond procedural adjustments to embrace genuine mutual respect and avoid cultural dominance or relativism.

Environmental sustainability has emerged as a pressing ethical priority, with climate research carrying significant policy and intergenerational implications. Integrating ecological responsibility into research means balancing scientific progress with sustainable practices. The rise of artificial intelligence and automation also challenges traditional ideas of human agency and accountability in research. As AI systems advance, frameworks must ensure human oversight and meaningful ethical control. Human participation in research is evolving with digital engagement, remote studies, and immersive technologies, creating both new opportunities and risks. Consent models, vulnerability protections, and participant care require constant adaptation. Data governance is another central concern as large scale collection and sharing call for frameworks that balance openness with protection of individual and collective rights.

Key priorities for the future include adaptive governance that keeps pace with social and technological change, stakeholder and community engagement in ethical decisions, and greater international cooperation. Education and training must move beyond compliance to foster critical reasoning and problem solving. The integration of diverse ethical perspectives, including biomedical, environmental, and data ethics, is vital. Ultimately, the future of research ethics depends on sustained commitment, proactive governance, and genuine collaboration to ensure that research advances both human welfare and global responsibility.

REFERENCES

Beaulieu, A., & Estalella, A. (2012). Rethinking Research Ethics for Mediated Settings. *Information Communication and Society*, *15*(1), 23–42. DOI: 10.1080/1369118X.2010.535838

Behr, K.-M., Nosper, A., Klimmt, C., & Hartmann, T. (2005). Some Practical Considerations of Ethical Issues in VR Research. *Presence (Cambridge, Mass.)*, *14*(6), 668–676. DOI: 10.1162/105474605775196535

Benatar, S. R., & Singer, P. A. (2000). *A new look at international research ethics*. DOI: 10.1136/bmj.321.7264.824

Blažun Vošner, H., Železnik, D., Kokol, P., Vošner, J., & Završnik, J. (2017). Trends in nursing ethics research: Mapping the literature production. *Nursing Ethics*, *24*(8), 892–907. DOI: 10.1177/0969733016654314 PMID: 27364534

Brall, C., Schröder-Bäck, P., & Maeckelberghe, E. (2019). Ethical aspects of digital health from a justice point of view. *European Journal of Public Health*, *29*(Supplement_3), 18–22. DOI: 10.1093/eurpub/ckz167 PMID: 31738439

Davies, S. E. (2020). The introduction of research ethics review procedures at a university in South Africa: Review outcomes of a social science research ethics committee. *Research Ethics*, *16*(1–2), 1–26. DOI: 10.1177/1747016119898408

Denisov, E. I., & Ильич, Д. Э. (2019). Robots, artificial intelligence, augmented and virtual reality: Ethical, legal and hygienic issues. *Hygiene and Sanitation*, *98*(1), 5–10. DOI: 10.18821/0016-9900-2019-98-1-5-10

Faden, R. R., Klag, M. J., Kass, N. E., & Krag, S. S. (2002). On the Importance of Research Ethics and Mentoring. *The American Journal of Bioethics*, *2*(4), 50–51. Advance online publication. DOI: 10.1162/152651602320957565 PMID: 12762925

Fischer, B. A.IV. (2006). A Summary of Important Documents in the Field of Research Ethics. *Schizophrenia Bulletin*, *32*(1), 69–80. DOI: 10.1093/schbul/sbj005 PMID: 16192409

Grigis, A., Beretta, G., Borry, P., & Sanchini, V. (2025). Vulnerability in research ethics: A systematic review of policy guidelines and documents. *PLoS One*, *20*(7), e0327086. DOI: 10.1371/journal.pone.0327086 PMID: 40591537

Hicks, J. (2023). The future of data ownership: An uncommon research agenda. *The Sociological Review*, *71*(3), 544–560. DOI: 10.1177/00380261221088120

Hilgartner, S., & Brandt-Rauf, S. I. (1994). Data Access, Ownership, and Control: Toward Empirical Studies of Access Practices. *Knowledge (Beverly Hills, Calif.), 15*(4), 355–372. DOI: 10.1177/107554709401500401

Komić, D., Marušić, S. L., & Marušić, A. (2015). Research Integrity and Research Ethics in Professional Codes of Ethics: Survey of Terminology Used by Professional Organizations across Research Disciplines. *PLoS One, 10*(7), e0133662. DOI: 10.1371/journal.pone.0133662 PMID: 26192805

Lange, M. M., Rogers, W., & Dodds, S. (2013). Vulnerability in Research Ethics: A Way Forward. *Bioethics, 27*(6), 333–340. DOI: 10.1111/bioe.12032 PMID: 23718774

Lee, S.-H. (2021). Changing Policy Trends in Research Ethics. *Health Policy and Management, 145–147*. Advance online publication. DOI: 10.4332/KJHPA.2021.31.2.145

Levine, R. J. (1979). Clarifying the Concepts of Research Ethics. *The Hastings Center Report, 9*(3), 21–26. DOI: 10.2307/3560793 PMID: 478836

Madani, M., Ghasemzadeh, N., Dizani, A., Gharamaleki, A. F., & Larijani, B. (2020). Policy considerations to achieve practical ethics: Closing the gap between ethical theory and practic. *Journal of Medical Ethics and History of Medicine, 13*, 8. DOI: 10.18502/jmehm.v13i8.4075 PMID: 33117501

Mauthner, N. S., & Parry, O. (2013). Open Access Digital Data Sharing: Principles, Policies and Practices☆. *Social Epistemology, 27*(1), 47–67. DOI: 10.1080/02691728.2012.760663

Mikk, K. A., Sleeper, H. A., & Topol, E. J. (2017). The Pathway to Patient Data Ownership and Better Health. *Journal of the American Medical Association, 318*(15), 1433–1434. DOI: 10.1001/jama.2017.12145 PMID: 28973063

Milne, C. (2005). Overseeing Research: Ethics and the Institutional Review Board. *Forum Qualitative Sozialforschung / Forum: Qualitative. Social Research, 6*(1). Advance online publication. DOI: 10.17169/fqs-6.1.531

Mirchev, M., Mircheva, I., & Kerekovska, A. (2020). The Academic Viewpoint on Patient Data Ownership in the Context of Big Data: Scoping Review. *Journal of Medical Internet Research, 22*(8), e22214. DOI: 10.2196/22214 PMID: 32808934

Nebeker, C., Torous, J., & Bartlett Ellis, R. J. (2019). Building the case for actionable ethics in digital health research supported by artificial intelligence. *BMC Medicine, 17*(1), 137. DOI: 10.1186/s12916-019-1377-7 PMID: 31311535

Perakslis, E. D. (2018). Using digital health to enable ethical health research in conflict and other humanitarian settings. *Conflict and Health, 12*(1), 23. DOI: 10.1186/s13031-018-0163-z PMID: 29785204

Sachs, G. A., & Siegler, M. (1993). Teaching scientific integrity and the responsible conduct of research. *Academic Medicine, 68*(12), 871–875. https://journals.lww.com/academicmedicine/abstract/1993/12000/Teaching_scientific_integrity_and_the_responsible.1.aspx. DOI: 10.1097/00001888-199312000-00001 PMID: 8259954

Schuppli, C. A., & Fraser, D. (2007). Factors influencing the effectiveness of research ethics committees. *Journal of Medical Ethics, 33*(5), 294–301. DOI: 10.1136/jme.2005.015057 PMID: 17470508

Shen, F. X., Silverman, B. C., Monette, P., Kimble, S., Rauch, S. L., & Baker, J. T. (2022). An Ethics Checklist for Digital Health Research in Psychiatry: Viewpoint. *Journal of Medical Internet Research, 24*(2), e31146. DOI: 10.2196/31146 PMID: 35138261

Soehartono, A. M., Yu, L. G., & Khor, K. A. (2022). Essential signals in publication trends and collaboration patterns in global Research Integrity and Research Ethics (RIRE). *Scientometrics, 127*(12), 7487–7497. DOI: 10.1007/s11192-022-04400-y PMID: 35755633

Stommel, W., & de Rijk, L. (2021). Ethical approval: None sought. How discourse analysts report ethical issues around publicly available online data. *Research Ethics, 17*(3), 275–297. DOI: 10.1177/1747016120988767

Strosberg, M. A., Gefenas, E., & Famenka, A. (2014). Research Ethics Review: Identifying Public Policy and Program Gaps. *Journal of Empirical Research on Human Research Ethics; JERHRE, 9*(2), 3–11. DOI: 10.1525/jer.2014.9.2.3 PMID: 24782068

Thakur, S., & Lahiry, S. (2019). Research ethics in the modern era. *Indian Journal of Dermatology, Venereology and Leprology, 85*(4), 351. DOI: 10.4103/ijdvl.IJDVL_499_18 PMID: 31115359

Torous, J., & Nebeker, C. (2017). Navigating Ethics in the Digital Age: Introducing Connected and Open Research Ethics (CORE), a Tool for Researchers and Institutional Review Boards. *Journal of Medical Internet Research, 19*(2), e6793. DOI: 10.2196/jmir.6793 PMID: 28179216

Wiltshire, D., & Alvanides, S. (2022). Ensuring the ethical use of big data: Lessons from secure data access. *Heliyon, 8*(2), e08981. Advance online publication. DOI: 10.1016/j.heliyon.2022.e08981 PMID: 35243099

Zimmer, M. (2018). Addressing Conceptual Gaps in Big Data Research Ethics: An Application of Contextual Integrity. *Social Media + Society*, *4*(2), 2056305118768300. DOI: 10.1177/2056305118768300

Compilation of References

AAMC. (2008). *Protecting Patients, Preserving Integrity, Advancing Health: Accelerating Implementation of COI Policies in Human Subjects Research.* https://www.aamc.org

Abbasi, P., Yoosefi-Lebni, J., Jalali, A., Ziapour, A., & Nouri, P. (2021). Causes of the plagiarism: A grounded theory study. *Nursing Ethics*, *28*(2), 282–296. DOI: 10.1177/0969733020945753 PMID: 32909912

Abdalla, H. B. (2022). A brief survey on big data: Technologies, terminologies and data-intensive applications. *Journal of Big Data*, *9*(1), 107. DOI: 10.1186/s40537-022-00659-3

Aboud, B. O., Zacarias, T. H., & Oliver, C. E. E. (2025). Advancing smart public administration: Challenges and benefits of artificial intelligence. *Urban Governance*, *1*(1), 1–10. DOI: 10.1016/j.ugj.2025.06.003

Adelakun, B. O., Onwubuariri, E. R., Adeniran, G. A., & Ntiakoh, A. (2024). Enhancing fraud detection in accounting through AI: Techniques and case studies. *Finance & Accounting Research Journal*, *6*(6), 978–999. DOI: 10.51594/farj.v6i6.1232

Adzima, K. (2020). Examining online cheating in higher education using traditional classroom cheating as a guide. *Electronic Journal of e-Learning*, *18*(6), 476–493. DOI: 10.34190/JEL.18.6.002

Aguinis, H., & Henle, C. A. (2004). Ethics in research. *Handbook of research methods in industrial and organizational psychology*, 34-56.

Aita, M., & Richer, M.-C. (2005). Essentials of research ethics for healthcare professionals. *Nursing & Health Sciences*, *7*(2), 119–125. DOI: 10.1111/j.1442-2018.2005.00216.x PMID: 15877688

Akbar, A. (2018). Defining plagiarism: A literature review. *Ethical Lingua: Journal of Language Teaching and Literature*, *5*(1), 31–38. DOI: 10.30605/ethicallingua.v5i1.750

Akhtar, S. (2014). The ethics of conducting research with Muslim participants. *Research Ethics*, *10*(3), 138–149.

Akpa-Inyang, F., & Chima, S. C. (2021). … values and beliefs regarding informed consent and limitations of the principle of respect for autonomy in African communities: A cross-cultural qualitative study. *BMC Medical Ethics*, *22*(1), 111. Advance online publication. DOI: 10.1186/s12910-021-00678-4 PMID: 34391415

Alahmad, G., Al-Jumah, M., & Dierickx, K. (2017). Confidentiality, informed consent and children's participation in research involving stored tissue samples: Interviews with medical professionals from the Middle East. *Narrative Inquiry in Bioethics*, *7*(1), 71–79. PMID: 28713147

Alahmad, G., & Dierickx, K. (2012). What do Islamic institutional fatwas say about medical and research confidentiality and breach of confidentiality? *Developing World Bioethics*, *12*(2), 104–112. DOI: 10.1111/j.1471-8847.2012.00329.x PMID: 22702346

Al-Attas, S. M. N. (1993). *Islam and secularism*. International Institute of Islamic Thought and Civilization.

Alawad, A. A. (2021). Students' attitudes toward plagiarism in the interior design field. *Civil Engineering and Architecture*, *9*(4), 1179–1188. DOI: 10.13189/cea.2021.090419

Al-Bar, M. A., & Chamsi-Pasha, H. (2015). *Contemporary bioethics: Islamic perspective*. Springer. DOI: 10.1007/978-3-319-18428-9

Alkire, L., Bilgihan, A., Bui, M., Buoye, A. J., Dogan, S., & Kim, S. (2024). RAISE: Leveraging responsible AI for service excellence. *Journal of Service Management*, *35*(4), 490–511. Advance online publication. DOI: 10.1108/JOSM-11-2023-0448

AlShanti, A. M., Al-Azab, H. A. H., Humeedat, M. M., & AlQudah, M. Z. (2024). Exploring the evolution of creative accounting and external auditors: Bibliometric analysis. *Cogent Business & Management*, *11*(1), 2300500. Advance online publication. DOI: 10.1080/23311975.2023.2300500

Alshater, M. (2022). Exploring the role of artificial intelligence in enhancing academic performance: A case study of ChatGPT. *Social Science Research Network*, *1*(1), 1–10. https://papers.ssrn.com/sol3/papers.cfm?abstract_id=4312358

American Accounting Association. (n.d.). *Publications ethics for academic research* [Guide for authors of Behavioral Research in Accounting]. Retrieved August 2025, from https://aaahq.org/Research/Journals/Behavioral-Research-in-Accounting/Guide-for-Authors

American Institute of Certified Public Accountants. (n.d.). *AICPA Code of Professional Conduct*. Wikipedia. https://en.wikipedia.org/wiki/AICPA_Code_of_Professional_Conduct

American Society for Public Administration. (2025). *Code of ethics*. ASPA. https://www.aspanet.org/ASPA/ASPA/Code-of-Ethics/Code-of-Ethics.aspx

Andersen, M. B. (1994). Ethical considerations in the supervision of applied sport psychology graduate students. *Journal of Applied Sport Psychology*, *6*(2), 152–167. DOI: 10.1080/10413209408406291

Andersson, H., Svensson, A., Frank, C., Rantala, A., Holmberg, M., & Bremer, A. (2022). Ethics education to support ethical competence learning in healthcare: An integrative systematic review. *BMC Medical Ethics*, *23*(1), 29. DOI: 10.1186/s12910-022-00766-z PMID: 35305627

Andrew, J., & Baker, M. (2021). The general data protection regulation in the age of surveillance capitalism. [REMOVED HYPERLINK FIELD]. *Journal of Business Ethics*, *168*(3), 565–578. Advance online publication. DOI: 10.1007/s10551-019-04239-z

Angwin, J., Larson, J., Mattu, S., & Kirchner, L. (2016). *Machine bias*. ProPublica. https://www.propublica.org/article/machine-bias-risk-assessments-in-criminal-sentencing

Apoorv, R., Dahiya, A., Sreeram, U., Patil, B. R., Irish, I., Graziano, R., & Starner, T. (2020). Examinator: A Plagiarism Detection Tool for Take-Home Exams. *L@S 2020 - Proceedings of the 7th ACM Conference on Learning @ Scale*, 261–264. DOI: 10.1145/3386527.3406723

Ara, A., & Ara, A. (2024). Exploring the ethical implications of generative AI. In *Exploring the Ethical Implications of Generative AI*. IGI Global., DOI: 10.4018/979-8-3693-1565-1

Arar, K., & Saiti, A. (2022). Ethical leadership, ethical dilemmas and decision making among school administrators. *Equity in Education & Society*, *1*(1), 126–141. Advance online publication. DOI: 10.1177/27526461211070828

Ashworth, P., Bannister, P., & Thorne, P.Students on the Qualitative Researc. (1997). Guilty in whose eyes? University students' perceptions of cheating and plagiarism in academic work and assessment. *Studies in Higher Education*, *22*(2), 187–203. DOI: 10.1080/03075079712331381034

Association, W. M. (2013). Declaration of Helsinki: Ethical Principles for Medical Research Involving Human Subjects. *Journal of the American Medical Association*. PMID: 24141714

Aurey, X. (2020). *From the Martens Clause to Consent to Human Experimentations, the Legal Journey of the Judges during the Nuremberg Doctors' Trial*. Biolaw and International Criminal Law., DOI: 10.1163/9789004364424_015

Awasthi, S. (2019). Plagiarism and academic misconduct: A systematic review. *DESIDOC Journal of Library and Information Technology*, *39*(2), 94–100. DOI: 10.14429/djlit.39.2.13622

Bal, M., Stok, M., Bombaerts, G., Huijts, N., Schneider, P., Spahn, A., & Buskens, V. (2023). A fairway to fairness: Toward a richer conceptualization of fairness perceptions for just energy transitions. *Energy Research & Social Science*, *103*, 103213. DOI: 10.1016/j.erss.2023.103213

Barrientos, A., Cuadros, M., Alba, J., & Cruz, A. S. (2021). Implementation of a remote system for the supervision of online exams through the use of cameras with artificial intelligence. *Proceedings of the 2021 IEEE Engineering International Research Conference, EIRCON 2021*. DOI: 10.1109/EIRCON52903.2021.9613352

Bayaa Martin Saana, S. B., Ablordeppey, E., Mensah, N. J., & Karikari, T. K. (2016). Academic dishonesty in higher education: Students' perceptions and involvement in an African institution. *BMC Research Notes*, *9*(1), 234. Advance online publication. DOI: 10.1186/s13104-016-2044-0 PMID: 27112550

Beauchamp, T. L., & Childress, J. F. (2019). *Principles of biomedical ethics* (8th ed.). Oxford University Press.

Beaulieu, A., & Estalella, A. (2012). Rethinking Research Ethics for Mediated Settings. *Information Communication and Society*, *15*(1), 23–42. DOI: 10.1080/1369118X.2010.535838

Behr, K.-M., Nosper, A., Klimmt, C., & Hartmann, T. (2005). Some Practical Considerations of Ethical Issues in VR Research. *Presence (Cambridge, Mass.)*, *14*(6), 668–676. DOI: 10.1162/105474605775196535

Bekelman, J. E., Li, Y., & Gross, C. P. (2003). Scope and impact of financial conflicts of interest in biomedical research: A systematic review. *Journal of the American Medical Association, 289*(4), 454–465. DOI: 10.1001/jama.289.4.454 PMID: 12533125

Benatar, S. R., & Singer, P. A. (2000). *A new look at international research ethics.* DOI: 10.1136/bmj.321.7264.824

Bendorf, C. M., Yun, S. C., Kurath, G., & Hedrick, R. P. (2022). Comparative susceptibilities of selected California Chinook salmon and steelhead populations to isolates of L genogroup Infectious Hematopoietic Necrosis virus (IHNV). *Animals (Basel), 12*(13), 1733. DOI: 10.3390/ani12131733 PMID: 35804631

Berg, C. (2023). The Case for Generative AI in Scholarly Practice. SSRN *Electronic Journal, April,* 1–15. DOI: 10.2139/ssrn.4407587

Bernal Arellano, W. M., Galarza Parra, J. N., Villavicencio Reinoso, J. M., & Quito Ochoa, J. F. (2024). A study on academic dishonesty among English as a foreign language students. *Heliyon, 10*(13), e33876. DOI: 10.1016/j.heliyon.2024.e33876 PMID: 39050424

Bernardi, R. A., & Bean, D. F. (2006). The debate on ethics in accounting education: The views of our forgotten stakeholders. *The CPA Journal, 76*(7), 56–57. https://ssrn.com/abstract=2250592

Bernardi, R. A., & Bean, D. F. (2006). The growth of accounting research in ethics journals. In Jeffrey, C. (Ed.), *Research on professional responsibility and ethics in accounting* (Vol. 11, pp. 1–28). Emerald Group Publishing., DOI: 10.1016/S1574-0765(06)11003-1

Bero, L. (2017). Addressing bias and conflict of interest among biomedical researchers. *Journal of the American Medical Association, 317*(17), 1723–1724. DOI: 10.1001/jama.2017.3854 PMID: 28464166

Bertholf, R. L. (2001). Protecting human research subjects. *Annals of Clinical and Laboratory Science, 31*(1), 119–127. PMID: 11314861

Bérubé, J., & Gendron, Y. (2023). Developing ethical sensitivity in future accounting practitioners: The case of a dialogic learning for final-year undergraduates. *Journal of Business Ethics, 183*(3), 763–781. DOI: 10.1007/s10551-021-04947-5

Bester, J. C. (2020). Beneficence, interests, and wellbeing in medicine: What it means to provide benefit to patients. *The American Journal of Bioethics, 20*(3), 53–62. Advance online publication. DOI: 10.1080/15265161.2020.1714793 PMID: 32105204

Bhatt, B. (2022). Ethical complexity of social change: Negotiated actions of a social enterprise. *Journal of Business Ethics*, *177*(4), 743–762. Advance online publication. DOI: 10.1007/s10551-022-05100-6

Bird, S. J. (1996). The Role of Science Professionals in Teaching Responsible Research Conduct. *Bioscience*, *46*(10), 783–786. DOI: 10.2307/1312856

Bishop, D. V. M. (2020). The psychology of experimental psychologists: Overcoming cognitive constraints to improve research: The 47th Sir Frederic Bartlett Lecture. *Quarterly Journal of Experimental Psychology*, *73*(1), 1–19. Advance online publication. DOI: 10.1177/1747021819886519 PMID: 31724919

Bitter, C. C., Ngabirano, A. A., Simon, E. L., & Taylor, D. M. (2020). Principles of research ethics: A research primer for low-and middle-income countries. *African Journal of Emergency Medicine : Revue Africaine de la Médecine d'Urgence*, *10*, S125–S129. DOI: 10.1016/j.afjem.2020.07.006 PMID: 32837877

Blažun Vošner, H., Železnik, D., Kokol, P., Vošner, J., & Završnik, J. (2017). Trends in nursing ethics research: Mapping the literature production. *Nursing Ethics*, *24*(8), 892–907. DOI: 10.1177/0969733016654314 PMID: 27364534

Bloemraad, I., & Menjívar, C. (2022). Precarious times, professional tensions: The ethics of migration research and the drive for scientific accountability. *The International Migration Review*, *56*(1), 4–32. Advance online publication. DOI: 10.1177/01979183211014455 PMID: 36467777

Bloomberg Philanthropies. (2023). *State of cities: Generative AI in local governments*. https://cityaiconnect.jhu.edu/pdfs/Final-Gen-AI-In-Cities-Report_10.18.2023.pdf

Bouzanis, C. (2023). Kant, Immanuel. *The International Encyclopedia of Ethics*, 1–12.

Boyd, E. A., & Bero, L. A. (2006). Institutional policies on faculty conflicts of interest: A survey of medical school deans. *Journal of the American Medical Association*, *296*(9), 1105–1112. DOI: 10.1001/jama.296.9.1105

Brall, C., Schröder-Bäck, P., & Maeckelberghe, E. (2019). Ethical aspects of digital health from a justice point of view. *European Journal of Public Health*, *29*(Supplement_3), 18–22. DOI: 10.1093/eurpub/ckz167 PMID: 31738439

Brandt, A. M. (1978). Racism and research: The case of the Tuskegee Syphilis Study. *The Hastings Center Report*, *8*(6), 21–29. DOI: 10.2307/3561468 PMID: 721302

Bray, T. (2020). Military Information Technology Certification Training Addressing Implementation Procedures Against Cyber-Attacks: An Exploratory Case Study. [HTML]

Brazil, (2018). *Lei n° 13.709, de 14 de agosto de 2018. Lei Geral de Proteção de Dados Pessoais (LGPD)* [Law No. 13.709, of August 14, 2018. General Law for the Protection of Personal Data (LGPD)]. Diário Oficial da União. Retrieved from https://www.planalto.gov.br/ccivil_03/_ato2015-2018/2018/lei/l13709.htm

Brenner, E., Bleys, R. L., de Caro, R., Catereniuc, I., Chirculescu, A. R., Destrieux, C., & Osorio, M. T. V. (2023). The legal and ethical framework governing body donation in Europe–2nd update on current practice. *Annals of Anatomy-Anatomischer Anzeiger, 152195*. Advance online publication. DOI: 10.1016/j.aanat.2023.152195 PMID: 38042354

Bretag, T., & Green, M. (2014). The role of virtue ethics principles in academic integrity breach decision-making. *Journal of Academic Ethics*, *12*(3), 165–177. DOI: 10.1007/s10805-014-9209-z

Bretag, T., Harper, R., Burton, M., Ellis, C., Newton, P., Rozenberg, P., Saddiqui, S., & Van Haeringen, K. (2019). Contract cheating: A survey of Australian university students. *Studies in Higher Education*, *44*(11), 1837–1856. DOI: 10.1080/03075079.2018.1462788

Brink, C. B., & Lewis, D. I. (2023). The 12 RS framework as a comprehensive, unifying construct for principles guiding animal research ethics. *Animals (Basel)*, *13*(7), 1128. DOI: 10.3390/ani13071128 PMID: 37048384

Brockopp, J. E., & Eich, T. (2008). *Muslim medical ethics: From theory to practice*. University of South Carolina Press.

Brown, C., Spiro, J., & Quinton, S. (2020). The role of research ethics committees: Friend or foe in educational research? An exploratory study. *British Educational Research Journal*, *46*(4), 747–769. DOI: 10.1002/berj.3654

Bryson, J. J., & Winfield, A. F. T. (2017). Standardizing ethical design for artificial intelligence and autonomous systems. *Computer*, *50*(5), 116–119. DOI: 10.1109/MC.2017.154

Bulger, R. E., & Heitman, E. (2007). Expanding Responsible Conduct of Research Instruction across the University. *Academic Medicine*, *82*(9), 876–878. DOI: 10.1097/ACM.0b013e31812f7909 PMID: 17726396

Burgess, T., Rennie, S., & Moodley, K. (2023). Key ethical issues encountered during COVID-19 research: A thematic analysis of perspectives from South African research ethics committees. *BMC Medical Ethics*, *24*(1), 11. Advance online publication. DOI: 10.1186/s12910-023-00888-y PMID: 36793067

Caminiti, C., & De Fiore, L. (2023). Fake paper factories: Root cause solutions are needed. *Recenti Progressi in Medicina, 114*(3), 148–153. DOI: 10.1701/3981.39638 PMID: 36815417

Campbell, E. G., Weissman, J. S., Ehringhaus, S., Rao, S. R., Moy, B., Feibelmann, S., & Goold, S. D. (2007). Institutional academic-industry relationships. *Journal of the American Medical Association, 298*(15), 1779–1786. DOI: 10.1001/jama.298.15.1779 PMID: 17940234

Campbell, E. G., Weissman, J. S., Vogeli, C., Clarridge, B., Abraham, M., Marder, J. E., & Koski, G. (2007). Financial relationships between institutional review board members and industry. *The New England Journal of Medicine, 356*(18), 1811–1820. DOI: 10.1056/NEJMsa066498 PMID: 17135585

Capili, B., & Anastasi, J. K. (2024). Ethical research and the institutional review board: An introduction. *The American Journal of Nursing, 124*(3), 50–54. DOI: 10.1097/01.NAJ.0001008420.28033.e8 PMID: 38386835

Carberry, E. J., Engelen, P.-J., & Van Essen, M. (2018). Which firms get punished for unethical behavior? Explaining variation in stock market reactions to corporate misconduct. *Business Ethics Quarterly, 28*(2), 119–151. DOI: 10.1017/beq.2017.46

Carboni, J. L., Dickey, T., Moulton, S., O'keefe, S., O'leary, R., Piotrowski, S. J., & Sandfort, J. (2019). Start with the problem: Establishing research relevance with integrative public administration. *Perspectives on Public Management and Governance, 2*(4), 267–274. DOI: 10.1093/ppmgov/gvz021

Cavaliere, P., De Souza, D., Fenton, A. L., Giridharan, B., Gralla, C., Inshakova, N., & Zaharuk, G. (2020). *Academic misconduct and plagiarism: Case studies from universities around the world.* Rowman & Littlefield.

Centers for Disease Control and Prevention. (2024). About The Untreated Syphilis Study at Tuskegee. CDC. https://www.cdc.gov/tuskegee/about/index.html

Chen, H. L., & Wu, C. T. (2023). A digital role-playing game for learning: Effects on critical thinking and motivation. *Interactive Learning Environments, 31*(5), 3018–3030. Advance online publication. DOI: 10.1080/10494820.2021.1916765

Chesser, S., Porter, M. M., & Tuckett, A. G. (2020). Cultivating citizen science for all: Ethical considerations for research projects involving diverse and marginalized populations. [REMOVED HYPERLINK FIELD]. *International Journal of Social Research Methodology, 23*(5), 497–508. DOI: 10.1080/13645579.2019.1704355

Chou, C., Lee, I. J., & Fudano, J. (2024). The present situation of and challenges in research ethics and integrity promotion: Experiences in East Asia. [REMOVED HYPERLINK FIELD]. *Accountability in Research, 31*(6), 576–599. Advance online publication. DOI: 10.1080/08989621.2022.2155144 PMID: 36641764

Christensen, T., & Lægreid, P. (2025). Future directions of public administration research—Addressing fundamental issues and questions. *International Journal of Public Administration, 48*(5-6), 299–305. DOI: 10.1080/01900692.2025.2470006

Clark, J. M. (2017). The 3Rs in research: A contemporary approach to replacement, reduction and refinement. *British Journal of Nutrition, 120*(s1), S1–S7. DOI: 10.1017/S0007114517002227 PMID: 29081302

Code, N. (1947). Permissible medical experiments. In *Trials of war criminals before the Nuremberg Military Tribunals under Control Council Law No. 10* (Vol. 2, pp. 181–182). U.S. Government Printing Office.

Code, T. N. (1947). *Trials of War Criminals before the Nuremberg Military Tribunals under Control Council Law No. 10*. U.S. Government Printing Office.

Cohen, J. R., Pant, L. W., & Sharp, D. J. (1996). A methodological note on cross-cultural accounting ethics research. *The International Journal of Accounting, 31*(1), 55–66. DOI: 10.1016/S0020-7063(96)90013-8

Cohen-Mansfield, J. (2021). The rights of persons with dementia and their meanings. *Journal of the American Medical Directors Association, 22*(7), 1381–1385. Advance online publication. DOI: 10.1016/j.jamda.2021.03.007 PMID: 33844989

Collins, C., Dennehy, D., Conboy, K., & Mikalef, P. (2021). Artificial intelligence in information systems research: A systematic literature review and research agenda. *International Journal of Information Management, 60*, 1–13. DOI: 10.1016/j.ijinfomgt.2021.102383

Comas-Forgas, R., & Sureda-Negre, J. (2010). Academic plagiarism: Explanatory factors from students' perspective. *Journal of Academic Ethics, 8*(3), 217–232. DOI: 10.1007/s10805-010-9121-0

COMMISSION, U. G. (2018). *UNIVERSITY GRANTS COMMISSION (PROMOTION OF ACADEMIC INTEGRITY AND PREVENTION OF PLAGIARISM IN HIGHER EDUCATIONAL INSTITUTIONS) REGULATIONS, 2018. 2018*(D).

Connelly, L. M. (2014). Ethical considerations in research studies. *Medsurg Nursing, 23*(1), 54–56. https://go.gale.com/ps/i.do?id=GALE%7CA360608990&sid=googleScholar&v=2.1&it=r&linkaccess=abs&iss PMID: 24707669

Cooper, R. J., Gupta, M., Wilkes, M. S., & Hoffman, J. R. (2006). Conflict of interest disclosure policies and practices in peer-reviewed biomedical journals. *Journal of General Internal Medicine*, *21*(12), 1248–1252. DOI: 10.1111/j.1525-1497.2006.00598.x PMID: 17105524

COPE. (2021). *COPE Guidelines on Conflicts of Interest in Peer Review*. https://publicationethics.org

Coppola L, C. A. (2019). Biobanking in health care: evolution and future directions. *Journal of Translational Medicine*, 172. (2024). *FDA Rules and Regulations.* US Food and Drug Administration.

Copyright Office, U. S.U.S. Copyright Office. (2005). U.S. Copyright Office report on legal protection for databases. *Bulletin of the American Society for Information Science*, *24*(2), 25–29. DOI: 10.1002/bult.82

Corneli, A., Dombeck, C. B., McKenna, K., & Calvert, S. B. (2021). The Single IRB Review Process: Stakeholder Experiences and Recommendations for Guidance by the US Food and Drug Administration. *Ethics & Human Research*, *43*(3), 26. DOI: 10.1002/eahr.500092 PMID: 34019339

Cortina, J. M. (2020). On the whys and hows of quantitative research. *Journal of Business Ethics*, *167*(1), 19–29. https://idp.springer.com/authorize/casa?redirect_uri=https://link.springer.com/article/10.1007/s10551-019-04195-8&casa_token=pFPN_pEjrIEAAAAA:a7jsAzcP8Hm4W6wK6-3f6xraknWh930FRFSHrnaqzlPat7_SpNuyxuBIRici2x7ZxIJ86nCVCjJCbUdRvw. DOI: 10.1007/s10551-019-04195-8

Cosman, D. (2023). The Nuremberg Code on Human Experimentation. A Historical Overview. Romanian Journal of Artistic Creativity. [HTML]

Criado, J. I., & Gil-Garcia, J. R. (2019). Creating public value through smart technologies and strategies: From digital services to artificial intelligence and beyond. *International Journal of Public Sector Management*, *32*(5), 438–450. DOI: 10.1108/IJPSM-07-2019-0178

Criado, J. I., Sandoval-Almazán, R., & Gil-Garcia, J. R. (2025). Artificial intelligence and public administration: Understanding actors, governance, and policy from micro, meso, and macro perspectives. *Public Policy and Administration*, *40*(2), 173–184. DOI: 10.1177/09520767241272921

Crompton, H., & Burke, D. (2024). The Educational Affordances and Challenges of ChatGPT: State of the Field. *TechTrends*, *68*(2), 380–392. DOI: 10.1007/s11528-024-00939-0

Cunningham, J. B., & McCrum-Gardner, E. (2007). Power, effect and sample size using GPower: Practical issues for researchers and members of research ethics committees. *Evidence-Based Midwifery*, *5*(4), 132–137. https://go.gale.com/ps/i.do?p=HRCA&sw=w&issn=14794489&v=2.1&it=r&id=GALE%7CA172050741&sid=googleScholar&linkaccess=abs

Daar, A. S., & Al Khitamy, A. B. (2001). Bioethics for clinicians: 21. Islamic bioethics. *Canadian Medical Association Journal*, *164*(1), 60–63. PMID: 11202669

Darr, R. (2020). Virtues as qualities of character: Alasdair MacIntyre and the situationist critique of virtue ethics. [REMOVED HYPERLINK FIELD]. *The Journal of Religious Ethics*, *48*(1), 7–25. Advance online publication. DOI: 10.1111/jore.12297

Dastin, J. (2018, October 10). Amazon scraps secret AI recruiting tool that showed bias against women. *Reuters*. https://www.reuters.com/article/us-amazon-com-jobs-automation-insight-idUSKCN1MK08G

Davidson, M. D. (2021). How fairness principles in the climate debate relate to theories of distributive justice. *Sustainability (Basel)*, *13*(13), 7302. Advance online publication. DOI: 10.3390/su13137302

Davies, S. E. (2020). The introduction of research ethics review procedures at a university in South Africa: Review outcomes of a social science research ethics committee. *Research Ethics*, *16*(1–2), 1–26. DOI: 10.1177/1747016119898408

Day, M., Turner, G., & Drozdiak, N. (2019, April 11). Thousands of Amazon workers listen to Alexa users' conversations. *TIME*. https://time.com/5568815/amazon-workers-listen-to-alexa/

De Briyne, N., Vidović, J., Morton, D. B., & Magalhães-Sant'Ana, M. (2020). Evolution of the teaching of animal welfare science, ethics and law in European veterinary schools (2012–2019). *Animals (Basel)*, *10*(7), 1238. DOI: 10.3390/ani10071238 PMID: 32708281

De Sabbata, K. (2020). Dementia, treatment decisions, and the UN convention on the rights of persons with disabilities. A new framework for old problems. [REMOVED HYPERLINK FIELD]. *Frontiers in Psychiatry*, *11*, 571722. Advance online publication. DOI: 10.3389/fpsyt.2020.571722 PMID: 33240127

Declaration of Helsinki. (2013). *Ethical principles for medical research involving human subjects*. World Medical Association. (Original work published 1964)

Dempere, J., Modugu, K., Hesham, A., & Ramasamy, L. K. (2023). The impact of ChatGPT on higher education. *Frontiers in Education*, *8*, 1206936. Advance online publication. DOI: 10.3389/feduc.2023.1206936

Dempster, G., Sutherland, G., & Keogh, L. (2022). Scientific research in news media: A case study of misrepresentation, sensationalism, and harmful recommendations. *JCOM, Journal of Science Communication*, *21*(1), A06. DOI: 10.22323/2.21010206

Denisov, E. I., & Ильич, Д. Э. (2019). Robots, artificial intelligence, augmented and virtual reality: Ethical, legal and hygienic issues. *Hygiene and Sanitation*, *98*(1), 5–10. DOI: 10.18821/0016-9900-2019-98-1-5-10

Désiron, J. C., & Petko, D. (2023). Academic dishonesty when doing homework: How digital technologies are put to bad use in secondary schools. *Education and Information Technologies*, *28*(2), 1251–1271. DOI: 10.1007/s10639-022-11225-y PMID: 35912037

Devine, E. G., Pingitore, A. M., Margiotta, K. N., Hadaway, N. A., Reid, K., Peebles, K., & Hyun, J. W. (2021). Frequency of concealment, fabrication and falsification of study data by deceptive subjects. *Contemporary Clinical Trials Communications*, *21*, 100713. DOI: 10.1016/j.conctc.2021.100713 PMID: 33604482

Devlin, M., & Gray, K. (2007). In their own words: A qualitative study of the reasons Australian university students plagiarize. *Higher Education Research & Development*, *26*(2), 181–198. DOI: 10.1080/07294360701310805

Di Feliciantonio, C. (2024). 'I guess I really survived many crises': On the benefits of longitudinal ethnographic research. *Area*, *56*(1), e12886. Advance online publication. DOI: 10.1111/area.12886

Dorn, M. F., Haemapun, H., Jin, Z., Sgandar, T. O., Pillai, D., Sandeep, A., & Yilmaz, A. (2024, September 16). *Ethical considerations in big data analytics*. OxJournal. https://www.oxjournal.org/ethical-considerations-in-big-data-analytics

Drolet, M. J., Rose-Derouin, E., Leblanc, J. C., Ruest, M., & Williams-Jones, B. (2023). Ethical issues in research: Perceptions of researchers, research ethics board members and research ethics experts. *Journal of Academic Ethics*, *21*(2), 269–292. DOI: 10.1007/s10805-022-09455-3 PMID: 35975128

Du Sert, N. P., Ahluwalia, A., Alam, S., Avey, M. T., Baker, M., Browne, W. J., Clark, A., Cuthill, I. C., Dirnagl, U., Emerson, M., Garner, P., Holgate, S. T., Howells, D. W., Hurst, V., Karp, N. A., Lazic, S. E., Lidster, K., MacCallum, C. J., Macleod, M., & Würbel, H. (2020). Reporting animal research: Explanation and elaboration for the ARRIVE guidelines 2.0. *PLoS Biology*, *18*(7), e3000411. DOI: 10.1371/journal.pbio.3000411 PMID: 32663221

Eacersall, D., Pretorius, L., Smirnov, I., Spray, E., Illingworth, S., Chugh, R., . . . Howell, K. A. (2024). Navigating ethical challenges in generative AI-enhanced research: The ethical framework for responsible generative AI use. *arXiv preprint arXiv:2501.09021*.

Eaton, T. V., Hazel, M., Pyzoha, J. S., & Zhang, J. (2023). How do the accounting profession and higher education currently approach ethics training? *Issues in Accounting Education*, **•••**, 1–18. DOI: 10.2308/ISSUES-2023-104

Edmondson, A. (2023). Right Kind of Wrong: Why Learning to Fail Can Teach Us to Thrive. [HTML]

Edwards, J. R. (2020). The peaceful coexistence of ethics and quantitative research. *Journal of Business Ethics*, *167*(1), 31–40. https://www.jstor.org/stable/45386639. DOI: 10.1007/s10551-019-04197-6

Eilifsen, A., Messier, W. F., Glover, S. M., & Prawitt, D. F. (2014). *Auditing and assurance services: A systematic approach* (10th ed.). McGraw-Hill Education.

El Mestari, S. Z., Lenzini, G., & Demirci, H. (2024). Preserving data privacy in machine learning systems. *Computers & Security*, *137*, 1–10. DOI: 10.1016/j.cose.2023.103605

Elaad, E. (2022). Tunnel vision and confirmation bias among police investigators and laypeople in hypothetical criminal contexts. *SAGE Open*, *12*(2), 21582440221095022. Advance online publication. DOI: 10.1177/21582440221095022

Ellapen, T. J., & Paul, Y. (2016). Innovative sport technology through cross-disciplinary research: Future of sport science. *S.A. Journal for Research in Sport Physical Education and Recreation*, *38*(3), 51–59. https://www.ajol.info/index.php/sajrs/article/view/149803

Emanuel, E. J., Wendler, D., & Grady, C. (2000). What makes clinical research ethical? *Journal of the American Medical Association*, *283*(20), 2701–2711. DOI: 10.1001/jama.283.20.2701 PMID: 10819955

Engel, C. J. (2016). Ethical standards of scientific research for beginning accounting scholars. *Business and Economic Research*, *6*(1), 129–136. RePEc:mth:ber888:v:6:y:2016:i:1:p:129-136

Ercegovac, Z., & Richardson, J. V. (2004). Academic dishonesty, plagiarism included, in the digital age: A literature review. *College & Research Libraries*, *65*(4), 301–318. DOI: 10.5860/crl.65.4.301

Eshet, Y. (2025). Examining the dynamics of plagiarism: A comparative analysis before, during, and after the COVID-19 pandemic. *International Journal for Educational Integrity, 21*(1), 3. DOI: 10.1007/s40979-024-00178-z

Ess, C. M. (2020). Internet research ethics and social media. Handbook of research ethics and scientific integrity. https://doi.org/ [REMOVED HYPERLINK FIELD] DOI: 10.1007/978-3-030-16759-2_12

Etzel, E. F., & Skvarla, L. A. (2017). Ethical considerations in sport and performance psychology. In *Oxford Research Encyclopedia of Psychology*. https://oxfordre.com/psychology/display/10.1093/acrefore/9780190236557.001.0001/acrefore-9780190236557-e-141

European Commission. (2021). *Proposal for a regulation laying down harmonised rules on artificial intelligence (Artificial Intelligence Act).* https://eur-lex.europa.eu/legal-content/EN/TXT/?uri=CELEX%3A52021PC0206

European Commission. (2025). *AI Act*. EC. https://digital-strategy.ec.europa.eu/en/policies/regulatory-framework-ai

European Union. (2016). Regulation (EU) 2016/679 of the European Parliament and of the Council of 27 April 2016 on the protection of natural persons with regard to the processing of personal data and on the free movement of such data (General Data Protection Regulation). *Official Journal of the European Union, L119*, 1–88. https://eur-lex.europa.eu/eli/reg/2016/679/oj

Fadel, H. E. (2024). Ethics of clinical research: An Islamic perspective. *Journal of the Islamic Medical Association of North America, 56*(2), 45–53.

Fadel, M. (2019). The Islamic ethical framework for religious minority rights in research contexts. *The Journal of Religious Ethics, 47*(3), 430–458.

Faden, R. R., & Beauchamp, T. L. (1986). *A history and theory of informed consent*. Oxford University Press.

Faden, R. R., Klag, M. J., Kass, N. E., & Krag, S. S. (2002). On the Importance of Research Ethics and Mentoring. *The American Journal of Bioethics*, 2(4), 50–51. Advance online publication. DOI: 10.1162/152651602320957565 PMID: 12762925

Fagan, M. (2024). *AI for the people: Use cases for government*. Harvard Kennedy School. https://www.hks.harvard.edu/sites/default/files/centers/mrcbg/working.papers/M-RCBG%20Working%20Paper%202024-02_AI%20for%20the%20People.pdf

Fanelli, D. (2009). How many scientists fabricate and falsify research? A systematic review and meta-analysis of survey data. *PLoS One*, *4*(5), e5738. DOI: 10.1371/journal.pone.0005738 PMID: 19478950

Fardid, G., Akbari, A., & Ghafoori Nezhad, M. (2023). The Relationship between Dewey's Moral Contextualism and Kant's Deontological Ethics in Abortion. *International Journal of Ethics and Society*, *5*(1), 20–26. DOI: 10.52547/ijethics.5.1.4

Favaretto, M., De Clercq, E., Gaab, J., & Elger, B. S. (2020). First do no harm: An exploration of researchers' ethics of conduct in Big Data behavioral studies. *PLoS One*, *15*(11), e0241865. Advance online publication. DOI: 10.1371/journal.pone.0241865 PMID: 33152039

Favaretto, M., Shaw, D., De Clercq, E., Joda, T., & Elger, B. S. (2020). Big data and digitalization in dentistry: A systematic review of the ethical issues. *International Journal of Environmental Research and Public Health*, *17*(7), 2495. DOI: 10.3390/ijerph17072495 PMID: 32268509

Favela, L. H., & Amon, M. J. (2023). The ethics of human digital twins: Counterfeit people, personhood, and the right to privacy. *2023 IEEE 3rd International Conference on Digital Twins and Parallel Intelligence, DTPI 2023*. DOI: 10.1109/DTPI59677.2023.10365409

Felzmann, H., Sixsmith, J., O'Higgins, S., Chonnachtaigh, S. N., & Gabhainn, S. N. (2014). Ethical Review and Children's Research in Ireland. Centre of Bioethical Research and analysis (www.nuigalway.ie/cobra) and health Promotion Research Centre, National University of Ireland, Galway (www.nuigalway.ie/hprc)

Ferretti, A., Ienca, M., Hurst, S., & Vayena, E. (2020). Big data, biomedical research, and ethics review: New challenges for IRBs. [wiley.com]. *Ethics & Human Research*, *42*(5), 17–28. DOI: 10.1002/eahr.500065 PMID: 32937036

Floridi, L., & Chiriatti, M. (2020). GPT-3: Its nature, scope, limits, and consequences. *Minds and Machines*, *30*(4), 681–694. DOI: 10.1007/s11023-020-09548-1

Floridi, L., & Cowls, J. (2019). A unified framework of five principles for AI in society. *Harvard Data Science Review*, *1*(1). Advance online publication. DOI: 10.1162/99608f92.8cd550d1

Flynn, S. V. (2021). Research design for the behavioral sciences: An applied approach. [HTML]

Forbes Technology Council. (2024, May 24). How AI can unlock the power of unstructured data. *Forbes*. https://www.forbes.com/councils/forbestechcouncil/2024/05/24/how-ai-can-unlock-the-power-of-unstructured-data/

Fowler, D. S. (2023). Ethical implications of deceptive earnings management practices. *Business Ethics and Leadership*, *7*(3), 84–96. DOI: 10.61093/bel.7(3).84-96.2023

Galvão, A., Marques, C., & Ferreira, J. J. (2020). The role of entrepreneurship education and training programmes in advancing entrepreneurial skills and new ventures. [REMOVED HYPERLINK FIELD]. *European Journal of Training and Development*, *44*(6/7), 595–614. DOI: 10.1108/EJTD-10-2019-0174

Galvez, R. (2024, January 16). Data governance trends for 2024. *Precisely*. https://www.precisely.com/datagovernance/data-governance-trends-for-2024/

Gao, C. A., Howard, F. M., Markov, N. S., Dyer, E. C., Ramesh, S., Luo, Y., & Pearson, A. T. (2023). Comparing scientific abstracts generated by ChatGPT to real abstracts with detectors and blinded human reviewers. *NPJ Digital Medicine*, *6*(1), 75. Advance online publication. DOI: 10.1038/s41746-023-00819-6 PMID: 37100871

Gao, H., Kou, G., Liang, H., Zhang, H., Chao, X., Li, C.-C., & Dong, Y. (2024). Machine learning in business and finance: A literature review and research opportunities. *Financial Innovation*, *10*(1), 86. DOI: 10.1186/s40854-024-00629-z

Gao, X., Shen, H., & Duan, J. H. (2023). A Study on ChatGPT to Alleviate the Difficulties in Writing Scientific Technology Papers. *2023 International Conference on Cognitive Computing and Complex Data, ICCD 2023*, 118–122. DOI: 10.1109/ICCD59681.2023.10420820

Garg, R. K., Urs, V. L., Agarwal, A. A., Chaudhary, S. K., Paliwal, V., & Kar, S. K. (2023). Exploring the role of ChatGPT in patient care (diagnosis and treatment) and medical research: A systematic review. *Health Promotion Perspectives*, *13*(3), 183–191. DOI: 10.34172/hpp.2023.22 PMID: 37808939

Gelman, M. (2020). Kantian ethics—duty, reason, and the grounding of moral law. *Essays in the Philosophy of Humanism, 32*, Article 7. https://americanhumanist.org/what-we-do/publications/eph/journals/volume32/gelman/

Ghaly, M. (2016). Islamic bioethics in the twenty-first century. *Zygon*, *51*(3), 732–751.

Ghaly, M. (2021). Islamic perspectives on genomics and global bioethics. *Developing World Bioethics*, *21*(3), 105–113.

Ghosh, A. (2025). Navigating ethical dilemmas at the workplace: A study of Indian professional management accountants in business. *Journal of Human Values*, *0*(0), 09716858251352091. Advance online publication. DOI: 10.1177/09716858251352091

Glaser, J., Nouri, S., Fernandez, A., Sudore, R. L., Schillinger, D., Klein-Fedyshin, M., & Schenker, Y. (2020). Interventions to improve patient comprehension in informed consent for medical and surgical procedures: An updated systematic review. *Medical Decision Making, 40*(2), 119–143. https://doi.dox.org/10.1177/0272989X19896348. DOI: 10.1177/0272989X19896348 PMID: 31948345

Gong, J. (2017). Ethics in accounting: A decision-making approach. *Journal of Business Ethics, 142*(3), 621–623. DOI: 10.1007/s10551-016-3386-1

Goode, E. (1996). The ethics of deception in social research: A case study. *Qualitative Sociology, 19*(1), 11–33. DOI: 10.1007/BF02393246

Gostin, L. O. (1991). Ethical principles for the conduct of human subject research: Population-Based Research and Ethics. https://scholarship.law.georgetown.edu/facpub/764

Goyal, N., Chang, M., & Terry, M. (2024). Designing for Human-Agent Alignment: Understanding what humans want from their agents. *Conference on Human Factors in Computing Systems - Proceedings*. DOI: 10.1145/3613905.3650948

Grady, C. (2015). Institutional Review Boards: Purpose and Challenges. *Chest, 148*(5), 1148–1155. DOI: 10.1378/chest.15-0706 PMID: 26042632

Grigis, A., Beretta, G., Borry, P., & Sanchini, V. (2025). Vulnerability in research ethics: A systematic review of policy guidelines and documents. *PLoS One, 20*(7), e0327086. DOI: 10.1371/journal.pone.0327086 PMID: 40591537

Groves, M., & Mundt, K. (2021). A ghostwriter in the machine? Attitudes of academic staff towards machine translation use in internationalised Higher Education. *Journal of English for Academic Purposes, 50*, 100957. Advance online publication. DOI: 10.1016/j.jeap.2021.100957

Grzybowsk, A., Pawlikowska, K., & Lambert, C. (2024). A history of artificial intelligence. *Clinics in Dermatology, 1*(1), 1–28. DOI: 10.1016/j.clindermatol.2023.12.016

Guleria, A., Krishan, K., Sharma, V., & Kanchan, T. (2023). ChatGPT: Ethical concerns and challenges in academics and research. *Journal of Infection in Developing Countries, 17*(9), 1292–1299. DOI: 10.3855/jidc.18738 PMID: 37824352

Gullifer, J., & Tyson, G. A. (2010). Exploring university students' perceptions of plagiarism: A focus group study. *Studies in Higher Education, 35*(4), 463–481. DOI: 10.1080/03075070903096508

Gupta, M. (2025). Cross-cultural influences in international accounting: A comparative study. *International Journal of Academic Excellence and Research, 01*(01), 14–20. DOI: 10.62823/MGM/IJAER/01.01.37

Gupta, U. (2013). Informed consent in clinical research: Revisiting few concepts and areas. *Perspectives in Clinical Research, 4*(1), 26. DOI: 10.4103/2229-3485.106373 PMID: 23533976

Guragai, B., Hunt, N. C., Neri, M. P., & Taylor, E. Z. (2017). Accounting information systems and ethics research: Review, synthesis, and the future. *Journal of Information Systems, 31*(2), 65–81. DOI: 10.2308/isys-51265

Haines, D. (2017). Ethical considerations in qualitative case study research recruiting participants with profound intellectual disabilities. *Research Ethics, 13*(3–4), 219–232. DOI: 10.1177/1747016117711971

Harrasi, A. L. (2023). Developing a needs-based plagiarism management in second-language writing in a higher education institute: Practice-oriented research. *Instructional Science, 51*(6), 1079–1115. DOI: 10.1007/s11251-023-09628-6

Harrington, M. E. (2020). Research Ethics. In *The Design of Experiments in Neuroscience* (pp. 13–29). Cambridge University Press. DOI: 10.1017/9781108592468.003

Harriss, D. J., & Atkinson, G. (2009). International Journal of Sports Medicine – Ethical Standards in Sport and Exercise Science Research. *International Journal of Sports Medicine, 30*(10), 701–702. DOI: 10.1055/s-0029-1237378 PMID: 19809942

Harriss, D. J., Jones, C., & MacSween, A. (2022). Ethical Standards in Sport and Exercise Science Research: 2022 Update. *International Journal of Sports Medicine, 43*(13), 1065–1070. DOI: 10.1055/a-1957-2356 PMID: 36495253

Hassan, J., Shehzad, D., Habib, U., Aftab, M. U., Ahmad, M., Kuleev, R., & Mazzara, M. (2022). [Retracted] The Rise of Cloud Computing: Data Protection, Privacy, and Open Research Challenges—A Systematic Literature Review (SLR). *Computational Intelligence and Neuroscience, 2022*(1), 8303504. DOI: 10.1155/2023/9838129 PMID: 35712069

Hassan, S., Pandey, S., & Pandey, S. K. (2021). Should managers provide general or specific ethical guidelines to employees: Insights from a mixed methods study. *Journal of Business Ethics, 172*(3), 563–580. DOI: 10.1007/s10551-020-04442-3

Haven, T., Gopalakrishna, G., Tijdink, J., van der Schot, D., & Bouter, L. (2022). Promoting trust in research and researchers: How open science and research integrity are intertwined. *BMC Research Notes, 15*(1), 302. DOI: 10.1186/s13104-022-06169-y PMID: 36127719

Hayat, I. (2024). Islamic ethical perspectives on artificial intelligence and digital surveillance: Principles, challenges, and policy recommendations. *International Research Journal of Arts. Humanities and Social Sciences*, *2*(1), 12–24.

Head, G. (2020). Ethics in educational research: Review boards, ethical issues and researcher development. *European Educational Research Journal*, *19*(1), 72–83. Advance online publication. DOI: 10.1177/1474904118796315

Hébert, J. R., Satariano, W. A., Friedman, D. B., Armstead, C. A., Greiner, A., Felder, T. M., Coggins, T. A., Tanjasiri, S., & Braun, K. L. (2015). Fulfilling Ethical Responsibility: Moving Beyond the Minimal Standards of Protecting Human Subjects from Research Harm. *Progress in Community Health Partnerships, 9*(S), 41–50. https://doi.org/DOI: 10.1353/cpr.2015.0021

Heilferty, C. M. (2011). Ethical considerations in the study of online illness narratives: A qualitative review. *Journal of Advanced Nursing*, *67*(5), 945–953. DOI: 10.1111/j.1365-2648.2010.05563.x PMID: 21231955

Henrickson, M., Giwa, S., Hafford-Letchfield, T., Cocker, C., Mulé, N. J., Schaub, J., & Baril, A. (2020). Research ethics with gender and sexually diverse persons. *International Journal of Environmental Research and Public Health*, *17*(18), 6615. DOI: 10.3390/ijerph17186615 PMID: 32932832

Herreid, C. F. (1994). Case studies in science-A novel method of science education. *Journal of College Science Teaching*, *23*, 221–221.

Herrera, S. G., Porter, L., & Barko-Alva, K. (2020). Equity in school–parent partnerships: Cultivating community and family trust in culturally diverse classrooms. [HTML]

Hicks, D., Wouters, P., Waltman, L., de Rijcke, S., & Rafols, I. (2015). The Leiden Manifesto for research metrics. *Nature*, *520*(7548), 429–431. DOI: 10.1038/520429a PMID: 25903611

Hicks, J. (2023). The future of data ownership: An uncommon research agenda. *The Sociological Review*, *71*(3), 544–560. DOI: 10.1177/00380261221088120

High-Level Expert Group on Artificial Intelligence. (2019). *A definition of AI: Main capabilities and scientific disciplines*. European Commission. https://ec.europa.eu/newsroom/dae/document.cfm?doc_id=56341

Hilgartner, S., & Brandt-Rauf, S. I. (1994). Data Access, Ownership, and Control: Toward Empirical Studies of Access Practices. *Knowledge (Beverly Hills, Calif.)*, *15*(4), 355–372. DOI: 10.1177/107554709401500401

Hinton, G. E., Osindero, S., & Teh, Y. W. (2006). A fast learning algorithm for deep belief nets. *Neural Computation, 18*(7), 1527–1554. DOI: 10.1162/neco.2006.18.7.1527 PMID: 16764513

Hoekstra, F., Mrklas, K. J., Khan, M., McKay, R. C., Vis-Dunbar, M., Sibley, K. M., Nguyen, T., Graham, I. D., & Gainforth, H. L. (2020). A review of reviews on principles, strategies, outcomes and impacts of research partnerships approaches: A first step in synthesising the research partnership literature. [springer.com]. *Health Research Policy and Systems, 18*(1), 1–23. DOI: 10.1186/s12961-020-0544-9 PMID: 32450919

Holm, S. (2015). Informed Consent: Ethical and Legal Issues. *eLS* .

Ho, M. (2024). Relational tactics and trust in high-risk activism: Anonymity, preexisting ties, and bonding in Hong Kong's 2019–2020 protest. [HTML]. *International Journal of Comparative Sociology, 65*(4), 499–516. DOI: 10.1177/00207152231220524

Horn, L. (2013). Promoting responsible research conduct in a developing world academic context. *South African Journal of Bioethics and Law, 6*(1), 1. https://www.ajol.info/index.php/sajbl/article/view/90037. DOI: 10.7196/sajbl.256

Hosseini, M., Wieczorek, M., & Gordijn, B. (2022). Ethical issues in social science research employing big data. Science and Engineering Ethics. springer.com

Hota, P. K., Bhatt, B., & Qureshi, I. (2023). Institutional work to navigate ethical dilemmas: Evidence from a social enterprise. [HTML]. *Journal of Business Venturing, 38*(1), 106269. DOI: 10.1016/j.jbusvent.2022.106269

Howard, R. M. (1995). Plagiarisms, authorships, and the academic death penalty. *College English, 57*(7), 788–806. DOI: 10.58680/ce19959094

Hung, J., & Chen, J. (2023). The Benefits, Risks and Regulation of Using ChatGPT in Chinese Academia: A Content Analysis. *Social Sciences (Basel, Switzerland), 12*(7), 380. Advance online publication. DOI: 10.3390/socsci12070380

Huriye, A. Z. (2023). The ethics of artificial intelligence: Examining the ethical considerations surrounding the development and use of AI. *American Journal of Technology, 2*(1), 37–44. DOI: 10.58425/ajt.v2i1.142

Husband, G. (2020). Ethical data collection and recognizing the impact of semi-structured interviews on research respondents. Education Sciences. mdpi.com

Hussien, L. F., Alshaketheep, K., Al-Ahmed, H., Shajrawi, A., Alghizzawi, M., Zraqat, O., & Deeb, A. (2025). The impact of big data analytics on the sustainability reports quality. *Discover Sustainability, 6*(1), 776. DOI: 10.1007/s43621-025-01678-9

ICMJE. (2019). *Recommendations for the Conduct, Reporting, Editing, and Publication of Scholarly Work in Medical Journals.* https://www.icmje.org/recommendations/

IEEE Global Initiative on Ethics of Autonomous and Intelligent Systems. (2019). Ethically aligned design: A vision for prioritizing human well-being with autonomous and intelligent systems (Version 2). IEEE. https://standards.ieee.org/wp-content/uploads/import/documents/other/ead_v2.pdf

Ikezuruora, C. (2023). *Unveiling the truth: Why transparency is essential in the age of big data.* PrivacyEnd. https://www.privacyend.com/transparency-essential-age-big-data/

International Ethics Standards Board for Accountants. (2023, July 12). *IESBA emphasizes the critical importance of ethical behavior for all professional accountants.* International Federation of Accountants. https://www.ethicsboard.org/news-events/2023-07/iesba-emphasizes-critical-importance-ethical-behavior-all-professional-accountants

Introna, L., Hayes, N., Blair, L., & Wood, E. (2003). *Cultural attitudes towards plagiarism.* University of Lancaster.

Iqbal, M. (2007). *Science and Islam.* Greenwood Press. DOI: 10.5040/9798216011477

Isaak, J., & Hanna, M. J. (2018). User data privacy: Facebook, Cambridge Analytica, and privacy protection. *Computer, 51*(8), 56–59. DOI: 10.1109/MC.2018.3191268

Isbell, D. R., Brown, D., Chen, M., Derrick, D. J., Ghanem, R., Arvizu, M. N. G., Schnur, E., Zhang, M., & Plonsky, L. (2022). Misconduct and Questionable Research Practices: The Ethics of Quantitative Data Handling and Reporting in Applied Linguistics. *Modern Language Journal, 106*(1), 172–195. DOI: 10.1111/modl.12760

Islam, R., Shahan, A. M., & Rahman, A. (2024). Research ethics in public administration. In O. F. V. Feigenblatt & M, R. Islam. (Eds.), *Ethics in social science research* (pp. 155–166). Springer. DOI: 10.1007/978-981-97-9881-0_12

IV, B. A. (2006). A Summary of Important Documents in the Field of Research Ethics. *Schizophrenia Bulletin,* 69–80. PMID: 16192409

Ivanov, S. (2023). The dark side of artificial intelligence in higher education. *Service Industries Journal, 43*(15–16), 1055–1082. DOI: 10.1080/02642069.2023.2258799

Ivey, G. (2023). Interpreting hidden meaning in qualitative research interview data: Opportunities and challenges. [HTML]. *Qualitative Research in Psychology, 20*(1), 21–51. DOI: 10.1080/14780887.2022.2067509

Jain, A. K., Sahoo, S. R., & Kaubiyal, J. (2021). Online social networks security and privacy: comprehensive review and analysis. Complex & Intelligent Systems. springer.com

Jamnik, A. (2021). Rawls' Theory of Justice as Fairness. *Bogoslovska Smotra*, 1037–1058.

Jamrozik, E., Littler, K., Bull, S., Emerson, C., Kang, G., Kapulu, M., Rey, E., Saenz, C., Shah, S., Smith, P. G., Upshur, R., Weijer, C., & Selgelid, M. J. (2021). Key criteria for the ethical acceptability of COVID-19 human challenge studies: Report of a WHO Working Group. [sciencedirect.com]. *Vaccine*, *39*(4), 633–640. DOI: 10.1016/j.vaccine.2020.10.075 PMID: 33341309

Jejeniwa, T. O., Mhlongo, N. Z., & Jejeniwa, T. O.Temitayo Oluwaseun Jejeniwa-Noluthando Zamanjomane MhlongoTitilola Olaide Jejeniwa. (2024). The role of ethical practices in accounting: A review of corporate governance and compliance trends. *Finance & Accounting Research Journal*, *6*(4), 707–720. DOI: 10.51594/farj.v6i4.1070

Jereb, E., Perc, M., Lämmlein, B., Jerebic, J., Urh, M., Podbregar, I., & Šprajc, P. (2018). Factors influencing plagiarism in higher education: A comparison of German and Slovene students. *PLoS One*, *13*(8), e0202252. DOI: 10.1371/journal.pone.0202252 PMID: 30096189

Jereb, E., Urh, M., Jerebic, J., & Šprajc, P. (2018). Gender differences and the awareness of plagiarism in higher education. *Social Psychology of Education*, *21*(2), 409–426. DOI: 10.1007/s11218-017-9421-y

Jobin, A., Ienca, M., & Vayena, E. (2019). The global landscape of AI ethics guidelines. *Nature Machine Intelligence*, *1*(9), 389–399. DOI: 10.1038/s42256-019-0088-2

Jongsma, K. R. & Bredenoord, A. L. (2020). Ethics parallel research: an approach for (early) ethical guidance of biomedical innovation. BMC Medical Ethics. springer.com

Kalichman, M., & Plemmons, D. (2007). Reported goals for responsible conduct of research courses. *Academic Medicine*, *82*(9), 846–852. DOI: 10.1097/ACM.0b013e31812f78bf PMID: 17726389

Kalkstein, J. A. (2000). Moore v. Regents of the University of California Revisited. *Yale Symposium on Law and Technology*.

Kamali, M. H. (2008). *Principles of Islamic jurisprudence*. Islamic Texts Society.

Kamali, M. H. (2010). *Islamic law in the modern world: Malaysian perspectives*. Islamic Book Trust.

Kang, E., & Hwang, H.-J. (2020). The Consequences of Data Fabrication and Falsification among Researchers. *Journal of Research and Publication Ethics*, *1*(2), 7–10. DOI: 10.15722/jrpe.1.2.202009.7

Kang, E., & Hwang, H.-J. (2023). The importance of anonymity and confidentiality for conducting survey research. *Journal of Research and Publication Ethics*, *4*(1), 1–7. DOI: 10.15722/JRPE.4.1.202303.1

Karimova, G. S., Hoffmann, N. C., Heidbrink, L., & Hoffmann, S. (2020). Virtue ethics between east and west in consumer research: Review, synthesis and directions for future research. [HTML]. *Journal of Business Ethics*, *165*(2), 255–275. DOI: 10.1007/s10551-019-04321-6

Katz, J. (1972). *Experimentation with human beings: The authority of the investigator, subject, professions, and state*. Russell Sage Foundation. DOI: 10.2307/1227842

Keshta, I. & Odeh, A. (2021). Security and privacy of electronic health records: Concerns and challenges. Egyptian Informatics Journal. sciencedirect.com

Ketefian, S. (2015). Ethical considerations in research. Focus on vulnerable groups. *Investigacion y Educacion en Enfermeria*, *33*(1), 164–172. http://www.scielo.org.co/scielo.php?pid=S0120-53072015000100019&script=sci_arttext. DOI: 10.17533/udea.iee.v33n1a19 PMID: 26148168

Khaled, F., & Al-Tamimi, M. S. H. (2021). Plagiarism detection methods and tools: An overview. *Iraqi Journal of Science*, •••, 2771–2783.

Khalifa, M., & Albadawy, M. (2024). Using artificial intelligence in academic writing and research: An essential productivity tool. *Computer Methods and Programs in Biomedicine Update*, *5*, 11–19. DOI: 10.1016/j.cmpbup.2024.100145

Khuroo, M. S., & Sofi, A. A. (2020). The discovery of hepatitis viruses: Agents and disease. *Journal of Clinical and Experimental Hepatology*, *10*(4), 391–401. DOI: 10.1016/j.jceh.2020.04.006 PMID: 32655240

Kiani, A. K., Pheby, D., Henehan, G., Brown, R., Sieving, P., Sykora, P., Marks, R., Falsini, B., Capodicasa, N., Miertus, S., Lorusso, L., Dondossola, D., Tartaglia, G. M., Ergoren, M. C., Dundar, M., Michelini, S., Malacarne, D., Bonetti, G., Dautaj, A., & Bertelli, M. (2022). Ethical considerations regarding animal experimentation. *Journal of Preventive Medicine and Hygiene*, *63*(Suppl. 3), E255–E266. DOI: 10.15167/2421-4248/jpmh2022.63.2S3.2768 PMID: 36479489

Kieslich, K., Keller, B., & Starke, C. (2022). Artificial intelligence ethics by design. Evaluating public perception on the importance of ethical design principles of artificial intelligence. *Big Data & Society*, *9*(1), 1–10. DOI: 10.1177/20539517221092956

Kim, J. H., & Scialli, A. R. (2011). Thalidomide: The tragedy of birth defects and the effective treatment of disease. *Toxicological Sciences*, *122*(1), 1–6. DOI: 10.1093/toxsci/kfr088 PMID: 21507989

King, G. (2021). Towards a culture of care for ethical review: Connections and frictions in institutional and individual practices of social research ethics. *Social & Cultural Geography*, *24*(1), 104–120. DOI: 10.1080/14649365.2021.1939122 PMID: 36712287

Kitamura, F. C. (2023). ChatGPT Is Shaping the Future of Medical Writing But Still Requires Human Judgment. *Radiology*, *307*(2), e230171. Advance online publication. DOI: 10.1148/radiol.230171 PMID: 36728749

Klovdahl, A. S. (2005). Social network research and human subjects' protection: Towards more effective infectious disease control. *Social Networks*, *27*(2), 119–137. DOI: 10.1016/j.socnet.2005.01.006

Klykken, F. H. (2022). Implementing continuous consent in qualitative research. Qualitative Research. sagepub.com

Knöpfle, P., Haim, M., & Breuer, J. (2024). Key topic or bare necessity? How Research Ethics are Addressed and Discussed in Computational Communication Science. Publizistik. springer.com

Komić, D., Marušić, S. L., & Marušić, A. (2015). Research Integrity and Research Ethics in Professional Codes of Ethics: Survey of Terminology Used by Professional Organizations across Research Disciplines. *PLoS One*, *10*(7), e0133662. DOI: 10.1371/journal.pone.0133662 PMID: 26192805

Kotovich, J., & Oriol, M. (2023). Is ChatGPT 3 Safe for Students? In C. A., E. S., B. J., & M. B. (Eds.), *Lecture Notes in Computer Science (including subseries Lecture Notes in Artificial Intelligence and Lecture Notes in Bioinformatics)* (Vol. 14387, pp. 100–107). Springer Science and Business Media Deutschland GmbH. DOI: 10.1007/978-3-031-48639-5_8

Kranak, J. (2019). Kantian Deontology. In *Introduction to Philosophy: Ethics*. Rebus Press.

Kraus, K., Mikes, A., & Véliz, C. (2024). Bringing morality back in: Accounting as moral interlocutor in reflective equilibrium processes. *Accounting, Organizations and Society*, *113*, 101570. Advance online publication. DOI: 10.1016/j.aos.2024.101570

Krimsky, S. (2003). *Science in the Private Interest: Has the Lure of Profits Corrupted Biomedical Research?* Rowman & Littlefield.

Krimsky, S., & Rothenberg, L. S. (2001). Conflict of interest policies in science and medical journals: Editorial practices and author disclosures. *Science and Engineering Ethics*, *7*(2), 205–218. DOI: 10.1007/s11948-001-0041-7 PMID: 11349360

Kue, J., Szalacha, L. A., Happ, M. B., Crisp, A. L., & Menon, U. (2017). Culturally Relevant Human Subjects Protection Training: A case study in Community-Engaged Research in the United States. *Journal of Immigrant and Minority Health*, *20*(1), 107–114. DOI: 10.1007/s10903-017-0548-x PMID: 28239756

Kuziemski, M., & Misuraca, G. (2020). AI governance in the public sector: Three tales from the frontiers of automated decision-making in democratic settings. *Telecommunications Policy*, *44*(6), 1–10. DOI: 10.1016/j.telpol.2020.101976 PMID: 32313360

Kuzlu, M., Xiao, Z., Sarp, S., Catak, F. O., Gurler, N., & Guler, O. (2023). The Rise of Generative Artificial Intelligence in Healthcare. *12th Mediterranean Conference on Embedded Computing, MECO 2023*. DOI: 10.1109/MECO58584.2023.10155107

Kyngäs, H., Kääriäinen, M., & Elo, S. (2020). The trustworthiness of content analysis. The application of content analysis in nursing science research, 41-48. [HTML]

Labrecque, L. I., Markos, E., Swani, K., & Peña, P. (2021). When data security goes wrong: Examining the impact of stress, social contract violation, and data type on consumer coping responses following a data breach. [HTML]. *Journal of Business Research*, *135*, 559–571. DOI: 10.1016/j.jbusres.2021.06.054

Laird, L. D., de Marrais, J., & Barnes, L. L. (2007). Portraying Islam and Muslims in MEDLINE: A content analysis. *Social Science & Medicine*, *65*(12), 2425–2439. DOI: 10.1016/j.socscimed.2007.07.029 PMID: 17767988

Lancaster, T., & Clarke, R. (2016). Contract cheating: The outsourcing of assessed student work. Handbook of academic integrity, 639-654.

Lane, S. (2024, October 14). *The new era of ethical challenges for professional accountants*. ACCA Professional Insights. https://www.accaglobal.com/us/en/professional-insights/global-profession/ethical-challenges.html

Lange, M. M., Rogers, W., & Dodds, S. (2013). Vulnerability in Research Ethics: A Way Forward. *Bioethics*, *27*(6), 333–340. DOI: 10.1111/bioe.12032 PMID: 23718774

Largent, E. A. (2016). Recently proposed changes to legal and ethical guidelines governing human subjects research. *Journal of Law and the Biosciences*, *3*(1), 206–216. DOI: 10.1093/jlb/lsw001 PMID: 27774244

Laryeafio, M. N., & Ogbewe, O. C. (2023). Ethical consideration dilemma: Systematic review of ethics in qualitative data collection through interviews. *Journal of Ethics in Entrepreneurship and Technology, 3*(2), 94–110. https://www.emerald.com/insight/content/doi/10.1108/JEET-09-2022-0014/. DOI: 10.1108/JEET-09-2022-0014

Lategan, L. O. K. (2012). The Building of a Responsible Research Community: The Role of Ethics. *The Journal of Research Administration, 43*(1), 85–97. https://eric.ed.gov/?id=EJ976743

Laurijssen, S. J., van der Graaf, R., van Dijk, W. B., Schuit, E., Groenwold, R. H., Grobbee, D. E., & de Vries, M. C. (2022). When is it impractical to ask informed consent? A systematic review. [sagepub.com]. *Clinical Trials, 19*(5), 545–560. DOI: 10.1177/17407745221103567 PMID: 35775421

Lee, K. (2016). *Artificial intelligence, automation, and the economy*. The White House. https://obamawhitehouse.archives.gov/blog/2016/12/20/artificial-intelligence-automation-and-economy

Lee, S.-H. (2021). Changing Policy Trends in Research Ethics. *Health Policy and Management, 145–147*. Advance online publication. DOI: 10.4332/KJHPA.2021.31.2.145

Lemke, A. A., Esplin, E. D., Goldenberg, A. J., Gonzaga-Jauregui, C., Hanchard, N. A., Harris-Wai, J., Ideozu, J. E., Isasi, R., Landstrom, A. P., Prince, A. E. R., Turbitt, E., Sabatello, M., Schrier Vergano, S. A., Taylor, M. R. G., Yu, J.-H., Brothers, K. B., & Garrison, N. A. (2022). Addressing underrepresentation in genomics research through community engagement. [cell.com]. *American Journal of Human Genetics, 109*(9), 1563–1571. DOI: 10.1016/j.ajhg.2022.08.005 PMID: 36055208

Lens, (2025). *Scholar search: Animal research ethics and welfare standard* https://www.lens.org/lens/search/scholar/list?q=Animal%20research%20ethics,%20welfare%20standards

Lens. (2025). Retrieved, 2025, from https://www.lens.org/

Levine, C., Faden, R., Grady, C., Hammerschmidt, D., Eckenwiler, L., & Sugarman, J. (2004). A response to commentators on "The limitations of 'Vulnerability' as a protection for human research participants.". *The American Journal of Bioethics, 4*(3), W32. DOI: 10.1080/15265160490508954 PMID: 16192138

Levine, R. J. (1979). Clarifying the Concepts of Research Ethics. *The Hastings Center Report, 9*(3), 21–26. DOI: 10.2307/3560793 PMID: 478836

Levine, R. J. (1988). *Ethics and regulation of clinical research* (2nd ed.). Urban & Schwarzenberg.

Li, B., Chen, Q., Lin, J., Li, S., & Yen, J. (2024). Assessing GPT-4 Generated Abstracts: Text Relevance and Detectors Based on Faithfulness, Expressiveness, and Elegance Principle. In T. Y. & S. Y. (Eds.), *Communications in Computer and Information Science: Vol. 2017 CCIS* (pp. 165–180). Springer Science and Business Media Deutschland GmbH. DOI: 10.1007/978-981-97-0837-6_12

Li, M., Enkhtur, A., Cheng, F., & Yamamoto, B. A. (2024). Ethical Implications of ChatGPT in Higher Education: A Scoping Review. *Journal of Interdisciplinary Studies in Education, 13*(1), 55–68. DOI: 10.32674/jise.v13i1.6072

Littlewood, K., & Mellor, D. (2016). Changes in the welfare of an injured working farm dog assessed using the five domains model. *Animals (Basel), 6*(9), 58. DOI: 10.3390/ani6090058 PMID: 27657140

Lo, B., & Field, M. J. (2009). *Conflict of Interest in Medical Research, Education, and Practice*. National Academies Press.

Lo, C. K. (2023). What Is the Impact of ChatGPT on Education? A Rapid Review of the Literature. *Education Sciences, 13*(4), 410. Advance online publication. DOI: 10.3390/educsci13040410

Loland, S. (2009). The Ethics of Performance-Enhancing Technology in Sport. *Journal of the Philosophy of Sport*. https://www.tandfonline.com/doi/abs/10.1080/00948705.2009.9714754

Lowe, K., & Ridgway, A. (2023). The keepers of secrets: Ethics and the emotional labour of working with privileged populations during criminological research. In *Ethical Dilemmas in International Criminological Research* (pp. 113–129). Routledge. [HTML] DOI: 10.4324/9781003241515-9

Luger, T. M., Hamilton, A. B., & True, G. (2020). Measuring community-engaged research contexts, processes, and outcomes: a mapping review. The Milbank Quarterly. wiley.com

Lundh, A., Lexchin, J., Mintzes, B., Schroll, J. B., & Bero, L. (2017). Industry sponsorship and research outcome. *Cochrane Database of Systematic Reviews, 2017*(2). Advance online publication. DOI: 10.1002/14651858.MR000033.pub3 PMID: 28207928

Lynch, H. F., Eriksen, W., & Clapp, J. T. (2022). "We measure what we can measure": Struggles in defining and evaluating institutional review board quality. [HTML]. *Social Science & Medicine, 292*, 114614. DOI: 10.1016/j.socscimed.2021.114614 PMID: 34861569

Lynn, J. (2004). When does quality improvement count as research? Human subject protection and theories of knowledge. *BMJ Quality & Safety*, *13*(1), 67–70. DOI: 10.1136/qshc.2002.002436 PMID: 14757803

MacArthur Clark, J. A., & Sun, D. (2020). Guidelines for the ethical review of laboratory animal welfare People's Republic of China National Standard GB/T 35892-2018 [Issued 6 February 2018 Effective from 1 September 2018] [wiley.com]. *Animal Models and Experimental Medicine*, *3*(1), 103–113. DOI: 10.1002/ame2.12111 PMID: 32318667

Macfarlane, B. (2009). *Researching with Integrity: The Ethics of Academic Enquiry*. Routledge.

Macfarlane, B. (2017). The ethics of multiple authorship: Power, performativity and the gift economy. *Studies in Higher Education*, *42*(7), 1194–1210. DOI: 10.1080/03075079.2015.1085009

Macfarlane, B., Zhang, J., & Pun, A. (2014). Academic integrity: A review of the literature. *Studies in Higher Education*, *39*(2), 339–358. DOI: 10.1080/03075079.2012.709495

Macrina, F. L. (2007). Scientific Societies and Promotion of the Responsible Conduct of Research: Codes, Policies, and Education. *Academic Medicine*, *82*(9), 865–869. DOI: 10.1097/ACM.0b013e31812f7e58 PMID: 17726393

Madak, P. R. (1994). Ethical Considerations When Using Qualitative Methods in Evaluation. *The Canadian Journal of Program Evaluation*, *9*(2), 1–13. DOI: 10.3138/cjpe.9.001

Madani, M., Ghasemzadeh, N., Dizani, A., Gharamaleki, A. F., & Larijani, B. (2020). Policy considerations to achieve practical ethics: Closing the gap between ethical theory and practic. *Journal of Medical Ethics and History of Medicine*, *13*, 8. DOI: 10.18502/jmehm.v13i8.4075 PMID: 33117501

Madan, R., & Ashok, M. (2023). AI adoption and diffusion in public administration: A systematic literature review and future research agenda. *Government Information Quarterly*, *40*(1), 1–13. DOI: 10.1016/j.giq.2022.101774

Maddi, A., & Teixeira da Silva, J. A. (2024). Beyond authorship: Analyzing contributions in PLOS ONE and the challenges of appropriate attribution. *Journal of Data and Information Science*, *9*(3), 88–115. DOI: 10.2478/jdis-2024-0015

Mahabeer, P., & Pirtheepal, T. (2019). Assessment, plagiarism and its effect on academic integrity: Experiences of academics at a university in South Africa. *South African Journal of Science, 115*(11–12). Advance online publication. DOI: 10.17159/sajs.2019/6323

Mansoor, F., Ameen, K., & Arshad, A. (2024). An exploratory study of university librarians' perceptions on causes and deterrents of plagiarism: A Pakistani perspective. *Global Knowledge, Memory and Communication, 73*(4–5), 634–649. DOI: 10.1108/GKMC-04-2022-0074

Manti, S., & Licari, A. (2018). How to obtain informed consent for research. *Breathe (Sheffield, England), 14*(2), 145–152. DOI: 10.1183/20734735.001918 PMID: 29875834

Manton, A., Wolf, L. A., Baker, K. M., Carman, M. J., Clark, P. R., Henderson, D., & Zavotsky, K. E. (2014). Ethical Considerations in Human Subjects research. *Journal of Emergency Nursing: JEN, 40*(1), 92–94. DOI: 10.1016/j.jen.2013.11.001 PMID: 24384483

Mardawi, Z., Seguí-Mas, E., & Tormo-Carbó, G. (2021). Rethinking the accounting ethics education research in the post-COVID-19 context. *Cogent Business & Management, 8*(1), 1984627. Advance online publication. DOI: 10.1080/23311975.2021.1984627

Martínez-Padilla, L., Escribano, F., Tizon-Couto, D., & Gil-Borrelli, C. (2023). The 3Rs principle in animal experimentation: A legal review in Europe, the Americas, and Asia. *Animals (Basel), 13*(1), 1–16. DOI: 10.3390/ani13010077 PMID: 38200732

Martin, K. D., & Zimmermann, J. (2024). Artificial intelligence and its implications for data privacy. *Current Opinion in Psychology, 58*, 1–10. DOI: 10.1016/j.copsyc.2024.101829 PMID: 38954851

Maryon, T., Dubre, V., Elliott, K., Escareno, J., Fagan, M. H., Standridge, E., & Lieneck, C. (2022). COVID-19 Academic Integrity Violations and Trends: A Rapid Review. *Education Sciences, 12*(12), 901. Advance online publication. DOI: 10.3390/educsci12120901

Masic, I. (2014). Plagiarism in Scientific Research and Publications and How to Prevent It. *Materia Socio-Medica, 26*(2), 141. DOI: 10.5455/msm.2014.26.141-146 PMID: 24944543

Mason, G. J. (2023). Animal welfare research is fascinating, ethical, and useful—But how can it be more rigorous? *BMC Biology, 21*(1), 302. Advance online publication. DOI: 10.1186/s12915-023-01793-x PMID: 38155349

Mauthner, N. S., & Parry, O. (2013). Open Access Digital Data Sharing: Principles, Policies and Practices☆. *Social Epistemology*, *27*(1), 47–67. DOI: 10.1080/02691728.2012.760663

McAlpine, J. B., Ferreira, D., Pauli, N. E., Gafner, S., & Pauli, G. F. (2023). The Ethics of Publishing Biomedical and Natural Products Research. *Journal of Natural Products*, *86*(9), 2228–2237. DOI: 10.1021/acs.jnatprod.3c00165 PMID: 37638654

McCandless, S., Bishu, S. G., Gomez Hernandez, M., Paredes Eraso, E., Sabharwal, M., Santis, E. L., & Yates, S. (2022). A long road: Patterns and prospects for social equity, diversity, and inclusion in public administration. *Public Administration*, *100*(1), 129–148. DOI: 10.1111/padm.12830

McCarthy, J., Minsky, M. L., Rochester, N., & Shannon, C. E. (1956). *A proposal for the Dartmouth summer research project and artificial intelligence.* https://home.dartmouth.edu/about/artificial-intelligence-ai-coined-dartmouth

McDonald, B. D.III, Hall, J. L., O'Flynn, J., & Van Thiel, S. (2022). The future of public administration research: An editor's perspective. *Public Administration*, *100*(1), 59–71. DOI: 10.1111/padm.12829

McFee, G. (2014). Ethical considerations. In *Research methods in sports coaching* (pp. 98–108). Routledge., https://www.taylorfrancis.com/chapters/edit/10.4324/9780203797549-12/ethical-considerations-graham-mcfee DOI: 10.4324/9780203797549-10

McGowan, W. (2020). 'If you didn't laugh, you'd cry': Emotional labour, reflexivity and ethics-as-practice in a qualitative fieldwork context. Methodological innovations. sagepub.com

McIntosh, L. D., & Vitale, C. H. (2023). *Safeguarding scientific integrity: Examining conflicts of interest in the peer review process.* ArXiv preprint arXiv:2308.04297. https://doi.org//arXiv.2308.04297DOI: 10.48550

McLeod, S. (2025). Stanford Prison Experiment. *Simply Psychology*. megítélésére, K.-F. S. (2008). Influence of the Nuremberg physicians' trials—beginning a new era in the ethical judging of human experiments. *Orvosi Hetilap*, 195–201.

McNamee, M. J., Olivier, S., & Wainwright, P. (2006). *Research Ethics in Exercise, Health and Sports Sciences*. Routledge., DOI: 10.4324/9780203966853

McNamee, M., Anderson, L. C., Borry, P., Camporesi, S., Derman, W., Holm, S., Knox, T. R., Leuridan, B., Loland, S., Frias, F. J. L., Lorusso, L., Malcolm, D., McArdle, D., Partridge, B., Schramme, T., & Weed, M. (2025). Sport-related concussion research agenda beyond medical science: Culture, ethics, science, policy. *Journal of Medical Ethics*, *51*(1), 68–76. DOI: 10.1136/jme-2022-108812 PMID: 36868564

Med, D. G., Nohl, S., Wolf, C., Rupert, Y., Rimkus, L., Ehlers, J., Breuckmann, F., & Sellmann, T. (2023). A SWOT (Strengths, Weaknesses, Opportunities, and Threats) Analysis of ChatGPT in the Medical Literature: Concise Review. *Journal of Medical Internet Research*, *25*, e49368. Advance online publication. DOI: 10.2196/49368 PMID: 37865883

Mellor, D. (2017). Operational details of the five domains model and its key applications to the assessment and management of animal welfare. *Animals (Basel)*, *7*(8), 60. DOI: 10.3390/ani7080060 PMID: 28792485

Mergel, I., Edelmann, N., & Haug, N. (2019). Defining digital transformation: Results from expert interviews. *Government Information Quarterly*, *36*(4), 1–10. DOI: 10.1016/j.giq.2019.06.002

Mertens, D. M. (2021). Transformative research methods to increase social impact for vulnerable groups and cultural minorities. International Journal of Qualitative Methods. sagepub.com

Mignolo, W. D. (2020). *On decoloniality: Concepts, analytics, praxis*. Duke University Press. DOI: 10.1515/9780822371779

Mikk, K. A., Sleeper, H. A., & Topol, E. J. (2017). The Pathway to Patient Data Ownership and Better Health. *Journal of the American Medical Association*, *318*(15), 1433–1434. DOI: 10.1001/jama.2017.12145 PMID: 28973063

Milanović, L. Ran\djelović, N., Živković, D., & Savić, Z. (2021). Ethics and sports, ethics in sports, sports ethics: Aspects of consideration of different authors. *Physical Education and Sport through the Centuries, 8*(2), 16–32. https://scindeks.ceon.rs/Article.aspx?artid=2335-05982102016M

Milne, C. (2005). Overseeing Research: Ethics and the Institutional Review Board. *Forum Qualitative Sozialforschung / Forum: Qualitative. Social Research*, *6*(1). Advance online publication. DOI: 10.17169/fqs-6.1.531

Mintz, S. M. (2021). *Accounting Ethics Education*. The CPA Journal. https://www.cpajournal.com/2021/10/19/accounting-ethics-education/

Miranda-Rodríguez, R. A., Sánchez-Nieto, J. M., & Ruiz-Rodríguez, A. K. (2024, May). Effectiveness of intervention programs in reducing plagiarism by university students: A systematic review. [). Frontiers Media SA.]. *Frontiers in Education*, *9*, 1357853. DOI: 10.3389/feduc.2024.1357853

Mirchev, M., Mircheva, I., & Kerekovska, A. (2020). The Academic Viewpoint on Patient Data Ownership in the Context of Big Data: Scoping Review. *Journal of Medical Internet Research*, *22*(8), e22214. DOI: 10.2196/22214 PMID: 32808934

Miron, J., Wilson, R., Freeman, J., & Sears, K. (2022). Academic integrity in upper year nursing students' work-integrated settings. *International Journal for Educational Integrity*, *18*(1), 16. Advance online publication. DOI: 10.1007/s40979-022-00107-y

Misra, D. P., Ravindran, V., & Agarwal, V. (2018). Integrity of authorship and peer review practices: Challenges and opportunities for improvement. *Journal of Korean Medical Science*, *33*(46), e287. DOI: 10.3346/jkms.2018.33.e287 PMID: 30416407

Misselbrook, D. (2013). Duty, Kant, and deontology. *The British Journal of General Practice*, *63*(609), 211. DOI: 10.3399/bjgp13X665422 PMID: 23540473

Miteu, G. D. (2024). Ethics in scientific research: A lens into its importance, history, and future. *Annals of Medicine and Surgery (London)*, *86*(5), 2395–2398. DOI: 10.1097/MS9.0000000000001959 PMID: 38694399

Mittelstadt, B. D., Allo, P., Taddeo, M., Wachter, S., & Floridi, L. (2016). The ethics of algorithms: Mapping the debate. *Big Data & Society*, *3*(2), 1–21. DOI: 10.1177/2053951716679679

Mixon, K. (2024). The role of virtue ethics in modern moral dilemmas. *International Journal of Philosophy*, *3*(4), 14–28. https://www.researchgate.net/publication/382246077_The_Role_of_Virtue_Ethics_in_Modern_Moral_Dilemmas#:~:text=Ultimately%2C%20virtue%20ethics%20was%20found,more%20just%20and%20sustainable%20world. DOI: 10.47941/ijp.2094

Mökander, J., & Floridi, L. (2023). Operationalising AI governance through ethics-based auditing: An industry case study. *AI and Ethics*, *3*(3), 451–468. DOI: 10.1007/s43681-022-00171-7 PMID: 35669570

Mökander, J., Morley, J., Taddeo, M., & Floridi, L. (2021). Ethics-based auditing of automated decision-making systems: Nature, scope, and limitations. [springer.com]. *Science and Engineering Ethics*, *27*(4), 44. DOI: 10.1007/s11948-021-00319-4 PMID: 34231029

Molnar, G., & Jozsef, C. (2022). AI Based Plagiarism Checking : Ease of use and appli7cable system for teachers to find similarities in students' assessments. *CANDO-EPE 2022 - Proceedings: IEEE 5th International Conference and Workshop in Obuda on Electrical and Power Engineering*, 187–192. DOI: 10.1109/CANDO-EPE57516.2022.10046379

Moore, Z. E. (2003). Ethical dilemmas in sport psychology: Discussion and recommendations for practice. *Professional Psychology, Research and Practice*, *34*(6), 601–610. https://psycnet.apa.org/fulltext/2003-10053-004.html. DOI: 10.1037/0735-7028.34.6.601

Moriña, A. (2021). When people matter: The ethics of qualitative research in the health and social sciences. [HTML]. *Health & Social Care in the Community*, *29*(5), 1559–1565. DOI: 10.1111/hsc.13221 PMID: 33170531

Morocco-Clarke, A., Sodangi, F. A., & Momodu, F. (2024). The implications and effects of ChatGPT on academic scholarship and authorship: A death knell for original academic publications? *Information & Communications Technology Law*, *33*(1), 21–41. DOI: 10.1080/13600834.2023.2239623

Moya, B. A. (2024). Academic Integrity Policies in Hispanic South American Higher Education: Status and Recommendations. In *Springer International Handbooks of Education: Vol. Part F2304* (pp. 911–933). Springer Nature. DOI: 10.1007/978-3-031-54144-5_122

Moya, B. A., & Eaton, S. E. (2024). Academic Integrity Policy Analysis of Chilean Universities. *Journal of Academic Ethics*, *22*(4), 639–663. Advance online publication. DOI: 10.1007/s10805-024-09515-w

Mphahlele, A., & McKenna, S. (2019). The use of turnitin in the higher education sector: Decoding the myth. *Assessment & Evaluation in Higher Education*, *44*(7), 1079–1089. DOI: 10.1080/02602938.2019.1573971

Mulenga, R., & Shilongo, H. (2024). Academic integrity in higher education: Understanding and addressing plagiarism. *Acta Pedagogia Asiana*, *3*(1), 30–43. DOI: 10.53623/apga.v3i1.337

Munhall, P. L. (1988). Ethical Considerations in Qualitative Research. *Western Journal of Nursing Research*, *10*(2), 150–162. DOI: 10.1177/019394598801000204 PMID: 3394317

Mustafa, Y. (2014). Islam and the four principles of medical ethics. *Journal of Medical Ethics*, *40*(7), 479–483. DOI: 10.1136/medethics-2012-101309 PMID: 23975951

Muthu, S. S. Miguel Angel Gardetti, & Subramanian Senthilkannan Muthu (Eds.). (2016). *Environmental footprints and eco-design of products and processes*. Springer. https://doi.org/DOI: 10.1007/978-981-10-0245-8

Muto, H. (2022). *The Holocaust as an Inflection Point in the Development of Research Ethics*. Springer eBooks.

Nagai, H., Nakazawa, E., & Akabayashi, A. (2022). The creation of the Belmont Report and its effect on ethical principles: a historical study. Monash bioethics review. springer.com

Nagai, H. N. E., Nakazawa, E., & Akabayashi, A. (2022). The creation of the Belmont Report and its effect on ethical principles: A historical study. *Monash Bioethics Review, 40*(2), 157–170. DOI: 10.1007/s40592-022-00165-5 PMID: 36357708

Nannoni, E., & Mancini, C. (2024). Toward an integrated ethical review process: An animal-centered research framework for the refinement of research procedures. *Frontiers in Veterinary Science, 11*, 1343735. Advance online publication. DOI: 10.3389/fvets.2024.1343735 PMID: 38694478

NASEM. (2017). *Fostering Integrity in Research*. National Academies Press., DOI: 10.17226/21896

National Institute of Standards and Technology. (2023). *Artificial Intelligence Risk Management Framework (AI RMF 1.0) (NIST AI 100-1)*. U.S. Department of Commerce., DOI: 10.6028/NIST.AI.100-1

National Institute of Standards and Technology. (2025). *Artificial intelligence*. NIST. https://www.nist.gov/artificial-intelligence

National Research Act, Pub. L. No. 93-348, 88 Stat. 342 (1974).

Nawroth, C., & Krause, E. T. (2022). The academic, societal and animal welfare benefits of open science for animal science. *Frontiers in Veterinary Science, 9*, 810989. Advance online publication. DOI: 10.3389/fvets.2022.810989 PMID: 35498731

Nebeker, C., Torous, J., & Bartlett Ellis, R. J. (2019). Building the case for actionable ethics in digital health research supported by artificial intelligence. *BMC Medicine, 17*(1), 137. DOI: 10.1186/s12916-019-1377-7 PMID: 31311535

Neupane, N. (2023). Ins and outs of plagiarism in academia: Reconnoitring concepts, causes, and consequences. *Janapriya Journal of Interdisciplinary Studies, 12*(1), 95–105. DOI: 10.3126/jjis.v12i1.62244

Newman, P. A., Guta, A., & Black, T. (2021). Ethical considerations for qualitative research methods during the COVID-19 pandemic and other emergency situations: Navigating the virtual field. [sagepub.com]. *International Journal of Qualitative Methods, 20*, 16094069211047823. DOI: 10.1177/16094069211047823

Nickerson, R. S. (1998). Confirmation bias: A ubiquitous phenomenon in many guises. *Review of General Psychology*, *2*(2), 175–220. DOI: 10.1037/1089-2680.2.2.175

Niel, S., Florence, A., John, A., & Berky, B. (2024). Investigating the effectiveness of data visualization in communicating financial information. *Journal of Financial Reporting and Analysis*, *9*(2), 45–62.

Nielsen, B. L., & Rodenburg, T. B. (2025). Ethical approval of studies involving humans and other animals. *Animal Welfare (South Mimms, England)*, *34*, e12. Advance online publication. DOI: 10.1017/awf.2025.5 PMID: 40028438

NIH. (2020). *Financial Conflict of Interest*. https://grants.nih.gov/grants/policy/coi/

Nilsson, N. J. (2009). *The quest for artificial intelligence*. Cambridge University Press. DOI: 10.1017/CBO9780511819346

Noordt, C., & Tangi, L. (2023). The dynamics of AI capability and its influence on the public value of AI within public administration. *Government Information Quarterly*, *40*(4), 1–10. DOI: 10.1016/j.giq.2023.101860

Nørreklit, L., Nørreklit, H., Cinquini, L., & Mitchell, F. (2024). Accounting for a better world: Towards a conceptual framework to enable corporate reporting to contribute to the sustainability of the good life. *Meditari Accountancy Research*, *32*(5), 1608–1640. DOI: 10.1108/MEDAR-05-2023-2015

Nurunnabi, A. S. M. (2014). Institutional Review Board (IRB): Its role and responsibility in making research ethical. *Bangladesh Journal of Bioethics*, *5*(1), 5–10. https://banglajol.info/index.php/BIOETHICS/article/view/18442. DOI: 10.3329/bioethics.v5i1.18442

Nussbaum, M. (1999). Virtue Ethics: The Misleading Category. *The Journal of Ethics*, *3*(3), 163–201. DOI: 10.1023/A:1009877217694

O'Connell, J. (2016). Networked participatory online learning design and challenges for academic integrity in higher education. *International Journal for Educational Integrity*, *12*(1), 4. Advance online publication. DOI: 10.1007/s40979-016-0009-7

O'Neil, C. (2016). *Weapons of math destruction: How big data increases inequality and threatens democracy*. Crown.

O'Sullivan, L. O., Crowley, R., McAuliffe, É., & Doran, P. (2020). Contributory factors to the evolution of the concept and practice of informed consent in clinical research: A narrative review. *Contemporary Clinical Trials*, *19*, 100634. DOI: 10.1016/j.conctc.2020.100634 PMID: 33024880

Obermeyer, Z., Powers, B., Vogeli, C., & Mullainathan, S. (2019). Dissecting racial bias in an algorithm used to manage the health of populations. *Science*, *366*(6464), 447–453. DOI: 10.1126/science.aax2342 PMID: 31649194

Office for Human Research Protections. (2018). *Federal policy for the protection of human subjects ("Common Rule")*. U.S. Department of Health and Human Services.

Okougbo, P. O., Okike, E. N., & Alao, A. (2021). Accounting ethics education and the ethical awareness of undergraduates: An experimental study. *Accounting Education*, *30*(3), 258–276. DOI: 10.1080/09639284.2021.1888135

Oliver, J. (2010). Ethical practice in sport psychology: Challenges in the real world. In *Routledge handbook of applied sport psychology* (pp. 60–68). Routledge., https://www.taylorfrancis.com/chapters/edit/10.4324/9780203851043-8/ethical-practice-sport-psychology-jonah-oliver

Organisation for Economic Co-operation and Development. (2019). *OECD principles on AI*. https://www.oecd.org/going-digital/ai/principles/

Organization for Economic Co-operation and Development. (2019). *AI principles*. OECD. https://www.oecd.org/en/topics/sub-issues/ai-principles.html

Otley, D. T., & Berry, A. J. (1994). Case study research in management accounting and control. *Management Accounting Research*, *5*(1), 45–65. DOI: 10.1006/mare.1994.1004

Owusu, G. M. Y., & Korankye, G. (2023). The state of ethical decision-making research in accounting: A retrospective assessment from 1987 to 2022. *Business Ethics, the Environment & Responsibility*, *32*(2), 419–434. DOI: 10.1111/beer.12519

Pabel, A. S., & Akther, W. (2025). Navigating the ethical landscape: A systematic review of challenges in AI adoption within the accounting profession. *Multidisciplinary Reviews*, *8*(11), 2025357. DOI: 10.31893/multirev.2025357

Padela, A. I. (2013). Islamic bioethics: Between sacred law, lived experiences, and state authority. *Theoretical Medicine and Bioethics*, *34*(2), 65–80. DOI: 10.1007/s11017-013-9249-1 PMID: 23588855

Padela, A. I. (2018). Islamic bioethics and animal research: Principles, attitudes, and practices. *Cambridge Quarterly of Healthcare Ethics*, *27*(1), 121–132.

Padela, A. I., Malik, A. Y., Curlin, F., & De Vries, R. (2014). [Re]considering respect for persons in a globalizing world. *Developing World Bioethics*, *15*(2), 98–106. DOI: 10.1111/dewb.12045 PMID: 24720355

Panter, A. T. (2011). *Handbook of ethics in quantitative methodology*. Taylor & Francis., https://books.google.com/books?hl=en&lr=&id=DKgYxsYQDqsC&oi=fnd&pg=PP1&dq=ethics+in+quantitative+research+&ots=UNEt8NsWNC&sig=enO34T-SNnNmf4HHjOqK4__Iox4 DOI: 10.4324/9780203840023

Parija, S., & Mandal, J. (2014). Informed consent and research. *Tropical Parasitology*, *4*(2), 78. DOI: 10.4103/2229-5070.138533 PMID: 25250226

Park University. (2024, September 23). *Ethical considerations in accounting: Navigating the gray areas*. https://www.park.edu/blog/ethical-considerations-in-accounting/

Park, C. (2017). In other (people's) words: Plagiarism by university students-literature and lessons. In *Academic Ethics* (pp. 525–542). Taylor and Francis., DOI: 10.4324/9781315263465-42

Parker, M. (2023). *Burnout in Information Security: The Case of Healthcare*. Advanced Health Technology. [HTML]

Parks-Leduc, L., Mulligan, L., & Rutherford, M. A. (2021). Can ethics be taught? Examining the impact of distributed ethical training and individual characteristics on ethical decision-making. *Academy of Management Learning & Education*, *20*(1), 30–49. DOI: 10.5465/amle.2018.0157

Parry, O., & Mauthner, N. S. (2004). Whose data are they anyway? Practical, legal and ethical issues in archiving qualitative research data. *Sociology*, *38*(1), 139–152. DOI: 10.1177/0038038504039366

Patel, A., Bakhtiyari, K., & Taghavi, M. (2011). Evaluation of cheating detection methods in academic writings. *Library Hi Tech*, *29*(4), 623–640. DOI: 10.1108/07378831111189732

Patwardhan, B., Desai, A., Chourasia, A., Nag, S., & Bhatnagar, R. (2020). Good Academic Research Practices. *University Grants Commission, September*, 56. https://www.ugc.ac.in/e-book/UGC_GARP_2020_Good Academic Research Practices.pdf

Paunova, M. (2023). Principlism in Practice: Ethics in Survey Research. The Sage Handbook of Survey Development and Application. London ua: Sage, 54-80. [HTML]

Pecorari, D. (2003). Good and original: Plagiarism and patchwriting in academic second-language writing. *Journal of Second Language Writing*, *12*(4), 317–345. DOI: 10.1016/j.jslw.2003.08.004

Pennock, R. T. (2018). Beyond research ethics: How scientific virtue theory reframes and extends responsible conduct of research. In *Cultivating Moral Character and Virtue in Professional Practice*. Routledge. DOI: 10.4324/9781315182414-13

Pennycook, A. (1996). Borrowing others' words: Text, ownership, memory, and plagiarism. *TESOL Quarterly*, *30*(2), 201–230. DOI: 10.2307/3588141

Perakslis, E. D. (2018). Using digital health to enable ethical health research in conflict and other humanitarian settings. *Conflict and Health*, *12*(1), 23. DOI: 10.1186/s13031-018-0163-z PMID: 29785204

Perkins, M., & Roe, J. (2024). Academic publisher guidelines on AI usage: A ChatGPT supported thematic analysis. *F1000Research, 12*. DOI: 10.12688/f1000research.142411.2

Peterson, A., Karlawish, J., & Largent, E. (2021). Supported decision making with people at the margins of autonomy. [tandfonline.com]. *The American Journal of Bioethics*, *21*(11), 4–18. DOI: 10.1080/15265161.2020.1863507 PMID: 33372858

Petkov, C. I., Flecknell, P., Murphy, K., Basso, M. A., Mitchell, A. S., Hartig, R., & Thompson-Iritani, S. (2022). Unified ethical principles and an animal research 'Helsinki'declaration as foundations for international collaboration. [sciencedirect.com]. *Current Research in Neurobiology*, *3*, 100060. DOI: 10.1016/j.crneur.2022.100060 PMID: 36386861

Pham, V. A. T., Nguyen, L. A., Dellaportas, S., Phan, D. H. T., & Nguyen, Q. H. (2025). How does big data analytics impact accounting manipulation? *Accounting and Finance*, *65*(3), 2918–2934. Advance online publication. DOI: 10.1111/acfi.70024

Phyo, E. M., Lwin, T., Tun, H. P., Oo, Z. Z., Mya, K. S., & Silverman, H. (2023). Knowledge, attitudes, and practices regarding plagiarism of postgraduate students in Myanmar. *Accountability in Research*, *30*(8), 672–691. DOI: 10.1080/08989621.2022.2077643 PMID: 35686819

Pietilä, A. M., Nurmi, S. M., Halkoaho, A., & Kyngäs, H. (2020). Qualitative research: Ethical considerations. The application of content analysis in nursing science research, 49-69. [HTML]

Pietrzykowski, T. & Smilowska, K. (2021). The reality of informed consent: empirical studies on patient comprehension—systematic review. Trials. springer.com

Pietrzykowski, T. (2021). Ethical Review of Animal Research and the Standards of Procedural Justice: A European perspective. *Journal of Bioethical Inquiry*, *18*(3), 525–534. DOI: 10.1007/s11673-021-10111-5 PMID: 34283345

Pisani, A. R., Wyman, P. A., Mohr, D. C., Perrino, T., Gallo, C., Villamar, J., Kendziora, K., Howe, G. W., Sloboda, Z., & Brown, C. H. (2016). Human Subjects Protection and Technology in Prevention Science: Selected Opportunities and challenges. *Prevention Science*, *17*(6), 765–778. DOI: 10.1007/s11121-016-0664-1 PMID: 27220838

Pizzolato, D. & Dierickx, K. (2021). Stakeholders' perspectives on research integrity training practices: a qualitative study. BMC medical ethics. springer.com

Podschuweit, N. (2021). How ethical challenges of covert observations can be met in practice. Research Ethics. sagepub.com

Poje, T., & Groff, M. Z. (2021). Mapping ethics education in accounting research: A bibliometric analysis. *Journal of Business Ethics*, *179*(2), 451–472. DOI: 10.1007/s10551-021-04846-9 PMID: 34127872

Poje, T., Ličen, M., & Groff, M. Z. (2025). Who enrolls in accounting? Ethical predispositions of undergraduate accounting students. *Journal of Accounting Education*, *71*, 100966. Advance online publication. DOI: 10.1016/j.jaccedu.2025.100966

Popper, S. E., & McCloskey, K. (1995). Ethics in human experimentation: Historical perspectives. *Military Medicine*, *160*(1), 7–11. DOI: 10.1093/milmed/160.1.7 PMID: 7746436

Poulton, P. (2023). Being a teacher-researcher: Reflections on an insider research project from a virtues-based approach to research ethics. [HTML]. *Educational Action Research*, *31*(3), 575–591. DOI: 10.1080/09650792.2021.1962379

Powell, T. C. (2020). Can Quantitative Research Solve Social Problems? Pragmatism and the Ethics of Social Research. *Journal of Business Ethics*, *167*(1), 41–48. DOI: 10.1007/s10551-019-04196-7

Pradhan, T. K., & Kumar, A. (2023). Plagiarism in the Research Reports of Indian Doctoral Students: Causes and a Remedial Action Plan. *Center for Educational Policy Studies Journal*, *13*(3), 185–212. DOI: 10.26529/cepsj.1494

Pritchett, M., Ala'i-Rosales, S., Cruz, A. R., & Cihon, T. M. (2022). Social justice is the spirit and aim of an applied science of human behavior: Moving from colonial to participatory research practices. [springer.com]. *Behavior Analysis in Practice*, *15*(4), 1074–1092. DOI: 10.1007/s40617-021-00591-7 PMID: 34178290

Proctor, R. N. (2012). *Golden Holocaust: Origins of the Cigarette Catastrophe and the Case for Abolition*. University of California Press. DOI: 10.1525/9780520950436

PubMed. (2025). Retrieved, 2025, from https://pubmed.ncbi.nlm.nih.gov/

Pudasaini, S., Miralles-Pechuán, L., Lillis, D., & Salvador, M. L. (2024). Survey on plagiarism detection in large language models: The impact of chatgpt and gemini on academic integrity. *arXiv preprint arXiv:2407.13105*.

Putra, B. A. (2023). Complexities of the ethical dilemmas in qualitative international relations research: Research subjects, ethical codes, and constructing qualitative rigor. Social Sciences. mdpi.com

Queloz, M. & van Ackeren, M. (2024). Virtue Ethics and the Morality System. Topoi. springer.com

Radhika, D. (2012). Ethics in public administration. *Journal of Public Administration and Policy Research*, *4*(2), 23–31. DOI: 10.5897/JPAPR11.049

Rafiq, F., Awan, M. J., Yasin, A., Nobanee, H., Zain, A. M., & Bahaj, S. A. (2022). Privacy prevention of big data applications: A systematic literature review. [sagepub.com]. *SAGE Open*, *12*(2), 21582440221096445. DOI: 10.1177/21582440221096445

Rahman, S. A., Tuckerman, L., Vorley, T., & Gherhes, C. (2021). Resilient research in the field: Insights and lessons from adapting qualitative research projects during the COVID-19 pandemic. [sagepub.com]. *International Journal of Qualitative Methods*, *20*, 16094069211016106. DOI: 10.1177/16094069211016106

Ramadan, T. (2009). *Radical reform: Islamic ethics and liberation*. Oxford University Press.

Rani, R., & Sharma, R. K. (2012). Ethical Consideration in Research. *International Journal of Nursing Education*, *4*(1). http://ijone.org/scripts/IJONE%20Jan-June%202012.pdf#page=49

Rao, K. N., Mair, M., Arora, R. D., Dange, P., & Nagarkar, N. M. (2024). Misconducts in research and methods to uphold research integrity. *Indian Journal of Cancer*, *61*(2), 354–359. DOI: 10.4103/ijc.ijc_4_23 PMID: 39016268

Rattani, A., & Hyder, A. A. (2019). Developing an Islamic research ethics framework. *Journal of Religion and Health, 58*(1), 3–17.* https://doi.org/DOI: 10.1007/s10943-017-0508-8

Razack, H. I. A., Mathew, S. T., Saad, F. F. A., & Alqahtani, S. A. (2021). Artificial intelligence-assisted tools for redefining the communication landscape of the scholarly world. *Science Editing*, *8*(2), 134–144. DOI: 10.6087/kcse.244

Razaque, A., Frej, M. B. H., Alotaibi, B., & Alotaibi, M. (2021). Privacy preservation models for third-party auditor over cloud computing: A survey. Electronics. mdpi.com

Reijers, W., Wright, D., Brey, P., Weber, K., Rodrigues, R., O'Sullivan, D., & Gordijn, B. (2018). Methods for practising ethics in research and innovation: A literature review, critical analysis and recommendations. *Science and Engineering Ethics*, *24*(5), 1437–1481. DOI: 10.1007/s11948-017-9961-8 PMID: 28900898

Report, B. (1979). *Ethical principles and guidelines for the protection of human subjects of research. National Commission for the Protection of Human Subjects of Biomedical and Behavioral Research.* U.S. Department of Health, Education, and Welfare.

Research, T. N. (1979). *The Belmont Report*. U.S. Department of Health, Education, and Welfare.

Resnik, D. B. (2021). Standards of evidence for institutional review board decision-making. Accountability in research. nih.gov

Resnik, D. B. (2014). Data Fabrication and Falsification and Empiricist Philosophy of Science. *Science and Engineering Ethics*, *20*(2), 423–431. DOI: 10.1007/s11948-013-9466-z PMID: 23982326

Resnik, D. B. (2015). *Ethics of research with human subjects: Protecting people, advancing science, promoting trust*. Springer.

Resnik, D. B. (2018). *The Ethics of Research with Human Subjects: Protecting People, Advancing Science, Promoting Trust*. Springer. DOI: 10.1007/978-3-319-68756-8

Resnik, D. B., & Shamoo, A. E. (2017). Reproducibility and research integrity. *Accountability in Research*, *24*(2), 116–123. DOI: 10.1080/08989621.2016.1257387 PMID: 27820655

Resnik, D., & Hecking, J. (2020). Cultural variation in informed consent for clinical research participation. In *Handbook of Ethnography in Healthcare Research* (pp. 33–46). Routledge. [HTML] DOI: 10.4324/9780429320927-6

Rhim, J., Lee, J. H., Chen, M., & Lim, A. (2021). A deeper look at autonomous vehicle ethics: an integrative ethical decision-making framework to explain moral pluralism. Frontiers in Robotics and AI. frontiersin.org

Rhodes, R. (2005). Rethinking Research Ethics. *The American Journal of Bioethics*, *5*(1), 7–28. DOI: 10.1080/15265160590900678 PMID: 16036651

Richards, S. (2016). Academic integrity, plagiarism, and mercenary authorship in online learning environments. In N. J. & J. A. (Eds.), *Proceedings of the European Conference on e-Learning, ECEL* (Vols. 2016-Janua, pp. 575–578). Academic Conferences Limited. https://www.scopus.com/inward/record.uri?eid=2-s2.0-85000786211&partnerID=40&md5=caaa2f7156a580c7be5bd728a6ed469b

Robinson, K., McKenna, B., & Rooney, D. (2022). The relationship of risk to rules, values, virtues, and moral complexity: What we can learn from the moral struggles of military leaders. Journal of Business Ethics. springer.com

Rodgers, W., Murray, J. M., Stefanidis, A., Degbey, W. Y., & Tarba, S. Y. (2023). An artificial intelligence algorithmic approach to ethical decision-making in human resource management processes. [sciencedirect.com]. *Human Resource Management Review, 33*(1), 100925. DOI: 10.1016/j.hrmr.2022.100925

Roets, L. (2017). Protection of the human research participant: A structured review. *South African Medical Journal, 107*(10), 847–853. DOI: 10.7196/SAMJ.2017.v107i10.12533 PMID: 29022527

Rogers, J., & Breakey, H. (2023). *James Rest's four component model (FCM): A case for its central place in legal ethics*. Leading Works in Legal Ethics. [HTML]

Roig, M. (2010). Plagiarism and self-plagiarism: What every author should know. *Biochemia Medica, 20*(3), 295–300. DOI: 10.11613/BM.2010.037

Roka, Y. B. (2017). Plagiarism: Types, causes and how to avoid this worldwide problem. *Nepal Journal of Neuroscience, 14*(3), 2–6. DOI: 10.3126/njn.v14i3.20517

Romanelli, M. (2023). Advancing ethics within public administration. *Lumen Proceedings, 19*, 01-11. DOI: 10.18662/lumproc/gekos2022/01

Ross, J. S., Hill, K. P., Egilman, D. S., & Krumholz, H. M. (2008). Guest authorship and ghostwriting in publications related to rofecoxib: A case study of industry documents from rofecoxib litigation. *Journal of the American Medical Association, 299*(15), 1800–1812. DOI: 10.1001/jama.299.15.1800 PMID: 18413874

Rossouw, T. M., van Zyl, C., & Pope, A. (2014). Responsible conduct of research: Global trends, local opportunities. *South African Journal of Science, 110*(1–2), 1–6. http://www.scielo.org.za/scielo.php?script=sci_abstract&pid=S0038-23532014000100011&lng=en&nrm=iso&tlng=en. DOI: 10.1590/sajs.2014/20130103

Sachedina, A. (2009). *Islamic biomedical ethics: Principles and application*. Oxford University Press. DOI: 10.1093/acprof:oso/9780195378504.001.0001

Sachedina, A. (2012). Islam and biomedical ethics. In Kuhse, H., & Singer, P. (Eds.), *A companion to bioethics* (pp. 106–114). Wiley-Blackwell.

Sachs, G. A., & Siegler, M. (1993). Teaching scientific integrity and the responsible conduct of research. *Academic Medicine, 68*(12), 871–875. https://journals.lww.com/academicmedicine/abstract/1993/12000/Teaching_scientific_integrity_and_the_responsible.1.aspx. DOI: 10.1097/00001888-199312000-00001 PMID: 8259954

Saeed, S., Aamir, R., & Ramzan, M. (2011). Plagiarism and its implications on higher education in developing countries. *International Journal of Teaching and Case Studies, 3*(2/3/4), 123. DOI: 10.1504/IJTCS.2011.039552

Salauddin, N. (2023). Factors Influencing Plagiarism Among Faculty Members and Research Scholars in the Higher Education System. *DESIDOC Journal of Library and Information Technology, 43*(5), 364–374. DOI: 10.14429/djlit.43.05.19113

Saleh, S., Sambakunsi, H., Nyirenda, D., Kumwenda, M., Mortimer, K., & Chinouya, M. (2020). Participant compensation in global health research: A case study. [oup.com]. *International Health, 12*(6), 524–532. DOI: 10.1093/inthealth/ihaa064 PMID: 33165559

Sallam, M., Salim, N. A., Barakat, M., & Al-Tammemi, A. B. (2023). ChatGPT applications in medical, dental, pharmacy, and public health education: A descriptive study highlighting the advantages and limitations. *Narra J, 3*(1), e103. Advance online publication. DOI: 10.52225/narra.v3i1.103 PMID: 38450035

Samuel, G., & Wassenaar, D. (2025). Joint editorial: Informed consent and AI transcription of qualitative data. *Journal of Empirical Research on Human Research Ethics; JERHRE, 20*(1-2), 3–5. DOI: 10.1177/15562646241296712 PMID: 39600037

Sankar, P. (2022). Ethical and welfare issues in the use of animals in biomedical research. In Muthu, S. S. (Ed.), *Animal Welfare* (pp. 15–24). Springer., DOI: 10.1007/978-981-16-8508-8_2

Sartoretto, P., & Custódio, L. (2020). Dealing with ethical dilemmas in activist research on social movement media. Media Activist Research Ethics: Global Approaches to Negotiating Power in Social Justice Research, 51-66. [HTML]

Savulescu, J., Persson, I., & Wilkinson, D. (2020). Utilitarianism and the pandemic. Bioethics. wiley.com

Savur, S. G. (2022). Ethical Decision-Making—Synthesizing SK Chakraborty's Classification of Ethics with Levels of Moral Judgement and the Four-Component Model. In Global Perspectives on Indian Spirituality and Management: The Legacy of SK Chakraborty (pp. 107-121). Singapore: Springer Nature Singapore. [HTML]

Sayed, R. (2023). Ethical considerations in the use of big data for business analytics: A cross-industry perspective. *Journal of Innovations and Sustainability, 7*(4), 08. DOI: 10.51599/is.2023.07.04.08

Schreyer, M., Sattarov, T., & Borth, D. (2022). Federated and privacy-preserving learning of accounting data in financial statement audits. *Proceedings of the Third ACM International Conference on AI in Finance*, 105–113. DOI: 10.1145/3533271.3561674

Schuppli, C. A., & Fraser, D. (2007). Factors influencing the effectiveness of research ethics committees. *Journal of Medical Ethics, 33*(5), 294–301. DOI: 10.1136/jme.2005.015057 PMID: 17470508

Schwartz-Shea, P. (2021). Under threat? Methodological pluralism in public administration. *Public Performance & Management Review*, *44*(5), 975–1005. DOI: 10.1080/15309576.2019.1694547

Serpico, K. (2024). Defining institutional review board application quality: critical research gaps and future opportunities. Research Ethics. sagepub.com

Serpico, K. (2024). The Belmont Report doesn't need reform, our moral imagination does. Research Ethics. sagepub.com

Service, R. F. (2002). Bell Labs fires star physicist found guilty of falsifying data. *Science*, *298*(5595), 30–31. DOI: 10.1126/science.298.5595.30 PMID: 12364753

Shamoo, A. E., & Resnik, D. B. (2015). *Responsible Conduct of Research* (3rd ed.). Oxford University Press.

Sheikh, H., Prins, C., & Schrijvers, E. (2023). *Mission AI: The new system technology*. Springer. https://link.springer.com/book/10.1007/978-3-031-21448-6

Shen, F. X., Silverman, B. C., Monette, P., Kimble, S., Rauch, S. L., & Baker, J. T. (2022). An Ethics Checklist for Digital Health Research in Psychiatry: Viewpoint. *Journal of Medical Internet Research*, *24*(2), e31146. DOI: 10.2196/31146 PMID: 35138261

Shilton, K., Moss, E., Gilbert, S. A., Bietz, M. J., Fiesler, C., Metcalf, J., Vitak, J., & Zimmer, M. (2021). Excavating awareness and power in data science: A manifesto for trustworthy pervasive data research. [sagepub.com]. *Big Data & Society*, *8*(2), 20539517211040759. DOI: 10.1177/20539517211040759

Shorey, S. (2025). *AI and government workers: Use cases in public administration. Roosevelt Institute*. https://rooseveltinstitute.org/publications/ai-and-government-workers/

Siddhpura, A., & Siddhpura, M. (2020). Plagiarism, contract cheating and other academic misconducts in online engineering education: Analysis, detection and prevention strategies. In M. H., G. Y., O. Y., R. Ma.M.T., S. J., V. N., W. G., Y. M., & C.-U. L. L. (Eds.), *Proceedings of 2020 IEEE International Conference on Teaching, Assessment, and Learning for Engineering, TALE 2020* (pp. 112–119). Institute of Electrical and Electronics Engineers Inc. DOI: 10.1109/TALE48869.2020.9368311

Sidorkin, A. M. (2024). Embracing Chatbots in Higher Education: The Use of Artificial Intelligence in Teaching, Administration, and Scholarship. In *Embracing Chatbots in Higher Education: The Use of Artificial Intelligence in Teaching, Administration, and Scholarship*. Taylor and Francis., DOI: 10.4324/9781032686028

Simmons, J. P., Nelson, L. D., & Simonsohn, U. (2011). False-positive psychology: Undisclosed flexibility in data collection and analysis allows presenting anything as significant. *Psychological Science*, *22*(11), 1359–1366. DOI: 10.1177/0956797611417632 PMID: 22006061

Singer, E., & Levine, F. J. (2003). PROTECTION OF HUMAN SUBJECTS OF RESEARCH: RECENT DEVELOPMENTS AND FUTURE PROSPECTS FOR THE SOCIAL SCIENCES. *Public Opinion Quarterly*, *67*(1), 148–164. DOI: 10.1086/346013

Sivasubramaniam, S., Dlabolová, D. H., Kralikova, V., & Khan, Z. R. (2021). Assisting you to advance with ethics in research: An introduction to ethical governance and application procedures. *International Journal for Educational Integrity*, *17*(1), 14. Advance online publication. DOI: 10.1007/s40979-021-00078-6

Smith, L. T. (2021). *Decolonizing methodologies: Research and Indigenous peoples* (3rd ed.). Zed Books. DOI: 10.5040/9781350225282

Sng, B., Yip, C., & Han, N. (2016). Legal and ethical issues in research. *Indian Journal of Anaesthesia*, *60*(9), 684. DOI: 10.4103/0019-5049.190627 PMID: 27729698

Soehartono, A. M., Yu, L. G., & Khor, K. A. (2022). Essential signals in publication trends and collaboration patterns in global Research Integrity and Research Ethics (RIRE). *Scientometrics*, *127*(12), 7487–7497. DOI: 10.1007/s11192-022-04400-y PMID: 35755633

Soniya Billore, T. A. (2023). Self-regulation and goal-directed behavior: A systematic literature review, public policy recommendations, and research agenda. *Journal of Business Research.*

Sowden, C. (2005). Plagiarism and the culture of multilingual students in higher education abroad. *ELT Journal*, *59*(3), 226–233. DOI: 10.1093/elt/cci042

Sperling, D. (2022). "Could You Help Me Die?": On the Ethics of Researcher-Participant Relationship and the Limitations of Qualitative Research. International Journal of Qualitative Methods. sagepub.com

Stapleton, A. B., Hankes, D. M., Hays, K. F., & Parham, W. D. (2010). Ethical dilemmas in sport psychology: A dialogue on the unique aspects impacting practice. *Professional Psychology, Research and Practice*, *41*(2), 143a. https://psycnet.apa.org/record/2010-06890-009. DOI: 10.1037/a0017976

Steneck, N. H. (2006). Fostering integrity in research: Definitions, current knowledge, and future directions. *Science and Engineering Ethics*, *12*(1), 53–74. DOI: 10.1007/PL00022268 PMID: 16501647

Steneck, N. H., & Bulger, R. E. (2007). The History, Purpose, and Future of Instruction in the Responsible Conduct of Research. *Academic Medicine*, *82*(9), 829–834. DOI: 10.1097/ACM.0b013e31812f7d4d PMID: 17726385

Stewart, R. J., & Reider, B. (2016). The Ethics of Sports Medicine Research. *Clinics in Sports Medicine*, *35*(2), 303–314. DOI: 10.1016/j.csm.2015.10.009 PMID: 26832979

Stilgoe, J., Owen, R., & Macnaghten, P. (2020). Developing a framework for responsible innovation. In The Ethics of Nanotechnology, Geoengineering, and Clean Energy (pp. 347-359). Routledge. sciencedirect.com DOI: 10.4324/9781003075028-22

Stineman, M. G., & Musick, D. W. (2001). Protection of human subjects with disability: Guidelines for research. *Archives of Physical Medicine and Rehabilitation*, *82*(12), S9–S14. DOI: 10.1016/S0003-9993(01)65984-7 PMID: 11805914

Stommel, W., & de Rijk, L. (2021). Ethical approval: None sought. How discourse analysts report ethical issues around publicly available online data. *Research Ethics*, *17*(3), 275–297. DOI: 10.1177/1747016120988767

Strosberg, M. A., Gefenas, E., & Famenka, A. (2014). Research Ethics Review: Identifying Public Policy and Program Gaps. *Journal of Empirical Research on Human Research Ethics; JERHRE*, *9*(2), 3–11. DOI: 10.1525/jer.2014.9.2.3 PMID: 24782068

Sullivan, M., Kelly, A., & McLaughlan, P. (2023). ChatGPT in higher education: Considerations for academic integrity and student learning. *Journal of Applied Learning and Teaching*, *6*(1), 31–40. DOI: 10.37074/jalt.2023.6.1.17

Sutherland-Smith, W. (2008). *Plagiarism, the Internet, and student learning: Improving academic integrity*. Routledge. DOI: 10.4324/9780203928370

Svensson, R., & Broberg, G. (2013). Ethics and public health: The Vipeholm experiments. In *The history of medical ethics* (pp. 303–322). Springer.

Talyshinskii, A., Naik, N., Hameed, B. M. Z., Zhanbyrbekuly, U., Khairli, G., Guliev, B., Juilebø-Jones, P., Tzelves, L., & Somani, B. K. (2023). Expanding horizons and navigating challenges for enhanced clinical workflows: ChatGPT in urology. *Frontiers in Surgery, 10*, 1257191. Advance online publication. DOI: 10.3389/fsurg.2023.1257191 PMID: 37744723

Tamminen, K. A., Bundon, A., Smith, B., McDonough, M. H., Poucher, Z. A., & Atkinson, M. (2021). Considerations for making informed choices about engaging in open qualitative research. [tandfonline.com]. *Qualitative Research in Sport, Exercise and Health, 13*(5), 864–886. DOI: 10.1080/2159676X.2021.1901138

Taquette, S. R., & Borges da Matta Souza, L. M. (2022). Ethical dilemmas in qualitative research: A critical literature review. [sagepub.com]. *International Journal of Qualitative Methods, 21*, 16094069221078731. DOI: 10.1177/16094069221078731

Taylor, G., Cumming, D., Li, D., & Shi, J. (2025). The role of legal, regulatory and political reforms and institutions in influencing accounting and financial outcomes. *The British Accounting Review, 57*(3), 101647. DOI: 10.1016/j.bar.2025.101647

Taylor, J. S. (2024). Reassessing academic plagiarism. *Journal of Academic Ethics, 22*(2), orari211–230. DOI: 10.1007/s10805-023-09478-4

Tesla, Inc. (2023). *Tesla vehicle safety report Q1–Q4 2023*. Tesla Investor Relations. https://www.tesla.com/presskit

Thacker, P. D. (2005). Searching for transparency: Conflicts of interest cloud medical research. *Journal of the American Medical Association, 294*(6), 689–692. DOI: 10.1001/jama.294.6.689

Thakur, S., & Lahiry, S. (2019). Research ethics in the modern era. *Indian Journal of Dermatology, Venereology and Leprology, 85*(4), 351. DOI: 10.4103/ijdvl.IJDVL_499_18 PMID: 31115359

Thapa, C., & Camtepe, S. (2021). Precision health data: Requirements, challenges and existing techniques for data security and privacy. [PDF]. *Computers in Biology and Medicine, 129*, 104130. DOI: 10.1016/j.compbiomed.2020.104130 PMID: 33271399

Thomas SB, Q. S. (1991). The Tuskegee Syphilis Study, 1932 to 1972: implications for HIV education and AIDS risk education programs in the black community. *Am J Public Health*.

Thompson, D. F. (1993). Understanding financial conflicts of interest. *The New England Journal of Medicine*, *329*(8), 573–576. DOI: 10.1056/NEJM199308193290812 PMID: 8336759

Thorne LeClair, D., Ferrell, O. C., & Ferrell, L. (1997). Federal Sentencing Guidelines for Organizations: Legal, ethical, and public policy issues for international marketing. *Journal of Public Policy & Marketing*, *16*(1), 26–37. DOI: 10.1177/074391569701600104

Thottoli, M. M. (2024). Leveraging information communication technology (ICT) and artificial intelligence (AI) to enhance auditing practices. *Accounting Research Journal*, *37*(2), 134–150. DOI: 10.1108/ARJ-09-2023-0269

Tibi, B. (2009). *Islam's predicament with modernity: Religious reform and cultural change*. Routledge. DOI: 10.4324/9780203881552

Tindall, I. K., Fu, K. W., Tremayne, K., & Curtis, G. J. (2021). Can negative emotions increase students' plagiarism and cheating? *International Journal for Educational Integrity*, *17*(1), 25. Advance online publication. DOI: 10.1007/s40979-021-00093-7

Tonis, R. B.-M., Păun, D. G., Coelho, A. S., Urdes, L., Mihoreanu, L., & Martins, O. M. D. (2025). Ethical challenges using technology in sport competitions. *Intangible Capital*, *21*(3), 406–426. DOI: 10.3926/ic.3034

Topol, E. J. (2004). Failing the public health—Rofecoxib, Merck, and the FDA. *The New England Journal of Medicine*, *351*(17), 1707–1709. DOI: 10.1056/NEJMp048286 PMID: 15470193

Torous, J., & Nebeker, C. (2017). Navigating Ethics in the Digital Age: Introducing Connected and Open Research Ethics (CORE), a Tool for Researchers and Institutional Review Boards. *Journal of Medical Internet Research*, *19*(2), e6793. DOI: 10.2196/jmir.6793 PMID: 28179216

Torres, C. G., & Fons, I. T. (2018). Cyber-plagiarism in Spanish higher education: Exploring the causes in order to reach specific solutions. *Synergies Espagne, 11*, 195–212. https://www.scopus.com/inward/record.uri?eid=2-s2.0-85059908937&partnerID=40&md5=ba93a4a01eeaaf564061a3ba18f47dbb

Tortora, L. (2024). Beyond Discrimination: Generative AI Applications and Ethical Challenges in Forensic Psychiatry. *Frontiers in Psychiatry*, *15*, 1346059. Advance online publication. DOI: 10.3389/fpsyt.2024.1346059 PMID: 38525252

Tripathi, S., & Patel, D. (2021). Academic integrity: Looking beyond plagiarism. *Annals of Library and Information Studies*, *68*(2), 190–197. https://www.scopus.com/inward/record.uri?eid=2-s2.0-85108826989&partnerID=40&md5=53e0df4e6cebc6e399c5b3eab8289a03

Trondal, J. (2025). The multilevel administrative state and the future of public administration research. *International Journal of Public Administration*, *48*(5-6), 321–333. DOI: 10.1080/01900692.2024.2412273

Tsan, M. F. (2025). Impact of the revised Common Rule on enhancing human research subject protections and reducing researcher burdens [Research Article]. *Qeios*, *7*(3), TU6VRN.2. Advance online publication. DOI: 10.32388/TU6VRN.2

Tsosie, K. S., Claw, K. G., & Garrison, N. A. (2021). Considering "Respect for Sovereignty" beyond the Belmont report and the common rule: Ethical and legal implications for American Indian and Alaska native peoples. [tandfonline.com]. *The American Journal of Bioethics*, *21*(10), 27–30. DOI: 10.1080/15265161.2021.1968068 PMID: 34554064

Udo Schuklenk, S. P. (2013). *International Research Ethics. International Encyclopedia of Ethics*. Wiley.

United States. National Commission for the Protection of Human Subjects of Biomedical, & Behavioral Research. (1978). *The Belmont report: ethical principles and guidelines for the protection of human subjects of research (Vol. 2)*. The Commission. https://www.google.com/books/edition/The_Belmont_Report/cAqtFohwYhIC?hl=en&gbpv=0

Valkenburg, G., Dix, G., Tijdink, J., & de Rijcke, S. (2021). Expanding research integrity: A cultural-practice perspective. [springer.com]. *Science and Engineering Ethics*, *27*(1), 10. DOI: 10.1007/s11948-021-00291-z PMID: 33559767

van Stekelenburg, M. H. M. (2024). *Navigating on the Ethical Compass.: What it Means and Implies to Equip Students of Universities of Applied Sciences with an Ethical Compass* [PhD-Thesis - Research and graduation internal]. DOI: 10.5463/thesis.782

Vanclay, F., Baines, J. T., & Taylor, C. N. (2013). Principles for ethical research involving humans: Ethical professional practice in impact assessment Part I. *Impact Assessment and Project Appraisal*, *31*(4), 243–253. DOI: 10.1080/14615517.2013.850307

Vatamanu, A. F., & Tofan, M. (2025). Integrating artificial intelligence into public administration: Challenges and vulnerabilities. *Administrative Sciences*, *15*(4), 149–169. DOI: 10.3390/admsci15040149

Velmurugan, V. S. (2024). Types and Definitions of Plagiarism: An Overview. *Global Research Journal of Social Sciences and Management*, *2*(1), 13–18. DOI: 10.55306/GRJSSM.2024.2102

Vijayananthan, A. N. O., & Nawawi, O. (2008). The importance of Good Clinical Practice guidelines and their role in clinical trials. *Biomedical Imaging and Intervention Journal*, *4*(1). Advance online publication. DOI: 10.2349/biij.4.1.e5 PMID: 21614316

Vinson, J., Curtis, M., Conover, T., & Chui, L. (2020). Ethical relativism in accounting: A cross-cultural examination of the influence of culture and risk-taking propensity on ethical decision-making. *Journal of International Accounting, Auditing & Taxation*, *41*, 100350. DOI: 10.1016/j.intaccaudtax.2020.100350

Vitak, J., Proferes, N., Shilton, K., & Ashktorab, Z. (2017). Ethics Regulation in Social Computing Research: Examining the Role of Institutional Review Boards. *Journal of Empirical Research on Human Research Ethics; JERHRE*, *12*(5), 372–382. DOI: 10.1177/1556264617725200 PMID: 28831844

Vlad, M. D., & Vlad, S. (2021). The Evolution of Intellectual Property Concepts at National and International Level. *SCIENTIA MORALITAS-International Journal of Multidisciplinary Research*, *6*(1), 31–39.

VOSviewer. (https://www.vosviewer.com/) PubMed. *Scholar search: Animal research ethics and welfare standard* https://pubmed.ncbi.nlm.nih.gov/?term=Animal+research+ethics%2C+welfare+standards+

VOSviewer. https://www.vosviewer.com/

Wachter, S., Mittelstadt, B., & Floridi, L. (2017). Why a right to explanation of automated decision-making does not exist in the General Data Protection Regulation. *International Data Privacy Law*, *7*(2), 76–99. DOI: 10.1093/idpl/ipx005

Waddington, I. (2013). Researching the World of Professional Football. In *Doing Real World Research in Sports Studies*. Routledge.

Walker, W. (2007). Ethical considerations in phenomenological research. *Nurse Researcher*, *14*(3), 36–45. DOI: 10.7748/nr2007.04.14.3.36.c6031 PMID: 17494467

Wallace, S. E. & Miola, J. (2021). Adding dynamic consent to a longitudinal cohort study: a qualitative study of EXCEED participant perspectives. BMC medical ethics. springer.com

Waqas, M. (2025). Exploring the ethical framework of science: A critical study from the Islamic perspective. *Pakistan Journal of Islamic Philosophy*, *7*(1), 18–31.

Warner, C. H., Fortin, M., & Melkonian, T. (2024). When are we more ethical? A review and categorization of the factors influencing dual-process ethical decision-making. [HTML]. *Journal of Business Ethics, 189*(4), 843–882. DOI: 10.1007/s10551-022-05281-0

Watanabe, A., & Wiseman, S. M. (2023). A New Era in Surgical Research: The Evolving Role of Artificial Intelligence. *American Journal of Surgery, 226*(6), 923–925. DOI: 10.1016/j.amjsurg.2023.06.040 PMID: 37419728

Watson, J. C., II, & Etzel, E. F. (2019). *Ethical and legal issues in sport and performance psychology.* https://psycnet.apa.org/record/2019-05400-041

Watson, J. C.II, Harris, B. S., & Baillie, P. (2020). Ethical Issues Impacting the Profession of Sport Psychology. In *Handbook of Sport Psychology* (pp. 751–772). John Wiley & Sons, Ltd., DOI: 10.1002/9781119568124.ch36

Webster, S., Lewis, J., & Brown, A. (2013). Considerations in qualitative research. *Qualitative Research Practice: A Guide for Social Science Students and Researchers, 1*, 77. https://www.torrossa.com/gs/resourceProxy?an=5017584&publisher=FZ7200#page=102

Wendler, R. (2012). Human Subjects Protection: a source for Ethical Service-Learning practice. In *Michigan Journal of Community Service Learning* (pp. 29–39).

West, S. & Schill, C. (2022). Negotiating the ethical-political dimensions of research methods: a key competency in mixed methods, inter-and transdisciplinary, and co-production research. Humanities and Social Sciences Communications. nature.com

Wester, K. L., Deaton, J. D., & Jayamohan, H. (2023). *Ethics in Research.* Reimagining Research. [HTML]

Whitcomb, D., Battaly, H., Baehr, J., & Howard-Snyder, D. (2017). Intellectual humility: Owning our limitations. *Philosophy Compass, 12*(8), e12415. DOI: 10.1111/phc3.12415

White, M. G. (2020). Why human subjects research protection is important. In *Ochsner Journal* (Vol. 20, pp. 16–33). https://doi.org/DOI: 10.31486/toj.20.5012

White, T., Blok, E., & Calhoun, V. D. (2022). Data sharing and privacy issues in neuroimaging research: Opportunities, obstacles, challenges, and monsters under the bed. Human Brain Mapping. wiley.com

Wieringa, J., Kannan, P. K., Ma, X., Reutterer, T., Risselada, H., & Skiera, B. (2021). Data analytics in a privacy-concerned world. [sciencedirect.com]. *Journal of Business Research, 122*, 915–925. DOI: 10.1016/j.jbusres.2019.05.005

Williams, O., Sarre, S., Papoulias, S. C., Knowles, S., Robert, G., Beresford, P., Rose, D., Carr, S., Kaur, M., & Palmer, V. J. (2020). Lost in the shadows: Reflections on the dark side of co-production. [springer.com]. *Health Research Policy and Systems, 18*(1), 1–10. DOI: 10.1186/s12961-020-00558-0 PMID: 32380998

Wiltshire, D., & Alvanides, S. (2022). Ensuring the ethical use of big data: Lessons from secure data access. *Heliyon, 8*(2), e08981. Advance online publication. DOI: 10.1016/j.heliyon.2022.e08981 PMID: 35243099

Winterberg, C. A., Brummel, B. J., Ayres, T. B., & Kerr, A. J. (2021). Applying ethical dilemmas to professional ethics training and education: Practical implications and future directions. *Industrial and Organizational Psychology: Perspectives on Science and Practice, 14*(3), 325–330. DOI: 10.1017/iop.2021.73

Wittenberg, H. (2024). Turn-It-Off: Reflections on Technology, the Knowledge Commons and the Academic Plagiarism Industry. *Scrutiny2*. DOI: 10.1080/18125441.2024.2320892

World Health Organization (WHO). (2021). *WHO Handbook for Guideline Development*. https://www.who.int/publications

World Medical Association. (1964). *Declaration of Helsinki: Ethical principles for medical research involving human subjects.* https://www.wma.net/policies-post/wma-declaration-of-helsinki-ethical-principles-for-medical-research-involving-human-subjects/

World Medical Association. (2013). Declaration of Helsinki: Ethical principles for medical research involving human subjects. *Journal of the American Medical Association, 310*(20), 2191–2194. DOI: 10.1001/jama.2013.281053 PMID: 24141714

Wu, J.-S., Chien, T.-H., Chien, L.-R., & Yang, C.-Y. (2021). Using artificial intelligence to predict class loyalty and plagiarism in students in an online blended programming course during the covid-19 pandemic. *Electronics (Basel), 10*(18), 2203. Advance online publication. DOI: 10.3390/electronics10182203

Xu, A., Baysari, M. T., Stocker, S. L., Leow, L. J., Day, R. O., & Carland, J. E. (2020). Researchers' views on, and experiences with, the requirement to obtain informed consent in research involving human participants: A qualitative study. *BMC Medical Ethics, 21*(1), 93. Advance online publication. DOI: 10.1186/s12910-020-00538-7 PMID: 33008387

Xu, Y., Liu, X., Cao, X., Huang, C., Liu, E., Qian, S., Liu, X., Wu, Y., Dong, F., Qiu, C. W., Qiu, J., Hua, K., Su, W., Wu, J., Xu, H., Han, Y., Fu, C., Yin, Z., Liu, M., & Zhang, J. (2021). Artificial intelligence: A powerful paradigm for scientific research. *Innovation (Cambridge (Mass.))*, *2*(4), 1–10. DOI: 10.1016/j.xinn.2021.100179 PMID: 34877560

Yadav, B. (2024). Ethics in public administration: Why integrity matters? *Research Review International Journal of Multidisciplinary*, *9*(1), 282–286. DOI: 10.31305/rrijm.2024.v09.n01.035

Yang, L., Tian, M., Xin, D., Cheng, Q., & Zheng, J. (2024). AI-driven anonymization: Protecting personal data privacy while leveraging machine learning. *Applied and Computational Engineering*, *71*(1), 7–13. DOI: 10.54254/2755-2721/71/2024MA0053

Yechuri, S. (2024). Enhanced utility-driven data anonymization: Leveraging AI and machine learning for sensitive data privacy. *Journal of Artificial Intelligence General Science*, *1*(1), 229–232. DOI: 10.60087/jaigs.v1i1.229

Yin, R. K. (2018). *Case study research and applications: Design and methods* (6th ed.). Sage Publications.

Youmans, R. J. (2011). Does the adoption of plagiarism-detection software in higher education reduce plagiarism? *Studies in Higher Education*, *36*(7), 749–761. DOI: 10.1080/03075079.2010.523457

Zhang, W. (2024). Ethical dilemmas in accounting: A comprehensive analysis of professional ethics. *Academic Journal of Business & Management*, *6*(2). Advance online publication.

Zhang, X., Li, D., Wang, C., Jiang, Z., Ngao, A. I., Liu, D., Peters, M. A., & Tian, H. (2023). From ChatGPT to China' Sci-Tech: Implications for Chinese High er Education. *Beijing International Review of Education*, *5*(3), 296–314. DOI: 10.1163/25902539-05030007

Zhang, Y., Fu, J., Lai, J., Deng, S., Guo, Z., Zhong, C., Tang, J., Cao, W., & Wu, Y. (2024). Reporting of ethical considerations in qualitative research utilizing social media data on public health care: Scoping review. *Journal of Medical Internet Research*, *26*, e51496. https://www.jmir.org/2024/1/e51496/. DOI: 10.2196/51496 PMID: 38758590

Zimmer, M. (2018). Addressing Conceptual Gaps in Big Data Research Ethics: An Application of Contextual Integrity. *Social Media + Society*, *4*(2), 2056305118768300. DOI: 10.1177/2056305118768300

Zlateva, P., Steshina, L., Petukhov, I., & Velev, D. (2024). A Conceptual Framework for Solving Ethical Issues in Generative Artificial Intelligence. In T.-B. A.J., C.-A. E., & L.-G. D.A. (Eds.), *Frontiers in Artificial Intelligence and Applications* (Vol. 381, pp. 110–119). IOS Press BV. DOI: 10.3233/FAIA231182

Zuiderwijk, A., Chen, Y. C., & Salem, F. (2021). Implications of the use of artificial intelligence in public governance: A systematic literature review and a research agenda. *Government Information Quarterly*, *38*(3), 1–12. DOI: 10.1016/j.giq.2021.101577

Zuniga, D. H., Goyanes, M., & Durotoye, T. (2024). A scholarly definition of artificial intelligence (AI): Advancing AI as a conceptual framework in communication research. *Political Communication*, *41*(2), 317–334. DOI: 10.1080/10584609.2023.2290497

Zwitter, A. (2014). Big data ethics. *Big Data & Society*, *1*(2), 1–6. DOI: 10.1177/2053951714559253

Zyphur, M. J., & Pierides, D. C. (2017). Is quantitative research ethical? Tools for ethically practicing, evaluating, and using quantitative research. *Journal of Business Ethics*, *143*(1), 1–16. https://idp.springer.com/authorize/casa?redirect_uri=https://link.springer.com/article/10.1007/s10551-017-3549-8&casa_token=RBVoBlCDYocAAAAA:FloGK-8NKEx3MjMP8zzMpiDbaCAqRAoqA_SS0dsFqGAT1vl3sNAUpqvZWK-nqIVq6nv5KrbD8wgbwEtuWA. DOI: 10.1007/s10551-017-3549-8

About the Contributors

Janice E. Jones, PhD is the Director of the PhD program at Carroll University in Waukesha, WI. Dr. Jones has many years of experience teaching research to adults and served for 17 years in leadership positions working with adults who were earning a PhD in Leadership. As Professor of Leadership, Dr. Jones has taught courses in research methods, writing, diversity and leadership. Dr. Jones currently serves as the chair of the Institutional Review Board at Carroll University. Dr. Jones has published books on mixed methods research, and has published articles in qualitative, quantitative and mixed methodologies. Dr. Jones is an advocate for children in out of home care and is currently studying attachment processes between foster parents and infants in their care.

Mette Baran is an award-winning educator and author whose scholarship focuses on mixed methods research, higher education teaching and learning, and leadership. She has published books and refereed journal articles in education and business and has served on numerous doctoral dissertation committees as both chair and committee member. Dr. Baran previously served as a professor of educational leadership at Cardinal Stritch University, where she taught in the Doctoral Leadership Programs. She is currently a lecturer at Troy University, teaching management and leadership courses in the Sorrell College of Business. Her research interests include educational looping, research methods, leadership, and international studies.

***

Andi Asrifan, S.Pd, M.Pd, a distinguished figure in the realm of education, particularly in English Language Education, has made significant contributions through his extensive research, teaching, and leadership roles. As the Head of the Institute for Cooperation and International Affairs at Universitas Muhammadiyah Sidenreng Rappang, Dr. Asrifan's academic journey is marked by a rich tapestry of achievements, including a Doctor of Education in English Language Education.

His work spans various facets of education, language, and technology, reflecting the diverse topics covered in his publications and research endeavors. These include, but are not limited to, the impact of the COVID-19 pandemic on educational practices, methodologies in English as a Foreign Language (EFL) classrooms, and the integration of technology in teaching and learning processes. Dr. Asrifan's commitment to advancing educational standards and practices is further evidenced by his active participation in international conferences, webinars, and workshops across countries like India, the Philippines, and Pakistan. These events, organized by various institutions and associations, underscore his role as a global educator and thought leader. His dedication to professional development is also highlighted through his involvement in numerous courses and certifications in personal branding, public speaking, and literacy teaching. Moreover, Dr. Asrifan's leadership extends beyond academia into organizational and educational development, where he has served in various capacities, including as an external examiner for Ph.D. dissertations and a reviewer for scientific journals. His accolades, memberships in professional organizations, and roles as an ambassador for international associations speak volumes of his influence and commitment to fostering educational excellence and sustainable development goals globally.

Tashi Ongmu Bhutia is a distinguished academician in the field of Physical Education, currently serving at the Government B.Ed. College, Soreng, Sikkim. She began her academic journey at the prestigious Lakshmibai National Institute of Physical Education (LNIPE), Gwalior, where she completed both her Bachelor of Physical Education (B.P.Ed.) and Master of Physical Education (M.P.Ed.) with excellence. Driven by a passion for research and academic growth, she pursued her doctoral studies at the University of Delhi, earning a Ph.D. in Physical Education. Over the years, Dr. Bhutia has contributed significantly to promoting physical education, sports, and wellness through her teaching, mentorship, and scholarly pursuits. Her academic background and professional dedication reflect her commitment to advancing the standards of teacher education in Sikkim. Known for her expertise, discipline, and inspiring approach, she continues to shape the future of aspiring educators while fostering holistic development through physical education.

Suratman Dahlan was born in Bumi Rahmat Village, South Halmahera Regency, North Maluku. He completed his undergraduate degree in English Language Education (S.Pd.) at Universitas Khairun, Ternate. He subsequently earned his Master of Education (M.Pd.) in English Language Education from the National Campus of Universitas Sarjanawiyata Tamansiswa, Yogyakarta. Funding from the LPDP scholarship, he completed his doctoral studies (Dr) in the English Education Study Program, Faculty of Language and Literature, Universitas Negeri

Makassar. He is currently employed as a lecturer (Assistant Professor) in the English Language Education Study Program at Khairun University, Ternate. His academic interests include English grammar, translation, and Teaching English as a Foreign Language (TEFL). He is also the author of several books, including Simple English Grammar (for Beginners, University Students, and General Readers), published by Andi Publisher, Yogyakarta, in 2017, and Establishment of a New Pedagogical Paradigm through Artificial Intelligence Tools, published by Deepublish, Yogyakarta, in 2025. In addition to writing books, he actively conducts research and publishes scholarly articles in reputable national and international journals. In international academic activities, he recently participated as a paper presenter at the 23rd AsiaTEFL International Conference, hosted by The Education University of Hong Kong (EdUHK), held on 11–13 July 2025 in Hong Kong. Furthermore, on 13–14 December 2025, he was invited as a speaker at the International Conference on Innovation in Learning Instruction and Teacher Education (ILITE-4), organized by Hanoi National University of Education (HNUE), Hanoi, Vietnam.

Anindita Das is an esteemed Associate Professor at the Lakshmibai National Institute of Physical Education (LNIPE), Gwalior, with a distinguished career in teaching, research, and sports. Since joining LNIPE in 2007, she has been actively engaged in teaching core subjects such as Health Education, Physiology, and Football (practical). Dr. Das has made significant academic contributions through her authorship of numerous research papers in Scopus, patents, books and chapters in the field of physical education and sports sciences. She has published over twenty research articles in reputed national and international journals and has presented her work at a wide array of conferences. Notably, she presented papers at the Pre-Olympic Congress in Liverpool, UK (2012) and the World Congress on Kinanthropometry in Santiago, Chile (2018). A recipient of international training, Dr. Das completed a 3-month International Coaching Course (ICC) in General Conditioning at Semmelweis University, Budapest, Hungary, in 2009, further enriching her academic and practical expertise. Beyond academics, Dr. Das is a former athlete who has passionately contributed to women's football. She was a four-time member of her university's intervarsity women's football team and won a bronze medal in the All India Intervarsity Women Football Tournament in 2002. She also represented Madhya Pradesh in the National Women's Football Championship and played a key role when her team participated in the 1st Khelo India University Games (KIUG) held in Bhubaneswar, Odisha, in February 2020. Dr. Das continues to inspire students and peers alike with her commitment to excellence in physical education, research, and sports development.

Susanta Das is a seasoned educator with a proven record of teaching and mentoring diverse students, working at multiple institutions, and providing service to the various initiatives of organizations. He received his Ph.D. and M.A. degrees from Western Michigan University (WMU), USA, and M.Sc. degree from Banaras Hindu University (BHU), India all in Physics. He continued his research as a Marie-Curie post-doctoral fellow at Stockholm University, Stockholm, Sweden on beam diagnostics for the DESIREE (Double ElectroStatic Ion Ring ExpEriment) in the project DITANET (Diagnostic Techniques for particle Accelerators – a Marie-Curie initial training NETwork), at the Indian Institute of Science Education and Research-Kolkata, India on high-pressure physics as a Project Scientist-B, and at the University of Electro-Communications, Tokyo, Japan on ion-surface interactions as a Post-doctoral fellow. Throughout his career, Dr. Das worked and collaborated with many researchers, Ph.D., master, and visiting students from different countries. He further visited many countries to discuss research and an international conference participant (UK, Italy, Germany, Greece, Belgium, Bulgaria, Romania, Brazil etc.). He worked at several institutions before assuming his current role at Ajeenkya D Y Patil University. He received the Marie-Curie post-doctoral fellowship in Sweden, Science Academies' Summer Research Fellowship in India, Gowen Frostic Doctoral Fellowship, Department Graduate Research and Creative Scholar Award by WMU, and Leo R. Parpart Doctoral Fellowship by Dept. of Physics, WMU, among many others, throughout his academic journey. He co-authored several research articles in WoS/Scopus indexed international journals, conference proceedings, and scholarly book chapters and co-edited several books published by renowned international publishing houses (IGI-Global, Springer Nature, Nova Science Publishers, Wiley, Scrivener-Wiley, Notion Press). At present, Dr. Das continues to delve into cutting-edge fields, with a keen interest in nanotechnology, UN SDGs, and data science. His multifaceted contributions, spanning teaching, research, and administrative leadership, showcase a dedicated professional who is instrumental in advancing the vision and mission of the institutions he serves.

Adi F. Mahmud, S.Pd., M.Pd. is an associate Professor at English Language Education Department, Faculty of Education, Universitas Muhammadiyah Maluku Utara, Indonesia. He earned his Bachelor's degree in English Education from Universitas Khairun and completed his Master's degree in English Education at Universitas Negeri Makassar (UNM). He is currently pursuing a Doctoral degree in English Education at Universitas Negeri Makassar, Indonesia. In addition to his primary academic appointment, he has contributed as a visiting lecturer at Universitas Khairun and a health polytechnic institution in Ternate. His scholarly works include a book chapter and a reference book, alongside publications in national and international journals, including Scopus-indexed outlets.

Karuppasamy Govindasamy is a dedicated researcher in Physical Education and Sports Sciences at the Department of Sports, Recreation and Wellness, Symbiosis International (Deemed University), Hyderabad Campus, Telangana. He has authored over 104 publications in Scopus and Web of Science journals, along with patents, book chapters, and books. His research focuses on innovative training methodologies and the integration of theory and practice in sports science. Dr. Govindasamy is UGC-NET and JRF qualified, has received multiple Best Paper Presentation awards, and was honored with a national-level Young Researcher Award. He also serves as a reviewer for 13 top-ranking international journals.

Siriyama Kanthi Herath is an Associate Professor of Accounting at Clark Atlanta University's School of Business (CAU) and was a Renwick Faculty Fellow at New York University's Stern School of Business in 2022. She earned a Bachelor of Commerce Honors and an MBA from the University of Colombo in Sri Lanka, as well as a Ph.D. and a Master of Commerce Honors in Accounting from the University of Wollongong in Australia. She worked at the University of Lynchburg, Georgia State University and Georgia Institute of Technology before joining CAU. Dr. Herath worked in Australia at the University of Wollongong and Western Sydney University and in Sri Lanka at the University of Ruhuna and the University of Colombo. She has contributed over 70 publications to a diverse range of journals. She has published 17 book chapters with coauthors. She is a Distinguished Toastmaster and served as the President and served as an Area Director from 2020 to 2021 in Georgia, United States. She possesses the TESOL Professional Certificate from Arizona State University. Dr. Herath is the Editor-In-Chief of the Asian Journal of Economics and Business and Global Journal of Accounting and Economy Research.

Debajit Karmakar is a dedicated Ph.D. scholar in Physical Education Pedagogy at Lakshmibai National Institute of Physical Education (LNIPE), Gwalior. An accomplished researcher, Mr. Karmakar has authored numerous research papers, patents, and books, contributing significantly to sports education and physical fitness. His work focuses on advancing training methodologies and innovative approaches in physical education, aligning both theory and practice. He brings his passion for sports into his academic pursuits, making impactful contributions to sports science and education. Mr. Karmakar's achievements mark him as a promising figure in physical education and sports science.

Baljinder Kaur is an Assistant Professor in the Department of Library and Information Science at Punjabi University, Patiala. With extensive experience in library science, she has specialized in e-resources, web-based resources and services, knowledge management, media and information literacy, and library use and user studies. She has been an active member of several professional organizations, including the Indian Library Association and the Punjab Library Association. Dr. Kaur has a robust teaching portfolio, having taught various courses at the undergraduate and postgraduate levels, and has contributed significantly to research with numerous published papers and presentations at national and international conferences. Additionally, she has developed e-learning modules and has been involved in the writing and vetting of educational content for university courses.

Abhishek Kumar is a leading voice in the world of inclusive sports, celebrated for his dual role as a dedicated management professional and a pioneering coach. His work is a testament to the power of sports in transforming lives, particularly for athletes with intellectual disabilities. With a Master's in Physical Education and a Ph.D. from Delhi University, Abhishek's deep academic knowledge is matched by his extensive practical experience. As a two-time Gothia Cup winning coach in 2024 and 2025 for Special Olympics Bharat, he made history by leading India to its first-ever trophy in the special category of the world's largest soccer championship. His expertise in team management and athletic development has been instrumental in preparing athletes for global stages, including the 2023 Berlin World Games and the 2025 World Special Winter Games in Italy. Beyond his work in coaching, Abhishek's background spans managing sports academies and organizing professional boxing events, showcasing his commitment to developing comprehensive sports programs. His passion and leadership have cemented his reputation as a true champion for sports excellence.

Kritikka Kunte is a third-year undergraduate student pursuing a Bachelor of Technology in Artificial Intelligence and Data Science at the School of Engineering, Ajeenkya DY Patil University, Pune, Maharashtra, India. With a strong academic orientation toward data-driven problem solving, she has developed a focused interest in the domains of data analytics and data science, supported by both theoretical foundations and applied project experience. Throughout her academic journey, Kritikka has cultivated proficiency in core programming and analytical competencies, including Python, SQL, and Java, along with essential data science methodologies such as data preprocessing, exploratory data analysis (EDA), and machine learning techniques. She is proficient in industry-relevant tools and platforms, including Advanced Microsoft Excel, Matplotlib, Tableau, and Power BI, to derive insights from complex datasets and present them in a structured and interpretable manner. Her practical exposure includes the successful completion of projects such as Ocean Salinity Analysis, which involved analyzing environmental datasets to identify patterns and trends, and Speech Recognition and Translation, where she explored applied machine learning concepts in natural language processing. These projects reflect her ability to translate conceptual knowledge into functional, real-world solutions. Kritikka has further strengthened her technical profile through professional certifications, including the AWS re/Start Graduate credential from Amazon Web Services, a certification in Software Development, and Python training certified by Google. Her engagement in competitive and collaborative academic environments is demonstrated by her participation in the Smart India Hackathon (SIH) and by securing Second Runner-Up in an inter-collegiate Tableau competition, highlighting her analytical aptitude and visualization skills. With a growing interest in advanced transportation systems and data-centric engineering solutions, Kritikka brings a multidisciplinary perspective to the GHC Hyperloop Project Competition hosted at IIT Madras. She aspires to contribute meaningfully to innovative, technology-driven initiatives that integrate artificial intelligence, data science, and sustainable engineering to address complex global challenges.

Atisha Mandal is a dedicated Ph.D. scholar at LNIPE Gwalior, specializing in Yoga and Exercise Physiology. Ms. Mandal holds both an MPED and BPED from LNIPE, Gwalior, with notable academic achievements. Her research work has led to several publications in reputed journals, focusing on the effects of yoga and pranayama on physical health and balance. A passionate sportswoman, Ms. Mandal has represented her university in Yoga Championships multiple times, including in 2019. Her diverse skill set extends beyond yoga, as she is also trained in table tennis and swimming. She has been recognized with the prestigious UGC Junior Research Fellowship, further cementing her commitment to academia and physical education. Proficient in statistical tools like SPSS, she is actively contributing to research while maintaining a balance between her academic and professional roles.

Hj. Mardyawati was born on January 1, 1970, in Dabbare (Soppeng). She is a Senior Lecturer at UIN Alauddin, affiliated with the Faculty of Islamic Studies at Universitas Islam Makassar. She holds a Bachelor's degree in Islamic Education from Universitas Muslim Indonesia, a Master's degree in Islamic Education Management from the same university, and a Doctorate in Education and Teaching from Universitas Islam Negeri Alauddin Makassar. Throughout her career, Dr. Mardyawati has held several significant academic and administrative positions, including Head of the PGTKI/PGSDI Study Program at STAI Al-Gazali Makassar, Dean of the Faculty of Islamic Studies at UIM, and Vice Rector III for Student Affairs at UIM. She has also been actively involved in various community service projects, focusing on educational support and creative economic empowerment. Her research and publications cover topics such as the role of Islamic education in fostering harmonious families, work ethics from an Islamic perspective, and innovative teaching methods for memorizing the Quran. Dr. Mardyawati is also a member of several professional organizations, contributing her expertise to the development of Islamic education and community welfare in Sulawesi Selatan.

Krutika Patre, is an aspiring programmer and technology enthusiast, currently pursuing Engineering in Artificial Intelligence and Data Science. With a passion for learning, she is building experience in a variety of programming languages and technologies, including C, C++, Python, Data Structures (DS), HTML, CSS, and JavaScript, DBMS (My SQL). After learning this programming language, she has created interest in Full Stack Development. She had also attended an Arduino-Based Automation course at Dolphin Labs to improve my software knowledge. There, she learned how to work with sensors, microcontrollers, and circuit design. She has written review paper on Sustainable engineering on build environment. She is doing researching and learning many topics related to AI, Policy and governance and related to my interest. She has taken part in hackathons, workshops, and tech events to explore many fields. She is also working with a group on Hyperloop. Being part of this project has really helped her grow, especially in areas like teamwork, innovation, and systems thinking.

Ranita Rope, S.P., M.Sc. is a lecturer in Agribusiness at Universitas Muhammadiyah Maluku Utara (UMMU) and currently serves as Vice Rector II. She completed her undergraduate degree in Agricultural Cultivation at Universitas Khairun (Ternate), and earned her Master's and Doctoral degrees in Agricultural Economics from Universitas Gadjah Mada (UGM), Yogyakarta. Her research focuses on natural farming and agribusiness development, with publications in national and Scopus-indexed journals. Dr. Rope teaches courses in agricultural production economics, farm management, and agribusiness research methods. She has held key leadership roles, including Head of Agribusiness Study Program and Dean of the Faculty of Agriculture at UMMU.

Sunil Sankathala is pursuing B. Tech Computer Science and Engineering at Ajeenkya DY Patil University, with a diverse range of interests that bridge technology, sustainability, and innovation. His academic focus lies in computer science fundamentals like programming, data structures, algorithms, and software development. He actively supplements this with hands-on experience in web development and robotics. Beyond the core curriculum, He has a strong research inclination in emerging and impactful fields. His current research interests include Artificial Intelligence in Healthcare, bioplastics, nano bioplastics, e-waste management, and solid waste management—areas where he believes technology can make a meaningful difference in solving real-world problems. He has been publishing research articles and book chapters in reputed national and international journals and books. He is also part of an ambitious Hyperloop project, which involves interdisciplinary collaboration and cutting-edge engineering aimed at revolutionizing transportation. Working on this project has enhanced his skills in teamwork, innovation, and systems thinking. In the tech domain, he has developed proficiency in web development using technologies such as HTML, CSS, JavaScript, and React. His interests in robotics further reflect building smart, responsive systems that interact with the physical world. Throughout his course, he has helped strengthen problem-solving and collaborative skills. He is proficient in languages such as Python, C, C++, MySQL, Ubuntu, Machine Learning and familiar with frameworks and tools like React, Git. He actively participates in hackathons, coding competitions, and tech events to challenge himself and learn from peers. He is also a quick learner and highly adaptable, with an eagerness to stay updated with the latest trends in the tech world. He aspires to become a skilled software engineer and contribute to building innovative solutions that have a positive impact on society. The long-term goal is to work in a dynamic environment where knowledge, grow professionally, and make meaningful contributions to the field of technology. Driven by a spirit of exploration and sustainability, he aims to contribute to technological solutions that address both human and environmental needs. The long-term goal is to become a versatile engineer and researcher, working at the intersection of computer science, environmental innovation, and human well-being.

Daniel U. Scott, PhD, JD, MPA, was selected for inclusion in Marquis Who's Who, an honor recognizing individuals for their professional impact, accomplishments, and prominence. Dr. Scott has devoted more than 30 years to the legal and business professions, building extensive expertise across litigation, judicial service, and public administration. Since 2013, he has served as an administrative law judge for New York State, following earlier roles as a court attorney from 2004 to 2013 and a criminal litigator from 1996 to 2003. He also served as a judge advocate general in the New York Army National Guard and U.S. Army Reserve from 2005 to 2009. Dr. Scott holds a PhD and MPA from Liberty University, a JD from The University of Toledo, and a bachelor's degree from Daemen University. His professional credentials include certifications in mediation, conflict management, and social and behavioral research, along with memberships in several legal and professional organizations.

Meenu Sharma is currently working as a Professor in the Department of Public Administration at the Assam Royal Global University, Guwahati, Assam. She has 21 years of teaching experience. Her 60 research papers have been published in national and international journals. Her 25 chapters in edited books have been published. She is a life member of the Indian Institute of Public Administration,

New Public Administration Society of India, Indian Political Science Association, Indian Political Economy Association, Indian Public Administration Association, Agricultural Economics and Social Science Research Association (AESSRA), Indian Association for Women's Studies.

Jaswinder Pal Singh is a researcher at the Department of Library and Information Science, Punjabi University, Patiala, India, and is currently working as a University Librarian at Chitkara University, Punjab, India. He possesses a comprehensive 19-year tenure within the library field, accumulating extensive expertise and proficiency in various aspects of library management, including cataloging, information retrieval, collection development, and patron services. Throughout this substantial duration, he has demonstrated a deep commitment to fostering an enriching environment for patrons, leveraging his wealth of experience to enhance library operations and meet the diverse needs of users. His areas of research include Media and Information Literacy, Knowledge Sharing, Information retrieval, Knowledge Management in Libraries, Collections Development, Academic Integrity.

Devanshi Soni is currently pursuing a B.Tech. in Computer Engineering at the School of Engineering, Ajeenkya D Y Patil University. Her research interests are metal nanoparticles and their applications and the integration of Artificial Intelligence across various domains. Devanshi is particularly interested in exploring how metal nanoparticles can be combined with AI technologies to address practical challenges and develop innovative solutions. She aims to advance research in these fields and contribute to groundbreaking developments that tackle real-world issues.

Roopa U is an esteemed Assistant Professor at Dayananda Sagar College of Engineering, bringing over two decades of dedicated teaching experience in academia. Her academic journey includes the attainment of an M.Sc. and M.Phil. in Statistics, coupled with an MBA in Marketing. She is pursuing her Ph.D. from Visvesvaraya Technological University in the field of exploring the transformative impact of social media applications on higher education further enhancing her expertise and contributions to the field. Areas of Expertise • Business Analytics using Python, R, Excel, and SPSS tools. • Machine Learning, Quantitative Techniques, Operations Research, Business Mathematics. • Research Methodology and Statistical Analysis for interdisciplinary applications. • Curriculum design and facilitation of innovative teaching methods.

Aryan Vasa is a passionate and self-driven Bachelor of Technology (B.Tech.) student specializing in Computer Science Engineering at Ajeenkya DY Patil University, Pune. Currently in his third year, Aryan is committed to building a solid foundation in core technical domains, including web development, data science, and artificial intelligence. With an inquisitive mindset and a strong eagerness to explore diverse fields, he strives to balance his academic journey with hands-on learning and project development. Originally from Jamnagar, Gujarat, and having completed his schooling in Rajkot under the ICSE board, Aryan has always been drawn toward technology, logic, and creativity. He is actively engaged in learning full-stack development while also keeping a close eye on the evolving job market, particularly the shift toward AI and generative technologies. His academic curiosity extends to research in advanced topics such as Generative AI and Digital Synthetic Outputs, for which he is currently writing a comprehensive review article. Aryan has also been a dedicated athlete, having played basketball for over ten years. This experience has helped him develop teamwork, discipline, and a growth mindset. In addition, he enjoys writing analytical and review-based content, merging his technical knowledge with strong communication skills. Committed to personal growth, Aryan maintains a balanced lifestyle that includes regular workouts, general fitness, and a focus on healthy eating. While continuously learning, he aims to contribute meaningfully to India's tech innovation landscape and become a versatile engineer capable of solving real-world problems through thoughtful design and smart solutions. Whether developing web applications, studying emerging technologies, or sharing knowledge through articles and mentorship, Aryan Vasa stands out as a motivated learner and a future-ready technologist.

Index

C

D

E

F

G

H

I

J

M

N

O

P

R

S

T

V

www.ingramcontent.com/pod-product-compliance
Ingram Content Group UK Ltd.
Pitfield, Milton Keynes, MK11 3LW, UK
UKHW050843101026
15624UKWH00018B/466

9 798369 383025